```
QP 514.2 .D39

Davies, Julian.

Elementary biochemistry
```

```
QP 514.2 .D39

Davies, Julian.

Elementary biochemistry
```

**Anoka-Ramsey Community
College Learning Media Center
Coon Rapids, MN 55433**

QP 514.2 . D39

Elementary Biochemistry

Julian Davies
&
Barbara Shaffer Littlewood
University of Wisconsin, Madison

Elementary Biochemistry

An Introduction to the Chemistry of Living Cells

Prentice-Hall, Inc., Englewood Cliffs, New Jersey 07632

Library of Congress Cataloging in Publication Data
DAVIES, JULIAN.
 Elementary biochemistry.

 Includes bibliographies and index.
 1. Biological chemistry. I. Littlewood, Barbara
Shaffer, joint author. II. Title.
QP514.2.D39   574.1'92   78-11867
ISBN 0-13-252809-6

© 1979 by Prentice-Hall, Inc., Englewood Cliffs, N.J. 07632

All rights reserved. No part of this book may be reproduced in any form or by any means without permission in writing from the publisher.

Printed in the United States of America

10 9 8 7 6 5 4 3 2

Editorial/production supervision
 and interior design by Ian List
Chapter opening design by Janet Schmid
Cover design by Janet Schmid
Cover photograph by © Glen S. Heller
Manufacturing buyer: Phil Galea

Prentice-Hall International, Inc., *London*

Prentice-Hall of Australia Pty. Limited, *Sydney*

Prentice-Hall of Canada, Ltd., *Toronto*

Prentice-Hall of India Private Limited, *New Delhi*

Prentice-Hall of Japan, Inc., *Tokyo*

Prentice-Hall of Southeast Asia Pte. Ltd., *Singapore*

Whitehall Books Limited, *Wellington, New Zealand*

"Science is nothing but trained and organized common sense"

Thomas Henry Huxley (1825–1895)

Contents

Preface

1
Cells: the Basic Units of Life

1.1	Properties of living cells	1
1.2	Cell structure	2
1.3	The elemental composition of cells	4
1.4	The chemicals found in cells: water, organic compounds, trace elements	5
	SUMMARY	10
	PRACTICE PROBLEMS	10
	SUGGESTED READING	10

2
Protein Structure & Function

2.1 Proteins serve many biological functions	**12**
2.2 Amino acids are the building blocks of proteins	**15**
2.3 Each protein is a linear chain of amino acids held together by peptide bonds	**22**
2.4 Proteins have a preferred three-dimensional shape	**27**
2.5 Some proteins require metal ions or small organic molecules for biological activity	**37**
2.6 Destruction of the three-dimensional shape of a protein drastically alters its physical properties and reduces its ability to carry out its biological function	**39**
SUMMARY	**42**
PRACTICE PROBLEMS	**43**
SUGGESTED READING	**44**

3
Determining Protein Structure

3.1 The relationship between protein structure and function	**45**
3.2 Determining the complete structure of a protein involves four basic steps	**46**
3.3 To study the structure of a particular protein, that protein must be freed from all the other components of the cell	**47**
3.4 The amino acid composition of a protein is determined by breaking its peptide bonds and analyzing the resulting mixture of amino acids	**50**
3.5 Determining the linear sequence of the amino acids in a protein	**54**
3.6 The three-dimensional shape of a protein can be determined by x-ray crystallography	**59**
SUMMARY	**60**
PRACTICE PROBLEMS	**60**
SUGGESTED READING	**62**

4
Enzymes

4.1 A single enzyme molecule can be used over and over again to convert many molecules of substrate into product	**64**
4.2 Cells need enzymes to make reactions proceed at a fast rate	**64**
4.3 Each enzyme can catalyze only one type of metabolic reaction	**65**
4.4 Enzyme nomenclature	**65**
4.5 Enzymes promote chemical reactions by making them more energetically favorable	**67**
4.6 The catalytic ability of an enzyme depends on its three-dimensional shape	**68**
4.7 Enzyme assays	**71**
4.8 External factors affecting enzyme activity	**72**
4.9 Isoenzymes	**75**
4.10 Medical and industrial applications of enzymology	**77**
4.11 Enzyme activity is destroyed by inhibitors	**80**
4.12 Uses and misuses of enzyme inhibitors	**81**
SUMMARY	**84**
PRACTICE PROBLEMS	**85**
SUGGESTED READING	**87**

5
General Concepts in Metabolism

5.1 Every cellular reaction is one step in a metabolic pathway	**89**
5.2 Metabolic pathways are interconnected by branch-point compounds	**90**
5.3 Cells control metabolic reactions by regulating enzyme activity	**93**
5.4 Studying the metabolic maze	**94**
SUMMARY	**98**
PRACTICE PROBLEMS	**98**
SUGGESTED READING	**100**

6
Energy for Cell Work: the Role of ATP

6.1 ATP: the "middleman" between energy-releasing and energy-requiring processes	101
6.2 Energy-requiring processes: kinds of cell work	104
6.3 Muscle cells use ATP energy for mechanical work	106
6.4 Active transport: the Na^+–K^+ pump	109
6.5 The magnitude of the cellular energy cycle	111
SUMMARY	113
PRACTICE PROBLEMS	113
SUGGESTED READING	114

7
The Production of Energy & Intermediates from Glucose

7.1 Glycolysis. I. The energy-using steps of glucose breakdown	118
7.2 Glycolysis. II. The release of useful energy	121
7.3 How much energy can we get from glycolysis?	124
7.4 The fate of pyruvic acid under anaerobic conditions	124
7.5 The fate of pyruvic acid under aerobic conditions	127
SUMMARY	128
PRACTICE PROBLEMS	128
SUGGESTED READING	129

8
Degradation of Carbohydrates: Energy & Intermediates from Krebs Cycle & Electron Transport

8.1 Acetyl CoA—Molecule at the crossroads	130
8.2 Reactions of the Krebs cycle	131
8.3 The flow of intermediates into and out of the Krebs cycle	136
8.4 Energy from the Krebs cycle: ATP is formed by oxidative phosphorylation	139

8.5	Energy from glucose: the balance sheet	**142**
8.6	Cytochrome *c* and evolution	**142**
	SUMMARY	**145**
	PRACTICE PROBLEMS	**146**
	SUGGESTED READING	**147**

9
Carbohydrate Formation & Storage: How Cells Control Energy Supplies

9.1	Glucose is stored in polysaccharides	**149**
9.2	Monosaccharides and disaccharides	**151**
9.3	Gluconeogenesis: how glucose is made from noncarbohydrate precursors	**153**
9.4	Insulin and the control of sugar transport into cells	**154**
9.5	Glucagon and epinephrine control glycogen breakdown	**155**
9.6	Photosynthesis: the synthesis of carbohydrates using solar energy	**157**
9.7	From plants to animals	**160**
	SUMMARY	**160**
	PRACTICE PROBLEMS	**161**
	SUGGESTED READING	**162**

10
Lipids

10.1	The major lipids: fatty acids, glycerides, and sterols	**164**
10.2	Lipid metabolism in the whole animal	**167**
10.3	Fatty acids start and end as acetyl CoA	**168**
10.4	Cholesterol: from acetyl CoA to gallstones	**172**
10.5	Membranes and lipoproteins: structure and function	**174**
10.6	Many other lipids play specialized physiological roles	**177**
10.7	Lipids, diet, and heart disease	**182**
	SUMMARY	**183**
	PRACTICE PROBLEMS	**184**
	SUGGESTED READING	**185**

11
The Metabolism of Amino Acids & Other Nitrogen Compounds

11.1	The nitrogen cycle	**187**
11.2	Cells need a basic set of amino acids	**189**
11.3	The synthesis of nonessential amino acids from glycolytic and Krebs cycle intermediates and ammonia	**192**
11.4	The degradation of amino acids feeds the Krebs and urea cycles	**194**
11.5	Nitrogen enters most other biological compounds from amino acids	**197**
11.6	The aromatic amino acids are important intermediates in human metabolism	**200**
11.7	Purine and pyrimidine nucleotides	**202**
11.8	The biosynthesis of purine and pyrimidine nucleotides	**203**
11.9	The breakdown and excretion of purines and pyrimidines	**207**
	SUMMARY	**208**
	PRACTICE PROBLEMS	**209**
	SUGGESTED READING	**210**

12
Deoxyribonucleic Acid: the Genetic Material

12.1	The structure of DNA: single strands and the double helix	**211**
12.2	The DNA of higher organisms is organized into chromosomes	**214**
12.3	Semiconservative replication of DNA depends on Watson–Crick base pairing	**220**
12.4	The many enzymes of DNA replication, repair, and recombination	**221**
12.5	DNA replication in bacteria	**224**
12.6	Biochemistry meets genetics	**226**
12.7	Evidence that DNA is the genetic material	**227**
12.8	Genes code for proteins	**228**
12.9	The genetic code: a sequence of three bases signifies one amino acid	**231**
12.10	A look at an inherited metabolic disorder: sickle-cell anemia	**233**
	SUMMARY	**236**
	PRACTICE PROBLEMS	**237**
	SUGGESTED READING	**239**

13
Protein Synthesis: the Translation of Genetic Information into Protein Requires RNA

A.	*RIBONUCLEIC ACIDS, MOLECULES OF MANY USES*	**241**
13.1	The chemical composition and structure of RNA	242
13.2	The three-dimensional structure of RNA	245
13.3	Transcription: the synthesis of RNA on a DNA template	248
B.	*HOW PROTEINS ARE MADE— THE TRANSLATION SYSTEM IN CELLS*	**252**
13.4	An overview	252
13.5	Messenger RNA and the genetic code	252
13.6	Transfer RNA—the adapter molecule	254
13.7	The ribosomes—where the action is	255
13.8	Factors and their roles in protein synthesis	257
13.9	Getting it all together—the stages of protein synthesis	257
	SUMMARY	260
	PRACTICE PROBLEMS	261
	SUGGESTED READING	263

14
Regulation & Control: How Cells Turn Metabolic Reactions On & Off

14.1	Regulation at the level of transcription	266
14.2	Regulation at the level of translation: turning protein synthesis on and off	271
14.3	Post-translational control: the regulation of enzyme activity	272
14.4	Gene amplification: more genes mean more gene products	274
14.5	Hormones—chemical coordinators synchronize your cells	275
14.6	The mode of action of thyroid hormones	280
14.7	Pheromones	281
14.8	Compartmental regulation	281
	SUMMARY	282
	PRACTICE PROBLEMS	282
	SUGGESTED READING	284

15 Viruses: Cell's Natural Enemies

15.1	Viral structure	**286**
15.2	The life cycle of viruses: lysis versus lysogeny	**287**
15.3	Viruses and cancer	**293**
15.4	Copying the viral genome	**294**
15.5	Why we cannot cure viral infections	**296**
15.6	Viroids—science fiction in real life	**298**
	SUMMARY	**299**
	PRACTICE PROBLEMS	**299**
	SUGGESTED READING	**300**

16 Antimicrobial Agents

16.1	What are antibiotics?	**303**
16.2	Why are antibiotics effective?	**304**
16.3	Antibiotics that inhibit bacterial cell wall synthesis	**305**
16.4	Antibiotics that inhibit bacterial protein synthesis	**306**
16.5	Antibiotics that inhibit bacterial RNA synthesis	**307**
16.6	Antimetabolites	**308**
16.7	Bacterial resistance to antibiotics: bacteria fight back	**309**
	SUMMARY	**310**
	PRACTICE PROBLEMS	**311**
	SUGGESTED READING	**312**

Appendix: Important Concepts in Organic Chemistry

A.1 Covalent bonding	**313**
A.2 The shapes of organic compounds	**314**
A.3 The functional groups in organic compounds are responsible for their chemical reactivity	**316**
A.4 Oxidation-reduction reactions	**322**
TEXTBOOK REFERENCES	**324**
Index	**326**

Preface

This text is designed to introduce the principles of biochemistry and their applications to undergraduate students who desire a basic knowledge of biochemistry. The text is aimed at students in nursing, nutrition, agriculture, and other sciences for which a survey course in biochemistry is sufficient. However, we believe that students in fields other than science can and should comprehend the material.

The book is based on lectures given in an introductory biochemistry course at the University of Wisconsin. In working with the undergraduates in this course, we have found that many students view biochemistry as an "ivory tower" science which bears little or no relationship to everyday life, and that they cannot visualize how the material presented will ever be useful to them. In the hope of stifling the "Why are we learning this stuff?" refrain, we have emphasized the fact that biochemical principles underlie many common phenomena. For example, we include information about the enzymatic basis of inherited disease, drug therapy when it involves enzyme inhibition, and the biochemical basis of wine making, pickling, and other food preparation. We have tried to include explanations of topics currently being discussed in popular periodicals and newspapers.

Our experience has indicated that the beginning student often reacts negatively to biochemistry because the courses consist of the memorization of countless chemical formulas and the minute details of metabolic pathways. We believe this to be an inappropriate and stifling approach to the subject. In writing this text, we started with the premise that by emphasizing the principles and unity of biochemistry and by deemphasizing organic chemistry, biochemistry can be understood and appreciated by students who lack the background in chemistry, physics, and mathematics necessary to benefit from advanced biochemistry courses. In our efforts to emphasize the underlying principles and unity of biochemistry, there occur errors of generality, but we anticipate that these exceptions can be turned to advantage by a good lecturer.

The title of the book may be misleading, since biochemistry is not an elementary subject. However, its principles are simple and the most important feature of biochemistry is not the endless learning of pathways, but an appreciation of how molecules, pathways, cells, and organs interact in a controlled fashion to create and maintain the biochemistry that we call life.

Although we have tried to deemphasize the purely chemical aspects of biochemistry, we realize, all too well, that a sound knowledge of chemistry is essential to a complete comprehension of biochemical principles. We have tried to reduce the amount of chemistry that we use, without the intention of implying that chemistry is not needed. It is. Anyone who wants to attain a more complete understanding will require a firm background in organic chemistry. Even for courses using this text, students who have completed an introductory course in organic chemistry will be at an advantage. Those with limited backgrounds in the subject may wish to begin with the Appendix, which summarizes those principles of organic chemistry that are relevant to a study of cell metabolism.

It has also been our experience that teaching elementary biochemistry can be facilitated by the use of thought-provoking problems. To emphasize understanding rather than memorization, each chapter includes a problem set intended to stimulate the "thinking through" of biochemical principles and their applications.

This text, then, is intended as the basis for an undergraduate survey course in biochemistry. It is our hope that for some people this book will not be enough, but that it will serve as an introduction to some of the excellent advanced texts that are available at present and encourage students to delve into some subjects in more detail. We sincerely hope that students who use this text will complete their biochemistry course with "a good taste in their mouths" and a feeling that they have learned important information about the fascinating world inside living cells. Biochemistry may not be easy, but it *is* fun.

We are grateful to the many people who encouraged us in the writing of this textbook. Special thanks go to our families for their continuing patience and support, and to the students and teaching assistants in Biochemistry 201 who served as (unknowing) fodder for our ideas and have made this effort worthwhile. We thank Patricia Omilianowski for typing the manuscript and Melanie Loo and Leon LeVan for "debugging" the problem sets.

Madison, Wisconsin   JULIAN DAVIES
                                                    BARBARA SHAFFER LITTLEWOOD

1 Cells: the Basic Units of Life

Biochemistry is the study of the composition, the production, and the destruction of the chemical compounds found in living organisms. It encompasses all the chemical reactions carried out by cells and the relationships between these reactions. Biochemists work with a wide variety of biological materials, from the one-celled bacterium to the specialized multicellular tissues of man. Understanding this range of organisms would be a hopeless task except for the fact that the vast majority of biologically important reactions are common to all cells. The phrase "the unity of biochemistry" refers to the finding that the chemistry of all cells is remarkably similar.

1.1 Properties of living cells

The *cell* is the basic structural unit of all living organisms. Bacteria and blue-green algae, the smallest free-living organisms, consist of single, independent cells. Larger organisms—animals and plants—contain billions of cells organized into tissues with specialized functions. Before trying to understand the chemical basis of cellular phenomena, it is important to consider, in broad terms, those processes which characterize living cells. Although a precise definition of "life" is difficult, if not impossible, we *can* give a general description of the processes that distinguish living cells from the inanimate material found on the earth.

1. All cells are capable of metabolism; they can take up simple chemical compounds from their environment, modify them, and combine them into larger, more complex molecules. The newly synthesized compounds are then used to build the complicated structures necessary for cellular integrity and function. Metabolism is not a disorganized business; it is a tightly regulated network of reactions by which the cell produces (or degrades) thousands of different molecules in specific proportions, at specific rates, and often at specific times.

2. All cells are capable of energy transformation. During photosynthesis, green plant cells convert solar energy into the energy of the chemical bonds of sugar molecules. All cells degrade these sugars and store the released energy in the chemical bonds of "high-energy" compounds. Cells use their stored energy to perform cellular work such as biosynthesis, division, contraction, locomotion, and the transmission of electrical stimuli.

3. Cells are capable of self-directed growth and replication. By metabolic processes, cells increase their mass and at a critical time when all cellular constituents are present in the proper concentrations, cells divide, giving rise to two identical daughter cells. Each cell contains all the information necessary for perpetuating itself, and this information remains constant generation after generation.

1.2 Cell structure

Living cells carry out many different metabolic activities: the exchange of chemicals with the environment, the transformation of energy, the replication of informational material, and so on. To perform these tasks efficiently, an extraordinary system of internal organization has been developed. By (physically) grouping the chemical reactions concerned with specific metabolic activities, cells are able to perform and to regulate these activities with speed and efficiency.

Eucaryotic cells, the well-organized units of higher plants and animals, contain many different membrane-bound compartments, called *organelles* [Fig. 1.1(a)]. While the morphological appearance of these organelles may vary somewhat from cell to cell, their functions (Table 1.1) are identical in all cells. Eucaryotic cells always have a membrane-bound nucleus and usually have their DNA divided among several chromosomes. *Procaryotic cells* [Fig. 1.1(b)], those of lower organisms including bacteria and blue-green algae, are relatively small and lack most of the internal structures that characterize eucaryotic cells. Procaryotic cells have no membrane-enclosed nucleus and usually contain only one DNA molecule.

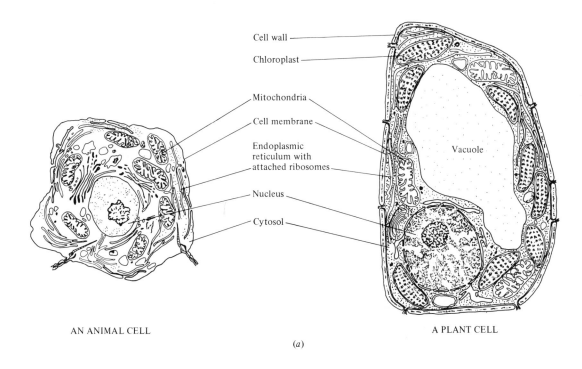

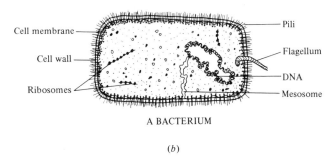

Figure 1.1 (*a*) Eucaryotic cells. (*b*) A procaryotic cell. [(*a*) After Robert D. Dyson, *Essentials of Cell Biology*, Allyn and Bacon, Inc., Boston, 1975, pp. 16, 17. (*b*) After Robert D. Dyson, *Essentials of Cell Biology*, Allyn and Bacon, Inc., Boston, 1975, p. 13.]

Biochemists have long been interested in both the chemical composition and the metabolic roles of the organelles of eucaryotic cells. This research has been greatly facilitated by two technical advances: the development of the electron microscope and of the ultracentrifuge. The high magnifying power of the electron microscope has enabled biologists to visualize individual structures within the intact cell. To determine the composition and function of subcellular organelles, it is necessary to prepare large quantities of each individual organelle free of contamination by other

TABLE 1.1 Structural Organization of Eucaryotic Cells

Subcellular organelle	Function
Cell membrane	Maintains cytoplasmic integrity; regulates exchange of chemicals with the environment
Cytosol	Contains water-soluble components of metabolic apparatus; most molecules diffuse freely through the cytosol
Nucleus	Storage, replication, and intracellular dispersal of the cell's informational material
Endoplasmic reticulum	Transport of metabolic products within the cell; may be involved in secretion
Ribosomes	Site of protein synthesis
Mitochondria	Site of oxidative reactions that transform energy into form used for cell work
Cell wall (absent from animal cells)	Protects the cell; maintains rigid cell shape
Lysosomes (in animal cells and protozoans)	Contain enzymes for intracellular digestion and destruction of foreign material; destruction of dead cells
Chloroplasts (algae and green plants only)	Site of photosynthesis

subcellular particles. Such a preparation is made by breaking the cells gently and separating the individual components by differential centrifugation, taking advantage of their different sizes, shapes, and densities. For this purpose, the high gravitational force and large capacity of the modern ultracentrifuge have proven invaluable.

1.3 The elemental composition of cells

The bulk of cellular material consists of the six elements that are found in organic molecules (Table 1.2). In the human body, for example, 99.4% of the atoms are accounted for by hydrogen, oxygen, carbon, and nitrogen. In addition, cells require small amounts of trace elements (Tables 1.2 and 1.5).

Clearly, cells are selective about which elements are used for metabolism; earthly life requires less than one-third of the 90 available elements. The evolutionary processes by which particular elements were selected for incorporation into living material are not understood. It can be argued that the capacity of carbon, nitrogen, hydrogen, and oxygen to form many combinations of covalent bonds makes these elements uniquely suited to be the material for the thousands of interconvertible organic compounds found in every cell. Although such arguments allow us to rationalize why water and not hydrogen sulfide, or why carbon and not silicon, are present in living forms on earth, it is quite possible that other planets may use alternative elements in "life."

Table 1.2 Chemical Elements Required by Cells

		Composition of human body (% of total number of atoms)
Elements found in organic molecules	Carbon (C)	9.5
	Hydrogen (H)	63.0
	Nitrogen (N)	1.4
	Oxygen (O)	25.5
	Phosphorus (P)	0.22
	Sulfur (S)	0.05
Trace elements required by all cells	Chlorine (Cl)	0.03
	Sodium (Na)	0.03
	Potassium (K)	0.06
	Magnesium (Mg)	0.01
	Calcium (Ca)	0.31
	Manganese (Mn)	All others less than 0.01
	Iron (Fe)	
	Cobalt (Co)	
	Copper (Cu)	
	Zinc (Zn)	
Trace elements required by certain cells	Boron (B)	
	Fluorine (F)	
	Silicon (Si)	
	Vanadium (V)	
	Chromium (Cr)	
	Selenium (Se)	
	Molybdenum (Mo)	
	Tin (Sn)	
	Iodine (I)	

1.4 The chemicals found in cells: water, organic compounds, trace elements

(a) WATER

Water is the most plentiful compound in all living cells; they contain about 70% water by weight. Most metabolic reactions are carried out in this aqueous environment. Not only is water the medium for cellular reactions, it is also a participant in many of these reactions. Water ionizes to yield hydrogen ions and hydroxyl ions:

$$H_2O \longrightarrow H^+ + OH^-$$

These ions are of critical importance as reactants in, or catalysts for, many metabolic

TABLE 1.3 Hierarchy of Organic Compounds in Cells

MONOMER		MACROMOLECULE		
Name	Elemental composition	Name (examples)	Monomer–monomer bond	Some cell organelles that include macromolecule
Amino acids	C, H, O, N (sometimes S)	Proteins (insulin, hemoglobin, all enzymes, actin, and myosin)	Peptide bond	Ribosomes, membranes, chromosomes, muscle fibers
Sugars	C, H, O	Polysaccharides (glycogen, starch, cellulose)	Glycosidic bond	Cell wall
Nucleotides	C, H, O, N, P	Nucleic acids (RNA, DNA)	Phosphodiester bond	Chromosomes, ribosomes
Fatty acids and glycerol	C, H, O	Lipids (triglycerides)	Ester bond	Membranes

Cells: The Basic Units of Life

reactions and in maintaining intracellular pH* near neutrality. The role of water in associations between molecules (bonding) is also significant.

(b) ORGANIC COMPOUNDS

Cells contain a great variety of organic molecules, which can be thought of as existing in a hierarchy of increasing molecular weight and complexity (Table 1.3). This hierarchy begins with the small molecules of low molecular weight such as amino acids and sugars; these small molecules can be combined to form very large *macromolecules*. Macromolecules are usually composed of only one type of small molecule, called a *monomer*, arranged in a definite, predetermined pattern and held together by a specific kind of chemical bond. The different types of macromolecules are, in turn, organized into more complex cell structures, including membranes, chromosomes, cell walls, contractile fibers, and ribosomes.

One such hierarchy of organic compounds is based on amino acids, small organic molecules with molecular weights of less than 200. Amino acids are the monomers for macromolecules called *proteins*; a chain of several hundred amino acids linked together by peptide bonds forms a protein. Proteins often combine with other types of macromolecules to yield complex structures such as ribosomes (proteins plus ribonucleic acids) or cell membranes (proteins plus lipids).

Where do cells obtain the materials to build the small molecules that are the basis for this molecular hierarchy? In general, cells cannot use carbon, nitrogen, and hydrogen in their elemental forms—organisms cannot incorporate pure carbon from coal, and only a few bacteria can use atmospheric nitrogen directly. Consequently, all cells require organic compounds as their sources of carbon, hydrogen, nitrogen, and oxygen. Cells can incorporate most of the small organic molecules (sugars, fatty acids, amino acids, and so on) present in the environment as a result of the decay of organic matter and excretion from other living organisms. Cells are capable of using only those organic compounds which are small enough to pass through the cell membrane; they are impermeable to (larger) macromolecules and therefore cannot metabolize them. (The fact that higher animals use macromolecules, such as starch and proteins, as food does not contradict this rule; macromolecules are digested into their constituent small molecules before they enter cells to be used in the intracellular metabolic chain.)

Although cells require certain quantities of organic material, this requirement is generally nonspecific; almost any organic compound will do since cells can interconvert most small organic molecules. For example, the sugar glucose can be converted into fatty acids, into the sugar ribose (which forms part of nucleotides, the monomers for RNA) and, given a suitable source of nitrogen, into amino acids. However, many cells do require a small number of specific organic compounds which they are incapable of synthesizing from other compounds present in the environment. These requirements vary from cell to cell, and from organism to organism; in human

---

* pH is a measure of the acidity of a solution, that is, the concentration of hydrogen ions (H^+).

Table 1.4 Specific Organic Compounds Needed in the Human Diet

Compound	Role in metabolism
Vitamins	
Thiamine (B$_1$)	Enzyme cofactors or precursors of cofactors
Riboflavin (B$_2$)	
Niacin	
Pyridoxal (B$_6$)	
Biotin	
Cobalamin (B$_{12}$)	
Ascorbic acid (C)	
Vitamin D	Ca^{2+} and PO$_4^{2-}$ absorption and utilization
Vitamin A	Precursor of visual pigment
Vitamin E	?
Vitamin K	Needed for blood clotting
Pantothenic acid	Precursor of coenzyme A
Essential amino acids	Monomers for protein synthesis
Polyunsaturated fatty acids	? (Precursors for prostaglandins, Chap. 10)

beings, for example, vitamins, some amino acids, and certain unsaturated fatty acids are essential for the maintenance of life (Table 1.4).

(c) TRACE ELEMENTS—RARE BUT CRUCIAL

Living cells require small quantities of between nine and 18 trace elements (Tables 1.4 and 1.5). Human beings require at least 14 such elements, probably the most familiar being iron. [Most people know this because of television commercials claiming that (iron-deficient) "tired blood" may cause personality problems.]

The roles of trace elements in cell metabolism are extremely varied. Sodium, potassium, and chloride ions are essential for the maintenance of the cell's water balance and osmotic pressure. Calcium, silicon, and magnesium are important for the production of skeletal structures such as bones and teeth. Many inorganic ions function as enzyme cofactors, a role that will be discussed in Chap. 4. The precise roles of tin and vanadium in animals, or for boron, vanadium, and selenium in plants, are not known, but the organisms do not grow and reproduce normally in their absence.

It is possible that some elements are missing from Table 1.5 because of experimental difficulties in establishing an essential requirement for an element, especially since minute quantities may be involved. The requirements for trace elements were first discovered when it was found that certain diseases of higher organisms could be cured by supplementing the diet with a specific element; for examples, dietary iron cures some anemias, iodine cures goiter. For other trace elements, such as manganese and chromium, no clinically important deficiency diseases have yet been recognized. The requirement for these elements has been established by showing that laboratory animals raised on diets lacking these elements did not develop normally; requirements

Table 1.5 Essential Trace Elements

Trace element	Metabolic role in all organisms	Examples of organism-specific metabolic role[a,b]
Chlorine	Cl^- is the principal cellular anion and is essential in maintaining water balance, osmotic pressure and acid–base balance	—
Potassium, Sodium	K^+ and Na^+ are the principal cellular cations and are essential in maintaining water balance, osmotic pressure, and acid–base balance	An: K^+ influences muscle activity
Magnesium	Enzyme cofactor	Pl: constituent of chlorophyll An: constituent of bone
Calcium	Enzyme cofactor	An: major component of bones, teeth, eggshells; needed for nerve and muscle function
Manganese	Enzyme cofactor	—
Iron	Enzyme cofactor	An: constituent of hemoglobin
Copper	Enzyme cofactor	Constituent of hemocyanin, the O_2 carrying protein of invertebrates
Zinc	Enzyme cofactor	An: constituent of insulin
Molybdenum	Enzyme cofactor	
Boron	—	Growth factor for lower plants
Fluorine	—	Growth factor in rats
Silicon	—	Necessary for development of feathers and skeleton in chickens; essential for hard structures of lower plants (diatoms)
Vanadium	—	Growth factor for rats and green algae
Chromium	—	An: promotes glucose utilization
Cobalt	—	Constituent of vitamin B_{12} needed for red blood cell formation in animals and for nitrogen fixation in certain bacteria
Selenium	—	An: essential for liver function Pl: growth factor
Tin	—	Growth factor for rats
Iodine	—	An: constituent of thyroid hormone

[a] An, higher animals; Pl, green plants.
[b] A growth factor is a substance essential for normal growth and development, to which a specific metabolic function has not yet been assigned.

for some other elements may well have been missed. In addition to omissions, Table 1.5 may be incomplete because elements essential in one case may, in fact, be of importance to many organisms. For example, silicon is presently known to be required for skeletal development in chicks only, but it could have a more universal role.

SUMMARY

Cells are the smallest units of living things which show the characteristics of life: the synthesis and degradation of organic molecules, energy transformation, and self-reproduction. Cells are classified as procaryotic (having a minimum of internal organization), or eucaryotic (containing many complex internal organelles).

Chemically, cells consist mainly of the elements in organic compounds: carbon, hydrogen, oxygen, and nitrogen (plus small amounts of sulfur and phosphorus). Cells take a limited number of small organic molecules from the environment and from them produce a large, but carefully balanced, collection of other small molecules: sugars, amino acids, fatty acids, and nucleotides. These are, in turn, used to construct large macromolecules: carbohydrates, proteins, lipids, and nucleic acids. Precise aggregation of macromolecules creates the necessary cell structures.

In addition to organic matter, cells require small quantities of several trace elements. Such trace elements play roles in the cell's water balance, act as enzyme cofactors, and/or are part of the supporting structures of higher animals. Many cells require specific organic compounds, the vitamins, to maintain their metabolic functions.

PRACTICE PROBLEMS

1. A large percentage of the elements present in the earth's crust are not required by living cells. Many of these nonessential elements can enter cells and inhibit metabolism. Name three such deleterious elements and describe briefly how man comes in contact with them.

2. Cells require small organic compounds as sources of the six elements found in macromolecules. Given glucose ($C_6H_{12}O_6$), what other elements found in organic compounds do cells need to build proteins? To build ribonucleic acid? To build fatty acids?

3. Which subcellular organelle would you isolate from rat liver if you wanted to study:
   (a) The incorporation of amino acids into proteins?
   (b) The incorporation of nucleotide bases into chromosomal nucleic acids?
   (c) The aerobic production of ATP?

4. A significant amount of nutritional "information" comes to us from television commercials and magazine advertisements. Describe three vitamin or trace-element preparations you have seen advertised. What is your personal reaction to these ads? To the products advertised? Is the information that is presented based on sound biochemical principles? (You may need to consult later chapters.)

SUGGESTED READING

BRACHET, JEAN, "The Living Cell," *Sci. American*, 205, No. 3, p. 50 (1961). A classic paper describing how cytological and biochemical techniques have been used to determine cell structure and function; many electron micrographs.

DEDUVE, CHRISTIAN, "The Lysosome," *Sci. American*, 208, No. 5, p. 64 (1963). Cell organelles are purified from cell homogenates by differential centrifugation.

DUDRICK, STANLEY J., and JONATHAN E. RHOADS, "Total Intravenous Feeding," *Sci. American*, 226, No. 5, p. 73 (1972). Patients survive several months with the infusion of a balanced solution containing sugar, amino acids, vitamins, and trace elements.

FRIEDEN, EARL, "The Chemical Elements of Life," *Sci. American*, 227, No. 1, p. 52 (1972). A comprehensive survey of the chemical composition of cells. The author theorizes on why carbon, hydrogen, oxygen, and nitrogen were selected as the major elements of cells. The article explains the roles of essential trace elements and how requirements for these elements are determined.

GILLIE, R. BRUCE, "Endemic Goiter," *Sci. American*, 224, No. 6, p. 92 (1971). Goiter is common in populations living where the iodine content of the soil is low. This ancient disorder has a long and interesting history.

HIRSCHHORN, NORBERT, and WILLIAM B. GREENOUGH III, "Cholera," *Sci. American*, 225, No. 2, p. 15 (1971). Cholera is characterized by the loss of body water and sodium and potassium salts; simple methods for replacing these substances can reduce the lethality from this disease.

MAZIA, DANIEL, "The Cell Cycle," *Sci. American*, 230, No. 1, p. 54 (1974). Describes the four phases of the life of a eucaryotic cell.

2 Protein Structure & Function

Proteins participate in every aspect of cell metabolism; all cell growth, development, and behavior depend upon these macromolecules. Different types of cells can be distinguished by the specific proteins they produce. Animal cells contain proteins not found in plant cells; nerve cells make different proteins from those made by muscle cells.

All proteins are high molecular weight polymers of amino acids. Yet, among proteins, there is enormous variation in size, shape, and chemical properties. This physical and chemical diversity among the hundreds of proteins in every cell allows each protein to perform its own unique function in cellular metabolism.

2.1 Proteins serve many biological functions

Each protein has its own "job" to do. It would be impractical (and boring) to list the thousands of known proteins and give their precise roles. To simplify matters, proteins are grouped according to their general biological function:

Enzymes are proteins that catalyze (speed up) the making and breaking of chemical bonds during metabolic reactions.

Structural proteins maintain the shape of a cell or organism. They protect the surface and provide supporting structures for soft organs in higher organisms.

Hormones are regulatory molecules produced by one cell to control the function and activity of other cells. Some hormones coordinate the metabolism of different cells and organs, while others trigger organ-specific differentiation (e.g., lactation). (*Not all hormones are proteins.*)

Antibodies are proteins that act as defense mechanisms in higher organisms. They protect higher organisms by inactivating potentially harmful foreign substances (which are often other proteins).

Contractile proteins are responsible for movement and locomotion of cells and animals.

Transport proteins carry small molecules from one place to another. They bring material across the cell membrane and, in higher animals, circulate in the blood, carrying metabolites between different tissues.

Toxins are small proteins excreted by certain bacteria, plants, and insects as a defense mechanism. They are poisonous to the "enemies" of the cells that produce them.

Storage proteins provide a reserve of amino acids for the next generation.

Many *proteins* are *identified by their association with other macromolecules*. Different kinds of macromolecules, loosely (noncovalently) joined together, form the functional and structural units of many subcellular organelles. For example, cell membranes are made up of *lipoproteins*—complexes between lipids and proteins. Ribosomes and chromosomes are made up of *nucleoproteins*—complexes between nucleic acids and proteins.

A few of the more familiar proteins in each category are listed in Table 2.1.

While certain classes of proteins, such as the nucleoproteins of ribosomes, are common to all cells, many proteins are found only in specific kinds of cells. In fact, differences in protein composition account for nearly all the functional differences between cells. The hemoglobin found only in red blood cells allows these cells to carry out oxygen transport. Contractile proteins are a major component of muscle cells, but they are absent from red blood cells or nerve cells. While all cells produce dozens of enzymes, certain enzymes are responsible for the unique characteristics of particular cells; for example, the iodine-requiring reactions that produce thyroxine are catalyzed by enzymes found only in cells of the thyroid gland.

Although all proteins are synthesized inside cells, some proteins are secreted from the producing cell to carry out their biological function in the extracellular environment. Bacterial toxins are good examples. In animals, extracellular fluids are rich in secreted proteins. Blood contains fibrinogen (needed for clotting), immunoglobulins, and many transport proteins; most of these plasma proteins are made by

TABLE 2.1 Representative Proteins in the Functional Classes

Functional class	Representative protein	
Enzymes	Chymotrypsin	Digestive enzyme; degrades proteins into short chains of amino acids
	Ribonuclease	Catalyzes the breakdown of ribonucleic acids
	Lactate dehydrogenase	Converts pyruvic acid into lactic acid; functions in working muscle
Structural proteins	Collagen	Water-insoluble framework on which calcium salts are deposited, forming bone and cartilage
	α-Keratin	Water-insoluble, nearly indestructible protein in hair, wool, nails
Hormones	Insulin	Regulates glucose utilization throughout the body
	Prolactin	Stimulates mammary glands to produce milk
Antibodies	Anti-Rh antibodies	Produced by Rh^- mother in response to Rh^+ fetus; destroys red blood cells of subsequent Rh^+ fetuses
	γG immunoglobulins	Group of high molecular weight serum proteins produced in response to foreign proteins; each can inactivate a particular foreign protein
Contractile proteins	Actin and myosin	Proteins that change position to produce contraction of muscle fibers
Transport proteins	Hemoglobin	In red blood cells; carries oxygen from lungs to the rest of the body
	Sulfate permease	Transports sulfate ions across the cell membrane into the cell interior
Toxins	*Vibrio cholerae* toxin	Causes severe loss of body fluids by way of the intestines; cholera victims die of dehydration
	Clostridium tetani toxin	Poisons the nerves of the spinal cord, causing many voluntary muscles to contract simultaneously (the spastic paralysis of tetanus)
Storage proteins	Ovalbumin	In egg white; nourishes the embryo before hatching
	Gliadin	In wheat seeds; nourishes the germinating embryo

liver cells. Digestive processes depend upon enzymes secreted from stomach and intestinal glands and from the pancreas. Digestive enzymes are also secreted by insectivorous plants! Secreted proteins have been obvious choices for biochemical study because these macromolecules are made in relatively large amounts and are naturally free of many cell components; they can be purified easily (Sec. 3.3).

Protein Structure and Function

We have chosen to classify proteins by their general function. If some other classification scheme had been selected—size, shape, or solubility—the proteins in Table 2.1 would have fallen into completely different groupings. Knowing the general biological function of a protein tells us little or nothing about the physical and chemical properties of that macromolecule; there is tremendous structural diversity among proteins with similar metabolic roles.

The common denominator among proteins is not their function; rather, it is that *all proteins are macromolecules made of amino acids*. Each protein contains characteristic proportions and arrangements of the amino acids.

2.2 Amino acids are the building blocks of proteins

(a) THE STRUCTURE OF AMINO ACIDS

Amino acids are low molecular weight organic compounds with the following generalized structure:

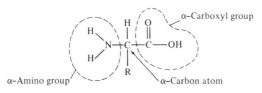

Structure of an amino acid

Amino acids contain a free carboxyl group (—$\overset{\overset{O}{\|}}{C}$—O—H) and a free amino group (—NH_2) attached to the α-carbon atom. The α-carbon is also attached to one hydrogen atom and to an organic substituent, the R group. The composition and structure of the R group varies from one amino acid to the next, so the R group gives each amino acid its unique physical and chemical properties.

Like other carbon-containing compounds, amino acids are not "flat," as the representation above implies. Rather, the four atoms attached to any carbon atom form a tetrahedron or pyramid around that carbon atom. The α-carbon, found in all amino acids, is therefore "buried" at the center of a pyramid with one of its four substituents (—H, —COOH, —NH_2, and —R) at each apex.

If you try to rearrange the substituents (i.e., place different groups at different points of the pyramid), you will see that there are only two arrangements which are not superimposable. These mirror-image arrangements are called the D and the L forms (isomers) of the amino acids:

$$\underset{\text{L-Amino acid}}{\text{H}-\underset{\underset{\text{COOH}}{|}}{\overset{\overset{\text{R}}{|}}{\text{C}}}-\text{NH}_2} \qquad \underset{\text{D-Amino acid}}{\text{H}_2\text{N}-\underset{\underset{\text{COOH}}{|}}{\overset{\overset{\text{R}}{|}}{\text{C}}}-\text{H}}$$

Only the L isomers of amino acids are found in proteins. With very few exceptions, only the L isomers participate in any metabolic reactions, since most enzymes recognize only this form.

The amino acids found in proteins are listed in Table 2.2. They are grouped according to the structure and charge properties of their R groups. *Neutral* amino

TABLE 2.2 Amino Acids Found in Proteins

Nature of R group at pH 7	Amino acid	Symbol	Structure at pH 7 (R group is shaded)
Neutral (uncharged), aliphatic	Alanine	Ala	$\text{H}_3\overset{+}{\text{N}}-\underset{\underset{\text{CH}_3}{|}}{\overset{\overset{\text{H}}{|}}{\text{C}}}-\overset{\overset{\text{O}}{\|}}{\text{C}}-\text{O}^-$
	Valine	Val	$\text{H}_3\overset{+}{\text{N}}-\underset{\underset{\underset{\text{CH}_3\quad\text{CH}_3}{\diagdown\diagup}}{\text{CH}}}{\overset{\overset{\text{H}}{|}}{\text{C}}}-\overset{\overset{\text{O}}{\|}}{\text{C}}-\text{O}^-$
	Leucine	Leu	$\text{H}_3\overset{+}{\text{N}}-\underset{\underset{\underset{\underset{\text{CH}_3\quad\text{CH}_3}{\diagdown\diagup}}{\text{CH}}}{\text{CH}_2}}{\overset{\overset{\text{H}}{|}}{\text{C}}}-\overset{\overset{\text{O}}{\|}}{\text{C}}-\text{O}^-$
	Isoleucine	Ile	$\text{H}_3\overset{+}{\text{N}}-\underset{\underset{\underset{\underset{\text{CH}_3}{|}}{\text{CH}_2}}{\text{HC}-\text{CH}_3}}{\overset{\overset{\text{H}}{|}}{\text{C}}}-\overset{\overset{\text{O}}{\|}}{\text{C}}-\text{O}^-$

TABLE 2.2 (continued)

Nature of R group at pH 7	Amino acid	Symbol	Structure at pH 7 (R group is shaded)
Neutral, aliphatic, sulfur-containing	Glycine	Gly	$H_3\overset{+}{N}-CH(H)-COO^-$
	Serine	Ser	$H_3\overset{+}{N}-CH(CH_2OH)-COO^-$
	Threonine	Thr	$H_3\overset{+}{N}-CH(CH(OH)CH_3)-COO^-$
	Methionine	Met	$H_3\overset{+}{N}-CH(CH_2CH_2SCH_3)-COO^-$
	Cysteine	Cys	$H_3\overset{+}{N}-CH(CH_2SH)-COO^-$
Neutral, aromatic	Phenylalanine	Phe	$H_3\overset{+}{N}-CH(CH_2C_6H_5)-COO^-$

TABLE 2.2 (continued)

Nature of R group at pH 7	Amino acid	Symbol	Structure at pH 7 (R group is shaded)
	Tyrosine	Tyr	
	Tryptophan	Trp	
Acidic (negative), aliphatic	Aspartic acid	Asp	
	Glutamic acid	Glu	
Basic (positive), aliphatic	Lysine	Lys	

TABLE 2.2 (continued)

Nature of R group at pH 7	Amino acid	Symbol	Structure at pH 7 (R group is shaded)
	Arginine	Arg	$H_3\overset{+}{N}-\underset{\underset{(CH_2)_3}{\vert}}{\overset{\overset{H}{\vert}}{C}}-\overset{\overset{O}{\Vert}}{C}-O^-$ with $(CH_2)_3-NH-C(NH_2)=\overset{+}{N}H_2$
	Histidine	His	$H_3\overset{+}{N}-\overset{\overset{H}{\vert}}{C}-\overset{\overset{O}{\Vert}}{C}-O^-$ with CH_2-imidazole ring
Imino acids, neutral	Proline	Pro	$H_2\overset{+}{N}-\overset{\overset{H}{\vert}}{C}-\overset{\overset{O}{\Vert}}{C}-O^-$ with pyrrolidine ring (CH_2, CH_2, CH_2)

acids have an R group that is always uncharged, regardless of the pH of the solution. *Acidic* and *basic* amino acids have R groups which are charged at neutral pH (i.e., the physiological pH of 7) [Sec. 2.2(c)]. The R groups of *aromatic* amino acids contain a five- or six-membered carbon ring; *aliphatic* amino acids have no such ring structure but have a straight or branched carbon chain. In *imino* acids, the nitrogen of the α-amino group has lost a hydrogen atom and, instead, is bonded to the third carbon of the R group.

In addition to the 18 amino acids listed in Table 2.2, proteins contain four amino acids derived from members of this set (Sec. 11.2). The four derived amino acids are glutamine (gln), asparagine (asn), hydroxyproline (hyp), and hydroxylysine (hyl).

It is not necessary to memorize the complete structures of all the amino acids. What *is* important is that you realize, first, that, with the exceptions of proline and hydroxyproline, all amino acids have an α-amino and an α-carboxyl group and, second, that amino acids differ because they have R groups with different chemical and physical properties.

(b) CHEMICAL REACTIONS OF AMINO ACIDS

The free amino and free carboxyl groups of amino acids take part in many chemical reactions. The linear or cyclic arrangements of —CH_2— groups found in the R groups are chemically quite inert. The biologically most important reaction of carboxyl and amino groups is the formation of amide bonds:

$$R-\underset{\text{Acid}}{\underset{\|}{\overset{O}{C}}-OH} + \underset{\text{Amine}}{H_2N-R'} \longrightarrow R-\underset{\text{Amide}}{\underset{\overset{|}{H}}{\overset{O}{\overset{\|}{C}}-N-R'}} + H_2O$$

In the special case where the amino and carboxyl groups both come from amino acids, the amide linkage is known as a *peptide bond*.

In the laboratory, it is often necessary to distinguish amino acids from other kinds of small molecules or to estimate the amount of an amino acid in a solution. Chemical reactions unique to the amino group are convenient for these purposes, since all amino acids contain an amino group, which is not present in several other classes of small molecules in cells such as sugars and fatty acids. An example of a reaction specific for amino groups is the ninhydrin reaction. When ninhydrin (a colorless organic compound) reacts with a molecule containing an amino group, a bright purple compound is produced. The intensity of the purple color is proportional to the concentration of amino groups in the solution, so this reaction can be used to estimate the amount of amino acid(s) in a solution.

The reaction between amino acids and fluoro-2,4-dinitrobenzene (FDNB) is another reaction of amino groups that is important in the determination of the structure of proteins (Chap. 3). FDNB reacts with the free amino group of an amino acid to produce a yellow compound, a dinitrophenyl amino acid (DNP-amino acid).

$$O_2N-\underset{\text{FDNB}}{\underset{}{\bigcirc}}\overset{NO_2}{-}F + H_2N-\underset{\underset{\text{Amino acid}}{R}}{\overset{H}{\overset{|}{C}}}-\overset{O}{\overset{\|}{C}}-OH \longrightarrow O_2N-\underset{\text{DNP-amino acid}}{\underset{}{\bigcirc}}\overset{NO_2}{-}N-\underset{\underset{}{H}\ \ R}{\overset{H}{\overset{|}{\underset{}{}}}-\overset{|}{C}-\overset{O}{\overset{\|}{C}}-OH} + HF$$

(c) CHARGE PROPERTIES OF AMINO ACIDS

The charge properties of amino acids are determined by the number of ionizable groups they contain. Free carboxyl and free amino groups are capable of losing and gaining hydrogen ions (H^+), thus altering the charge on the molecule. Whether or not this transfer of hydrogen ions occurs is a function of the hydrogen ion concentration in the solution surrounding the amino acid molecule. In a neutral solution (the physiological pH of 7), amino groups pick up an H^+ and become positively charged, while carboxyl groups lose an H^+ and are therefore negatively charged (see the diagram below). With the exception of proline, all amino acids contain a minimum of one free amino group and one free carboxyl group and have at least one positive and one negative charge at pH 7.

The charges on amino and carboxyl groups change as the pH of the solution is altered. In solutions of low pH, which contain a great excess of H^+ ions, the carboxyl group cannot ionize; it cannot give up its hydrogen ion to the solution. At low pH, therefore, the carboxyl group is uncharged. In solutions at high pH, where hydrogen ions are in low concentration, the carboxyl group can easily donate its H^+ to the surrounding solution, leaving a negative charge on the amino acid.

The amino group responds differently to changes in pH. At low pH, the amino group can easily attract a hydrogen ion from the excess of these ions in the surrounding solution, and this group becomes positively charged. At high pH, where the amino group can no longer accept an H^+, this part of the amino acid is uncharged. The pH dependence of the charges on carboxyl and amino groups is illustrated below.

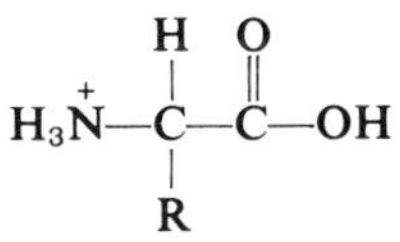

$$\underset{\substack{\text{In acidic solution}\\\text{(low pH)}}}{\overset{\overset{H}{|}\;\overset{O}{\|}}{\underset{\underset{R}{|}}{H_3\overset{+}{N}-C-C-OH}}} \qquad \underset{\substack{\text{In neutral solution}\\\text{(pH approx. 7)}}}{\overset{\overset{H}{|}\;\overset{O}{\|}}{\underset{\underset{R}{|}}{H_3\overset{+}{N}-C-C-O^-}}} \qquad \underset{\substack{\text{In basic solution}\\\text{(high pH)}}}{\overset{\overset{H}{|}\;\overset{O}{\|}}{\underset{\underset{R}{|}}{H_2N-C-C-O^-}}}$$

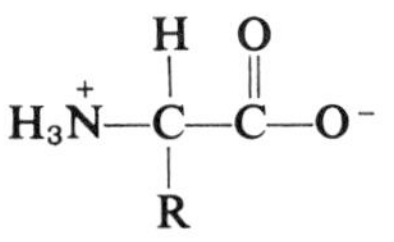

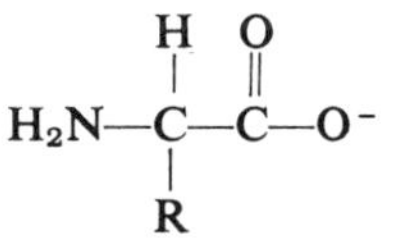

The charges on the α-amino and α-carboxyl groups are identical for all amino acids. But not all amino acids have the same *total* charge, because this depends on whether the R group of a particular amino acid can gain or lose hydrogen ions. The majority of amino acids contain R groups which are uncharged (neutral). It is principally five amino acids that have R groups which contribute to the total charge of a protein. Lysine and arginine have an amino group, histidine has an imidazole group, and glutamic acid and aspartic acid contain carboxyl groups. The amino and carboxyl groups on the R groups ionize in the same manner as the α-amino and α-carboxyl groups. The imidazole group, like the amino group, carries a $+1$ charge at low and neutral pH and is uncharged at high pH. The charge properties of lysine and aspartic acid in solution at different pHs are shown in Fig. 2.1.

The net charge on any amino acid at a given pH is the total of the positive and negative charges on the molecule (Table 2.3). In summary, at low pHs, amino acids bind the maximum number of hydrogen ions and are as positively charged as possible; at high pHs, amino acids bind the minimum number of hydrogen ions and are as negatively charged as possible.

Why is it important to learn about the charges of amino acids? First, in the laboratory, individual amino acids can be separated and identified on the basis of their charges at different pHs (Sec. 3.4). Second, the charges on the R groups of the amino acids play an important role in determining the three-dimensional shapes of proteins, since these charges limit the number of ways in which a chain of amino acids can fold over on itself (Sec. 2.4). And third, because of their ionizable R groups,

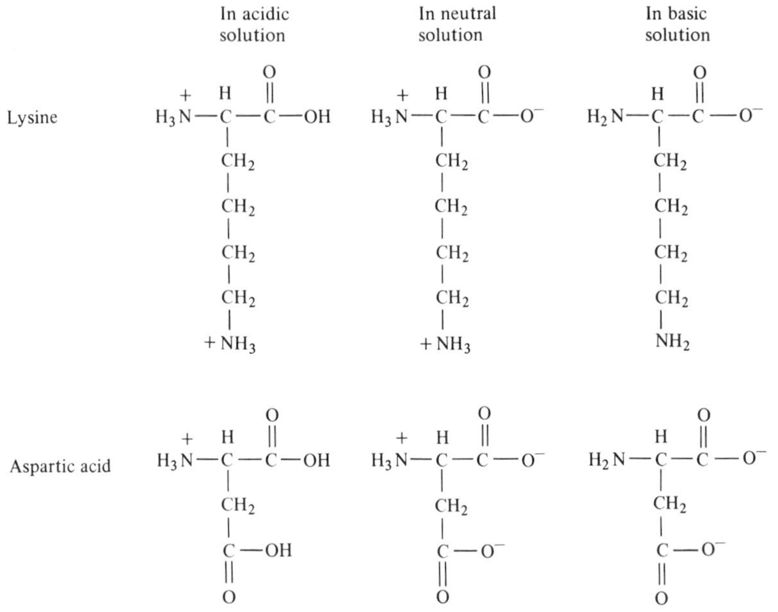

	In acidic solution	In neutral solution	In basic solution
Lysine	$\overset{+}{H_3N}-\underset{\underset{\underset{\underset{+NH_3}{|}}{\underset{CH_2}{|}}}{\underset{CH_2}{|}}}{\overset{H}{\underset{|}{C}}}-\overset{O}{\overset{\|}{C}}-OH$	$\overset{+}{H_3N}-\underset{\underset{\underset{\underset{+NH_3}{|}}{\underset{CH_2}{|}}}{\underset{CH_2}{|}}}{\overset{H}{\underset{|}{C}}}-\overset{O}{\overset{\|}{C}}-O^-$	$H_2N-\underset{\underset{\underset{\underset{NH_2}{|}}{\underset{CH_2}{|}}}{\underset{CH_2}{|}}}{\overset{H}{\underset{|}{C}}}-\overset{O}{\overset{\|}{C}}-O^-$
Aspartic acid	$\overset{+}{H_3N}-\underset{\underset{\underset{O}{\overset{\|}{C}-OH}}{\underset{CH_2}{|}}}{\overset{H}{\underset{|}{C}}}-\overset{O}{\overset{\|}{C}}-OH$	$\overset{+}{H_3N}-\underset{\underset{\underset{O}{\overset{\|}{C}-O^-}}{\underset{CH_2}{|}}}{\overset{H}{\underset{|}{C}}}-\overset{O}{\overset{\|}{C}}-O^-$	$H_2N-\underset{\underset{\underset{O}{\overset{\|}{C}-O^-}}{\underset{CH_2}{|}}}{\overset{H}{\underset{|}{C}}}-\overset{O}{\overset{\|}{C}}-O^-$

Figure 2.1 The pH dependence of the charge of lysine and aspartic acid.

TABLE 2.3 Net Charge on Amino Acids

	In acidic solution	In neutral solution	In basic solution
Amino acids with neutral R groups	+1	0	−1
Amino acids with —COOH in R groups	+1	−1	−2
Amino acids with —NH₂ in R groups	+2	+1	−1

proteins tend to absorb or release hydrogen ions as the pH of the solution bathing them changes, thus keeping the pH of the solution constant. The "buffering capacity" of hemoglobin and plasma proteins is important in keeping the blood pH at neutrality.

2.3 Each protein is a linear chain of amino acids held together by peptide bonds

Every protein consists of a unique linear sequence of amino acids joined together by peptide bonds. (This chain of amino acids is twisted and folded upon itself, giving the

Protein Structure and Function

protein molecule a complex three-dimensional structure.) Peptide bonds are formed between those parts of amino acids which are common to all amino acids: the α-carboxyl group of one amino acid is joined to the α-amino group of the next amino acid in the chain. Every time this "linking-up" process occurs, a water molecule is released. The R groups of the amino acids are *never* involved in the formation of peptide bonds; they remain free, extending out from the long, linear backbone of the chain. The formation of a peptide bond between two amino acid molecules is diagrammed below; R_1 and R_2 represent the R groups of any two different amino acids.

$$H_2N-\underset{R_1}{\underset{|}{C}}\underset{|}{\overset{H}{|}}-\underset{}{\overset{O}{\underset{||}{C}}}-OH + H_2N-\underset{R_2}{\underset{|}{C}}\underset{|}{\overset{H}{|}}-\underset{}{\overset{O}{\underset{||}{C}}}-OH \longrightarrow H_2N-\underset{R_1}{\underset{|}{C}}\underset{|}{\overset{H}{|}}-\underset{}{\overset{O}{\underset{||}{C}}}-N-\underset{R_2}{\underset{|}{C}}\underset{|}{\overset{H}{|}}-\underset{}{\overset{O}{\underset{||}{C}}}-OH + H_2O$$

α-Carboxyl group   α-Amino group   Peptide bond

Formation of a peptide bond

The α-carboxyl group of the R_2 amino acid can be linked to the α-amino group of a third amino acid, and so on. This "head-to-tail" joining of amino acids is repeated over and over until the linear chain contains the several hundred amino acids found in the average protein. Every protein, be it hemoglobin, ribonuclease, or insulin, has the same backbone of peptide bonds:

$$-N-\underset{R_1}{C}-C-N-\underset{R_2}{C}-C-N-\underset{R_3}{C}-C-N-\underset{R_4}{C}-C-N-\underset{R_5}{C}-C-N-\underset{R_6}{C}-C-$$

Backbone of peptide bonds

The formation of one possible peptide* containing glycine, lysine, aspartic acid, and alanine is shown in Fig. 2.2. It is clear that although alternative arrangements of these four amino acids, (e.g., alanine–aspartic acid–glycine–lysine) give a different structure in terms of the sequence of R groups, *the backbone of peptide bonds always remains the same.*

Peptide bonds are strong, covalent linkages which are not easily broken. They can be cleaved in enzymatic reactions, such as the action of pepsin in your stomach, or chemically, in the laboratory, by heating the protein with strong hydrochloric acid or alkali.

When amino acids are incorporated into a peptide chain, their α-amino and α-carboxyl groups are "tied up" in the peptide bonds. Such bonded groups are no longer chemically reactive. For example, they are no longer able to donate and accept hydrogen ions from the solution and they do not contribute to the charge on the protein. Each peptide chain—no matter how long it is—has only one free α-amino

---

* A peptide is a chain of amino acids joined by peptide bonds; the term can be used interchangeably with "protein" but usually refers to smaller molecules with only a few peptide bonds.

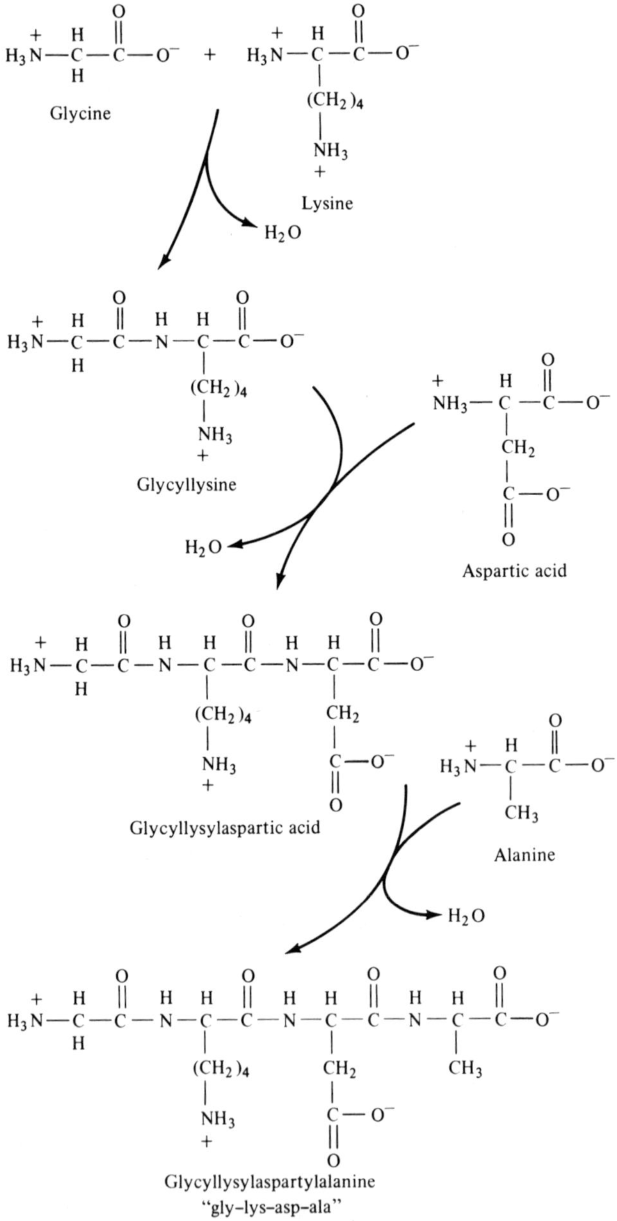

Figure 2.2 Formation of glycyllysylaspartylalanine.

group and one free α-carboxyl group: those belonging to the amino acids at the two ends of the macromolecule.

If every protein has the same backbone of peptide bonds, what makes proteins different from one another? Why is one protein suited to work as an enzyme while another serves as a hormone? Their unique properties come from the number of amino acids in each protein and from the order in which these amino acids occur in the peptide chain. For each specific protein, a fixed number of amino acids must be lined up in a particular sequence. Although every protein has the same backbone, the differences in the amino acid sequences mean that the nature and the order of the R groups projecting out from this backbone are peculiar to each protein. The R groups of the various amino acids have different physical and chemical properties: different sizes, different charges, different affinities for water, and so on. We shall see that the interactions between R groups, which become especially important as the peptide chain tries to fold upon itself, are defined by these physical and chemical properties. As an (nonbiological) example, try to imagine the differences between a protein containing 200 glycine molecules and one made up of 1000 lysine molecules. The "all-glycine" protein is much shorter and smaller, has only hydrogen atoms (as R groups) sticking out from the backbone, and, in solution at physiological pH, this protein has a net charge of zero. On the other hand, the "all-lysine" molecule has 1000 bulky R groups each carrying a $+1$ charge—and these positive charges tend to repel each other. Clearly, although both proteins have a backbone of peptide bonds, they are very different molecules.

The exact amino acid sequence of more than 100 proteins is known; determining such a sequence is an enormous task. The amino acid sequences of two proteins—insulin and ribonuclease—are shown in Fig. 2.3. These proteins have vastly different structures, which allow for their different cellular functions.

Because proteins are such large molecules, writing out every atom of their complete structure would be cumbersome and, therefore, a "shorthand" for writing amino acid sequences has been adopted. The peptide bond backbone, common to all proteins, is not drawn. Each amino acid is represented by its three-letter symbol (Table 2.2). The amino acid sequence of a protein is always written with the amino acid having a free α-amino group on the left end of the molecule; the free α-carboxyl group is at the right end.

Although proteins contain, at the most, only 22 different amino acids, the theoretical number of unique amino acid sequences that could be made from combinations of these amino acids is enormous. There are no known restrictions on the order of amino acids in proteins—any amino acid can be next to any other; nor are there limits on the number of copies of a particular amino acid in an individual protein. As a simple illustration of the possible number of different proteins, Fig. 2.2 shows only "one possible" peptide composed of alanine, lysine, aspartic acid, and glycine; if the sequence of amino acids is considered, there are 24 possible peptides containing just one molecule of each of these four amino acids. Biologically important proteins contain not four, but hundreds of amino acids. If all proteins contained 300 amino acids, there could be 22^{300} unique proteins. Of course, many sequences of

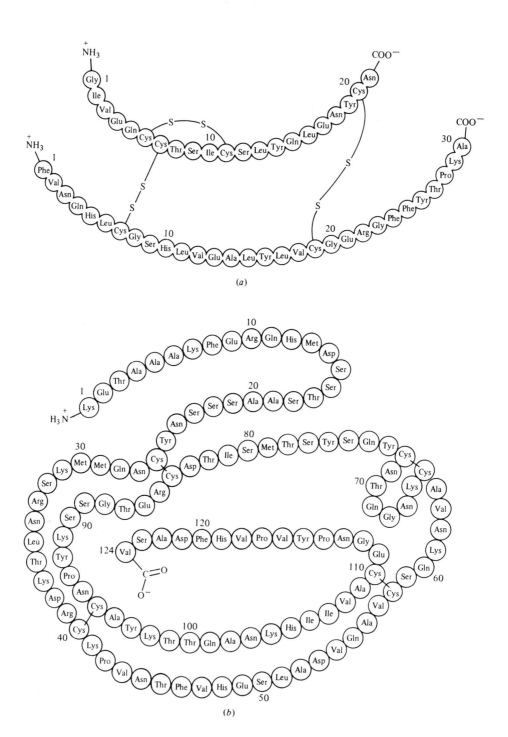

amino acids would have no biological activity, but this is not something that one can predict from the amino acid composition or even the amino acid sequence of a protein.

In order to have the physical and chemical properties necessary to perform a specific function, a protein must contain a specific sequence of amino acids. Obviously, cells do not line up amino acids randomly and then use only those sequences which result in a functional macromolecule. The hereditary material of the cell—the DNA—contains the information that tells the cell which amino acids to assemble, in what order, to make proteins that can serve particular metabolic functions. The mechanism by which the information in DNA is used by the cell to make proteins is discussed in detail in Chaps. 12 and 13.

2.4 Proteins have a preferred three-dimensional shape

We know that proteins are made up of amino acids joined together in a linear arrangement. But what shape does this "chain" have in the aqueous solution within a cell? Like string dropped into water, peptide chains do not remain rigidly stretched out in straight lines. Peptide chains bend and twist and fold over on top of themselves, attaining a complex three-dimensional shape. It is difficult to represent the three-dimensional nature of protein structure with a drawing. Try to imagine the arrangement of a 20-foot piece of string in an 8-ounce glass of water. Or, imagine that proteins resemble a long wire spring with the coils folded over on themselves. In both examples, the "backbone" is composed of only one material—string or wire—yet the possibilities for three-dimensional structure are almost infinite. In the case of proteins, the backbone is also homogeneous—one peptide bond after another—yet such macromolecules can be folded in a number of ways.

However unlike a string in water, proteins do not exist in random configurations. There are certain restrictions on how a polypeptide chain can fold and each specific protein has its own preferred three-dimensional structure. All hemoglobin molecules are folded in the same manner. Every molecule of the enzyme chymotrypsin has the same shape. Most important, it is only in this preferred configuration that a given protein can carry out its biological function. To carry oxygen, the peptide chain of hemoglobin *must* be folded in the preferred manner. The backbone of ribonuclease must attain a certain configuration for the macromolecule to bind to and break down nucleic acids. Although one can imagine that any polypeptide chain could fold in many ways, in fact, under physiological conditions, each active protein is folded up into its preferred, functional configuration (Figs. 2.4 through 2.6).

What factors determine the three-dimensional structure of a particular protein? The backbone of peptide bonds is common to all protein molecules, so this backbone could hardly be responsible for the huge variations in shape that are necessary if

---

Figure 2.3 Amino acid sequence of (*a*) porcine insulin and (*b*) bovine ribonuclease. [(*a*) After Ronald E. Chance, Robert M. Ellis, and William W. Bromer, *Science*, 161, p. 165 (1968). Copyright 1968 by the American Association for the Advancement of Science. (*b*) After Lubert Stryer, *Biochemistry*, W. H. Freeman and Company, San Francisco. Copyright 1975.]

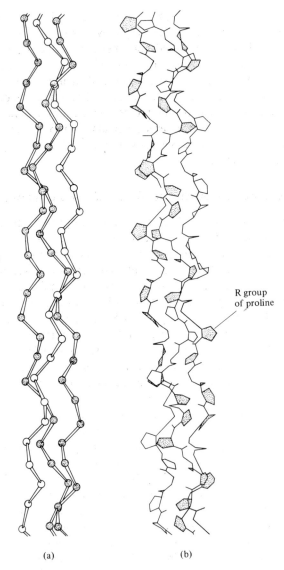

Figure 2.4 Three-dimensional structure of collagen. (*a*) A ball-and-stick model of the triple-stranded collagen helix. The balls indicate the positions of the α-carbon atoms. The strands are held together by hydrogen bonds between peptide bonds. (*b*) A skeletal model of the triple-stranded collagen helix. The repeating sequence here is -gly-pro-pro. The proline residues prevent the individual strands from folding into a more compact configuration. The proline R groups are on the outside; the glycine R groups are small enough to fit inside the coiled structure. (After Lubert Stryer, *Biochemistry*, W. H. Freeman and Company, San Francisco. Copyright 1975.)

proteins are to serve so many diverse functions. As we have pointed out before, the differences among proteins lie in the R groups which project out from the peptide backbone. It is largely the interactions of the R groups with each other and the interactions of R groups with water that determine the preferred three-dimensional shape of a protein.

There are a number of possible interactions between the R groups of different amino acids. Certain charged R groups are attracted to each other, while others repel each other. Some R groups are surrounded by water; others seek out a nonaqueous environment. Several R groups are large and bulky and prevent segments of the peptide backbone from coming in contact with each other. The combined effect of all these R-group interactions is to restrict the ways in which a particular protein molecule can be folded.

In general, the interactions between R groups involve the formation of weak, noncovalent bonds which have little strength compared to that of the peptide bond. However, hundreds of these weak interactions influence the three-dimensional shape of each protein, and their combined effect is to exert a considerable stabilizing force on the preferred three-dimensional shape of a protein. To make a simple analogy, each bond between two R groups is like a single thread in a piece of cloth: not very strong by itself, but many such threads together make a strong fabric.

(a) HYDROGEN BONDS AND THE α-HELIX

Before discussing interactions between R groups, it is important to consider the preferred configuration of the protein backbone—the string of peptide bonds. In other words, what is the shape of a protein with small, uncharged R groups that interact very little with each other, for example, polyalanine, a protein composed of a large number of residues of alanine? The answer is that such proteins coil up into a spring-shaped structure, an α-helix.

The α-helical structure is held together by hydrogen bonds between the —C=O group of one peptide bond and the —NH group of a second peptide bond. The oxygen of the —C=O group in a peptide bond has a slight negative charge and it tends to attract hydrogen ions, although it does not have the power to actually hold an H^+ and become charged. Conversely, the nitrogen of the —NH group of a peptide bond has a slight positive charge and tends to repel its hydrogen, although the nitrogen never really loses the hydrogen. Because of their tendencies to attract or repel hydrogen ions, there is a mutual attraction between —C=O and —NH groups. This attractive force is called a hydrogen bond:

$$\mathord{>}C\!\!=\!\!\overset{\delta-}{O}\text{---}\overset{\delta+}{H}\!\!-\!\!N\mathord{<}$$
Hydrogen bond

When a protein such as polyalanine with noninteracting R groups is placed in water, it arranges itself so as to maximize the number of hydrogen bonds between peptide bonds. Because of restrictions on how sharply a peptide bond can be bent,

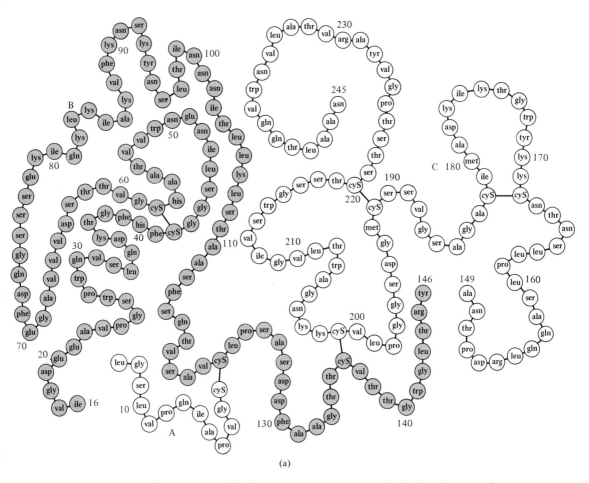

(a)

Figure 2.5 Structure of the digestive enzyme, chymotrypsin. Active chymotrypsin consists of three polypeptide chains: the A chain (amino acids 1–13), the B chain (amino acids 16–146) and the C chain (amino acids 149–245). The chains are held in position by hundreds of (individually) weak bonds between R groups: hydrogen bonds, ionic bonds, and hydrophobic bonds. As indicated in all three figures, the structure is also stabilized by four disulfide bridges. (a) The amino acid sequence of chymotrypsin. (b) Two models of the preferred three-dimensional shape of chymotrypsin—the tape model and the ball-and-stick model. In the latter the positions of the balls correspond to the positions of the α-carbons of the respective amino acids. Notice the α-helical region to the left on each model. Serine 195 and histidine 57 are part of the enzyme's active site (Chap. 5). [(a) After T. W. Goodwin, J. I. Harris, and B. S. Hartley (eds.), *The Structure and Activity of Enzymes*, Academic Press Ltd., London, 1964, p. 52. (b) After B. W. Matthews et al., *Nature*, 214, p. 652 (1967); after Robert M. Stroud, *Sci. American*, 231, p. 80 (1974).]

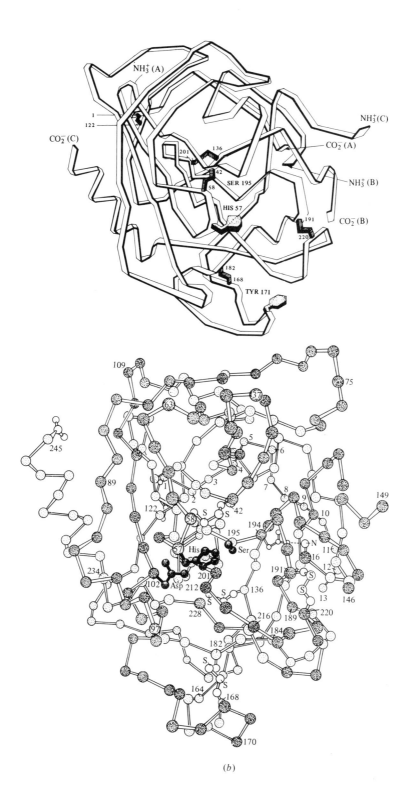

(b)

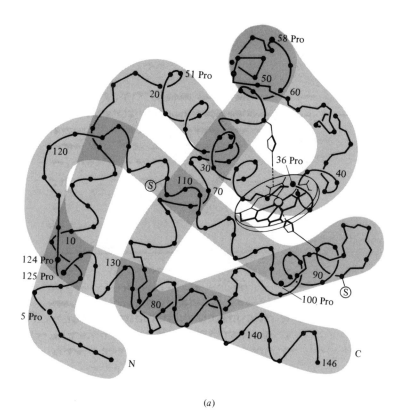

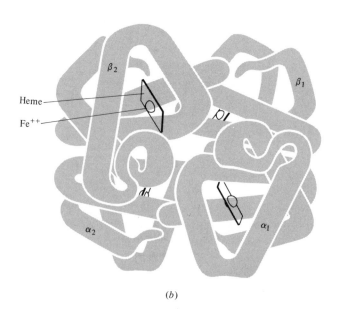

hydrogen bonds cannot form between adjacent peptide bonds. The closest possible intrachain hydrogen bonds are those linking the —NH of one peptide bond to the —C=O of the peptide bond three amino acids down the chain.

For a protein to contain the maximum number of hydrogen bonds, the —C=O and the —NH of each peptide bond must participate in (different) hydrogen bonds. When these groups are all bonded as shown in Fig. 2.7, the protein takes on the spring-shaped configuration of an α-helix. The hydrogen bonds run more or less parallel to the long axis of the helix, with each peptide bond involved in one hydrogen bond to the coil "above" it and one hydrogen bond to the coil "below" it. The R groups of the amino acids extend out from the α-helix; they are never in the center of the coil.

However, proteins in cells are more complicated than polyalanine; they contain all or nearly all of the amino acids, and as a consequence only certain segments of their peptide backbones are in a helical configuration. One reason is that imino acids, such as proline (Table 2.2), disrupt the helix because peptide bonds involving these imino acids cannot be bent sharply enough to fit into the tightly coiled α-helix. If our simple protein contained one proline molecule in the middle of the chain of alanine molecules, the protein would have the shape of a spring with one bent coil:

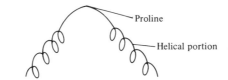

Proline destroys the α–helical configuration

This situation exists in hemoglobin; its preferred three-dimensional shape and therefore its ability to transport oxygen depends in part upon a specific proline residue that separates two helical portions of the peptide backbone. An extreme example of the effect of imino acids on the three-dimensional shape of a protein is found in the structural protein collagen; collagen contains a high proportion of glycine (about one-third), but also contains such a high proportion of proline (and hydroxyproline) that no segments of the polypeptide can form a normal α-helical structure (Fig. 2.4).

Figure 2.6 Three-dimensional structure of hemoglobin. (*a*) Three-dimensional model of the β-chain of hemoglobin. The solid line represents the twisting of the backbone of peptide bonds; some segments have the α-helical shape. The α-carbon atom of each amino acid residue is indicated by a solid circle. Notice that proline residues never occur in helical portions of the backbone. The disc on the right-hand side of the molecule represents the heme group, the organic cofactor in hemoglobin. The ball in the center of the heme group is an iron atom. (*b*) Active hemoglobin is a multimeric protein made up of 2α and 2β chains. Each chain carries a heme group. [(*a*) After M. F. Perutz, *The Hemoglobin Molecule.* Copyright November 1964 by Scientific American, Inc. All rights reserved. (*b*) After Richard E. Dickerson and Irving Geis, *The Structure and Action of Proteins,* W. A. Benjamin, Inc., Menlo Park, Calif., 1969.]

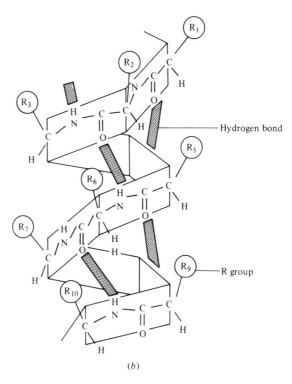

Figure 2.7 An α-helix. (*a*) Proteins are held in an α-helix by many hydrogen bonds between peptide bonds. Hydrogen bonding links the —NH of one peptide bond with the —C=O of the peptide bond three amino acids down the chain. (*b*) When all possible hydrogen bonding has occurred, the backbone of peptide bonds is wound into an α-helix. (After Robert Barker, *Organic Chemistry of Biological Compounds*, Prentice-Hall, Inc., Englewood Cliffs, N.J., 1971, p. 101.)

The helical configuration of the peptide backbone can also be distorted by the attractions and repulsions between R groups other than that of proline and by the interactions between R groups and water.

(b) THE DISULFIDE BRIDGE

The disulfide bridge is the one exception to the rule that the three-dimensional structure of a protein is stabilized by weak bonds. A disulfide bridge is a strong, covalent bond

Protein Structure and Function

formed between the sulfur atoms of the R groups of two cysteine molecules (Table 2.2). Since the participating cysteine molecules can be in any position along the polypeptide chain, such bridges can pull distant segments of the protein together. Several such bonds in a protein may introduce many folds into the structure.

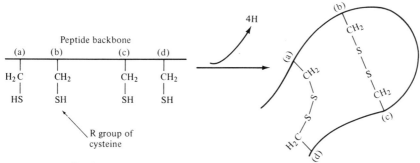

Disulfide bridges link distant segments of a protein

(c) INTERACTIONS BETWEEN CHARGED R GROUPS

Everyone knows that like charges repel each other and unlike charges attract. Because certain amino acids have R groups that are charged at physiological pH (Table 2.2), charge interactions are important factors in the stabilization of the preferred three-dimensional shape of a protein. R groups containing the negatively charged —COO$^-$ are attracted to R groups containing the positively charged —NH$_3^+$ group; this attraction is called an ionic (or salt) bond. For example, the negatively charged R group of glutamic acid can be attracted to the positively charged R group of lysine; this type of interaction can also link distant parts of a protein chain.

On the other hand, R groups carrying the same charge repel each other; glutamic acid R groups repel each other as well as repelling the negatively charged aspartic

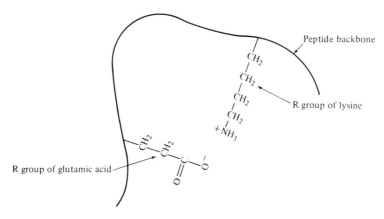

Ionic bonding between the R groups of glutamic acid and lysine

acid residues. In this manner, charged R groups, extending out from the polypeptide backbone, determine which segments of the chain can be in contact with each other and serve to stabilize contacts between different parts of the polypeptide chain.

(d) HYDROPHOBIC AND HYDROPHILIC INTERACTIONS

Hydrophilic and hydrophobic bonds refer to the interactions between the R groups of the various amino acids and water. "Hydro," of course, refers to water and "-philic" or "-phobic" refers to whether or not the particular R group likes being surrounded by water or prefers a nonaqueous environment. R groups that are charged or contain short carbon chains are hydrophilic—they try to surround themselves with water. Hydrophobic R groups—those which try to avoid water—include those containing an aromatic ring and those with many carbon atoms.

How do these interactions with water affect the three-dimensional structure of a protein? In the aqueous environment of a cell, the R groups projecting out from the polypeptide backbone arrange themselves according to their like or dislike of water. The hydrophilic R groups prefer to be on the outside of the protein, where they are completely surrounded by water. Hydrophobic R groups prefer to be next to each other rather than to be surrounded by water; by clustering tightly together, they form their own private, nonaqueous environment. The result of these interactions with water is that the polypeptide—whether a straight chain or in an α-helix—tends to form a sphere with the hydrophilic R groups sticking out from the sphere into the water, and the hydrophobic R groups huddled together in the center of the sphere. This arrangement of the R groups pulls certain segments of the polypeptide chain together while keeping other segments apart.

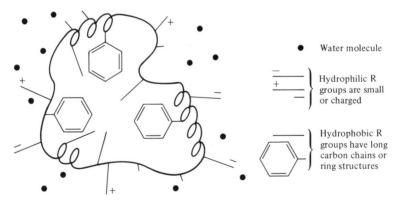

Cross-section of a protein in water

(e) INTERACTIONS BETWEEN POLYPEPTIDE CHAINS

Many biologically active proteins contain more than one polypeptide chain; these are called *multimeric proteins*. Their functionality depends upon each separate chain assuming a particular three-dimensional shape and then being arranged next to the other polypeptides in the proper relationship. Some multimeric proteins contain multiple copies of the same polypeptide chain; in others, the component chains have entirely different amino acid compositions. The interactions that hold separate chains together are the same interactions that stabilize the three-dimensional structure of a single-chain protein—only in this case the bonds link separate polypeptide chains as well as the segments of the individual macromolecules.

Multimeric proteins are found within all functional classes of proteins. Disulfide bridges link the two different polypeptide chains of insulin (Fig. 2.3). The four polypeptide chains of hemoglobin (two α and two β chains) are held together by hydrogen bonds, hydrophobic bonds and charge interactions between R groups. Immunoglobulins are composed of two "heavy" and two "light" polypeptide chains held in the preferred, active shape by several disulfide bridges.

Most structural proteins are multimeric and interchain bonding is essential for their activity. As mentioned previously, collagen molecules, because of their high content of proline and hydroxyproline, cannot form an α-helix and collagen molecules are nearly linear (Fig. 2.4). Several of these linear molecules line up side by side and are held in this position by hydrogen bonds, thus creating a fiber. Bundles of these fibers provide the rigid and stable structural matrix needed for bones and tendons.

2.5 Some proteins require metal ions or small organic molecules for biological activity

Many proteins require "helper" molecules to carry out their biological function. Often, what we call an active protein is, in reality, a complex between a polypeptide chain and a small, nonprotein molecule, called a *cofactor*. Alone, neither component of such a complex is biologically active; only together can they serve their metabolic function. Cofactors fall into two chemical groups: (a) metal ions and (b) small organic molecules. The cofactor is usually embedded in the three-dimensional "web" of the protein molecule. In most complexes, the small molecule is held in place by weak bonds, such as ionic bonds, between it and one or more of the R groups projecting out from the polypeptide chain. In some complexes, the cofactor is covalently linked to an R group.

As might be expected from the differences in their chemistry, the cofactors in different protein–cofactor complexes serve vastly different functions. The cofactor may influence the overall shape of the protein. In the case of enzymes or transport proteins, the cofactor may be important in binding or positioning the substrate (the molecule being acted upon) on the protein so that catalysis or transport can occur. In certain enzymes, the cofactor actually participates in the catalytic reaction as an acceptor or donor of electrons, hydrogen ions, or small organic groups, such as methyl groups or amino groups. In some cases, we simply do not understand why a particular cofactor is necessary.

(a) METAL ION COFACTORS

Superficially, there is little relationship between the function of a protein and the particular metal ion required for its activity. There is some correlation between the function of the protein and the charge on the ion; proteins with similar functions often require ions with the same charge. The oxygen-carrying protein in invertebrates, hemocyanin, requires copper ions, while hemoglobin, which serves a similar function in vertebrates, uses iron (Fig. 2.6). Like hemoglobin, the cytochromes, important in the energy conversions of mitochondria, need iron to function; in this case, the iron is an actual participant in the enzymatic reactions (Sec. 8.4). Carboxypeptidase, a digestive enzyme, and insulin, the hormone that regulates glucose metabolism, both contain zinc. The list of metal ions that serve as protein cofactors also includes magnesium, manganese, cobalt, and molybdenum. Does this list sound familiar? Because cells need metal ions to produce biologically active proteins and because cells must acquire these ions from their environment, the metal ions which act as protein cofactors make up a large portion of the trace elements which are essential for the maintenance of living cells (Table 1.5).

(b) SMALL ORGANIC COFACTORS

Small organic molecules often serve in protein–cofactor complexes which have enzymatic or transport activity. (Chapter 4 describes enzymes.) The small molecules are usually direct participants in the reaction being carried out, although, without the protein, they have no biological activity. Because they are direct participants in reactions, such molecules are often associated with several different proteins that carry out chemically similar reactions. For example, one such cofactor, nicotinamide adenine dinucleotide (NAD^+), is capable of accepting and donating hydrogen atoms and is a cofactor for enzymes that catalyze the transfer of hydrogen from one compound to another; a wide variety of enzymes using NAD^+ are essential for the processes by which cells transform the energy in foods into the forms used for cell growth and function (Chaps. 7 and 8). For future reference, the major organic cofactors and the types of reactions in which they function are listed in Table 2.4. The roles of these

TABLE 2.4 Organic Cofactors

Cofactor	Vitamin precursor	Type of reaction in which the cofactor functions
Thiamine pyrophosphate	Thiamine (B_1)	Decarboxylation
Flavin adenine dinucleotide (FAD)	Riboflavin (B_2)	Hydrogen transfer
Flavin mononucleotide (FMN)	Riboflavin (B_2)	Hydrogen transfer
Nicotinamide adenine dinucleotide (NAD^+)	Niacin	Hydrogen transfer
Pyridoxal phosphate	Pyridoxal (B_6)	Amino-group transfer
Biotin	Biotin	Carbon dioxide incorporation
5′-Deoxyadenosyl cobalamin	Cobalamin (B_{12})	Methyl-group transfer
Ascorbic acid	Ascorbic acid (C)	Hydroxylation
Heme	—	Oxygen transport; electron transport

Protein Structure and Function

and other small organic cofactors will be discussed in greater detail when particular enzymatic reactions are described.

Most of the small organic cofactors are vitamins (Table 2.4) or are made from vitamins by a very few chemical reactions within cells (i.e., the vitamin is the precursor of the cofactor). In other words, certain organisms, including human beings, are unable to produce the organic cofactors and must acquire these molecules as part of their diets.

2.6 Destruction of the three-dimensional shape of a protein drastically alters its physical properties and reduces its ability to carry out its biological function

We have emphasized that a protein must be in its preferred three-dimensional shape in order to carry out its biological function. Evidence for this relationship comes from studying what happens to an active protein when the weak bonds responsible for stabilizing its three-dimensional structure are destroyed while the strong peptide bonds are left intact. When their three-dimensional shape is destroyed, proteins lose the ability to carry out their normal biological functions; enzymes no longer act as catalysts, protein hormones no longer regulate metabolism, and membrane proteins no longer serve as permeability barriers. Destruction of the preferred three-dimensional shape of a protein, with the concomitant loss of biological activity, is called *denaturation.*

A wide variety of chemical reagents and physical conditions lead to protein denaturation: heating a protein in boiling water for a short time, placing a protein in a weakly acid solution, treating the protein with organic solvents, such as ethanol, whipping a protein in solution or adding detergents. These are not particularly drastic treatments—none will break peptide bonds—yet all lead to the destruction of the weak bonds stabilizing the overall configuration of a protein, and all lead to a loss of biological activity.

The correlation between protein denaturation and loss of biological activity is clearly illustrated by the consequences of heat denaturation on the toxicity of bacterial toxins. Most toxins lose their ability to cause disease when they are exposed to heat treatment. When heated to 80°C for a few minutes, the toxic protein excreted by *Clostridium botulinum* no longer causes fatal food poisoning in man. (For safety's sake, if you *have* to eat them, it is recommended that any suspected foods be heated to a full boil for 15 minutes before eating.) Heating also inactivates the bacterial toxins that cause diphtheria and tetanus; such heat denaturation is the basis for the preparation of vaccines against these particular diseases. Although incapable of producing disease, the denatured toxins retain their ability to induce active immunity in man (Chap. 15).

What occurs during denaturation? How do denaturing agents work? Although physicochemical studies clearly show that protein denaturation involves a transition from an ordered to a disordered state, the actual steps in this process are not well understood. Denaturing conditions may cause any or all of the following: breakage of hydrogen bonds, replacement of water molecules around the protein by a new solvent, alterations in the charges on the hydrophilic R groups on the outside of the molecule. Certain specific chemicals cause the breakage of disulfide bridges. Denaturation is difficult to study because it involves a rapid snowballing effect: once a small percentage of weak bonds are broken, the remaining weak bonds loosen up and the entire three-dimensional structure quickly collapses. In a short time, the structured protein completely unravels and R groups and segments of the peptide backbone that were previously in the interior hydrophobic region of the protein become exposed to the solvent. As their preferred configuration collapses, the protein molecules take on different, essentially random three-dimensional shapes.

Denaturation has other effects on proteins in addition to the destruction of their biological activity. Proteins that are very water-soluble in their preferred three-dimensional configuration often become essentially insoluble in water when their preferred configuration is lost. During denaturation, as their three-dimensional shape becomes more and more disorganized, many proteins coagulate and precipitate out of solution. This aspect of the denaturation process is familiar to anyone who is an active cook. Eggs become hardboiled because heating for 15 minutes or so at 100°C denatures the egg albumin, which becomes insoluble in the water of the egg white. Meringue shells are prepared by vigorously whipping egg whites until the proteins are denatured and "stiff." Sour milk may be prepared by adding a teaspoon of vinegar ($\sim 5\%$ acetic acid) to a cup of milk and letting the mixture stand at room temperature for a few minutes; this denatures one group of milk proteins, collectively called casein, which then precipitates—the milk curdles. Other "soured" dairy products are prepared, not by adding acid directly, but by growing acid-excreting bacteria—the "starter"—on different milk products. For example, yoghurt is prepared by growing lactic acid-producing bacteria in skim milk; as the concentration of acid increases, the casein is denatured and coagulates, giving yoghurt its semisolid form.

While all proteins can be denatured, specific proteins differ in their susceptibility to various denaturing conditions and reagents. This is to be expected since each individual protein depends on a unique number and combination of the possible weak bonds to stabilize its three-dimensional shape. For example, while the bacterial toxin causing botulism is totally destroyed by heating at 100°C for 15 minutes, at the same temperature it takes over 1 hour to inactivate the toxin from *Staphlococcus aureus*, which causes a less serious form of food poisoning.

A specific protein can often be isolated from a mixture of proteins by taking advantage of its individual denaturation characteristics. The commercial preparation of cheeses containing different milk proteins is an application of such selective denaturation. The acidification of milk by the combined action of lactic acid and the enzyme rennin denatures casein, which coagulates to produce the curd, which is the major ingredient of cottage cheese. The remaining casein-free solution—the whey—still contains a variety of acid-stable proteins which can be denatured by heating the acidic whey to 185°C. The coagulated whey proteins are collected, processed, and sold as Ricotta cheese, familiar to most of us as a basic ingredient of lasagna and other Italian delicacies.

Disulfide bridges are essential in stabilizing the three-dimensional shape of many proteins, including insulin (Fig. 2.3) and ribonuclease (Fig. 2.3). Breaking these covalent bonds unlinks segments of the polypeptide backbone and starts the denaturation process in motion. Disulfide bridges can be split by the chemical addition of a hydrogen atom (reduction) to each sulfur atom, thus releasing the R groups of the linked cysteine molecules. Mercaptoethanol is one chemical that can provide the hydrogen atoms necessary to break disulfide bridges. (Chemical reactions in which hydrogen *atoms* are covalently added to other atoms are called reduction reactions; only certain chemicals—*reducing agents*—are able to donate these hydrogen atoms. Do not confuse reduction with acidification, which involves the exchange of hydrogen *ions*.)

When ribonuclease (Fig. 2.3) is treated with mercaptoethanol in urea, its four disulfide bridges are broken and the protein is denatured. With its three-dimensional structure destroyed, the enzyme loses its ability to catalyze the breakdown of nucleic acids.

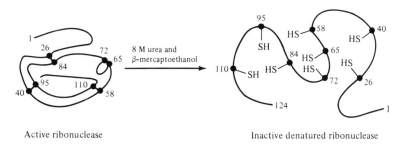

Active ribonuclease   Inactive denatured ribonuclease

(After Lubert Stryer, *Biochemistry*, W. H. Freeman and Company, San Francisco. Copyright 1975.)

In most instances, the denaturation of a protein is an irreversible process. There is no way to "un-hardboil" an egg or to "de-sour" milk. In a few specific cases, however, it is possible to get a denatured protein to return to its preferred three-dimensional shape and to regain its biological activity. This can be done for ribonuclease which has been denatured with mercaptoethanol; when a reagent is added that causes the disulfide bridges to form again, the protein takes on its preferred three-dimensional shape and regains enzymatic activity.

(a) THE DENATURATION OF MULTIMERIC PROTEINS

For those proteins whose active form consists of more than one polypeptide chain (multimeric proteins), denaturation may be thought of as a two-step process: first, the individual polypeptide chains are separated from each other and second, the three-dimensional shape of each individual protein is destroyed. Because the interchain linkages of a multimeric protein involve the same kinds of weak bonds found between segments of a single polypeptide chain, those denaturing agents which destroy the three-dimensional shape of a single chain protein also separate the individual polypeptide chains of a multimeric protein. For example, when hemoglobin (Fig. 2.6) is placed in a solution at an acid pH, the molecule dissociates completely into two α and two β chains, and then each individual chain loses its preferred three-dimensional shape.

Intermolecular disulfide bridges are essential for the integrity of many multimeric proteins, including keratin fibers, the structural units of hair and wool. The proteins found in keratin fibers are wrapped around each other to form "ropes" or fibers which are held in place by disulfide bridges between adjacent protein molecules. The destruction and re-forming of the intermolecular disulfide bridges of keratin is the basis for the process used to curl and straighten hair. If you have a "permanent," your hair is first treated with a reducing agent (that is what smells!), which destroys some of the intermolecular disulfide bridges. The hair is then arranged in the desired style and treated with a second chemical, an oxidizing agent, which promotes the formation of new disulfide bridges, locking the protein molecules into new positions. In other words, the newly formed bonds stabilize a realignment of the protein molecules, thus changing the shape of the keratin fibers and, in fact, of the hair itself.

The large keratin fibers, containing many proteins held together by disulfide bridges, are insoluble in water. For this reason, the digestive enzymes of most animals cannot degrade keratin fibers, and hair and wool cannot be used as a protein source. The clothes moth is unique in that it can attack and "eat" woolen fabrics; its gastric juices contain high concentrations of chemicals capable of breaking disulfide bridges, thereby releasing individual protein molecules from the keratin fibers. Once separated, the individual molecules are more soluble and susceptible to breakdown by digestive enzymes, thus providing the clothes moth with a source of amino acids.

SUMMARY

Proteins are a large group of functionally diverse macromolecules. Some proteins act as enzymes, others as hormones, and still others serve a structural or transport role. Each cell produces a characteristic mixture of proteins; differences in protein content explain most of the differences between cells.

Proteins are long chains of amino acids. All amino acids have the same general structure: $H_2N-\underset{R}{\overset{H}{\underset{|}{\overset{|}{C}}}}-COOH$. The differences between amino acids lie in the nature (size, shape, chemical reactivity, and charge) of their R groups. Peptide bonds can be formed between the amino group of one amino acid and the carboxyl group of a second amino acid; such "head-to-tail" joining of several hundred amino acids in a specific sequence produces the linear chain of amino acids characteristic of a given protein.

Protein Structure and Function

In metabolically active proteins, the linear amino acid chain is bent and folded upon itself to produce a complicated three-dimensional structure. This three-dimensional structure is stabilized, primarily, by weak interactions between amino acid R groups. For each protein, there is one preferred three-dimensional shape which must be maintained to ensure its biological activity; the denaturation of a protein destroys its ability to function in cell metabolism. In addition to having the proper configuration, some proteins require metal ions or organic cofactors for biological activity.

PRACTICE PROBLEMS

The structural formulas of the amino acids, given in Table 2.2, may be used to answer the following questions.

1. Complete the following chart:

	N-terminal	C-terminal	Net charge pH 1	pH 7	pH 12
Ala	—	—			
Glu	—	—			
Glu-ala-leu					
Ser-gly-met-phe					
Lys-lys-arg-val					
Tyr-lys-glu-asp					

2. Draw structures of peptides with the following properties:
   (a) A four-membered peptide with a net charge of +1 at pH 7.
   (b) Two different three-membered peptides with a net charge of −1 at pH 12 (basic pH).
   (c) A sulfur containing four-membered peptide that has a net charge of zero at pH 1 (acidic pH).
   Name the peptides using the three-letter symbols for the amino acids (Table 2.2).

3. One segment of hemoglobin has the amino acid sequence shown below. In the active protein, which of the R groups in the segment would you expect to find on the exterior of the hemoglobin molecule? Which on the interior? Would the entire segment be wound up in an α-helix?
   -ala-leu-glu-arg-met-phe-leu-ser-pro-thr-thr-lys-

4. Cofactor–protein complexes are held together by weak ionic bonds between the cofactor and the R groups on the protein. Which R groups might be responsible for binding positively charged metal ion cofactors such as Fe^+ or Mg^{2+}? Which R groups might bind pyridoxal phosphate, the negatively charged active form of vitamin B_6?

5. Suggest explanations for the following.
   (a) Insulin is given intraveneously rather than orally.
   (b) Enzyme detergents do not work in boiling water.
   (c) Cold water is recommended for removing blood stains from clothing.
   (d) Collagen fibers are unaffected by mercaptoethanol.

SUGGESTED READING

DICKERSON, RICHARD E., "The Structure and History of an Ancient Protein," *Sci. American*, 226, No. 4, p. 58 (1972). Good illustrations of the amino acid sequence and three-dimensional structure of cytochrome *c*, which carries electrons.

DICKERSON, RICHARD E., and IRVING GEIS, *The Structure and Action of Proteins*, Harper & Row, Inc., New York, 1969. This book comes with a stereo supplement and a viewer, which makes the drawings take on the biologically important "depth" dimension.

DOTY, PAUL, "Proteins," *Sci. American*, 197, No. 3, p. 173 (1957). Early studies on the α-helix and other aspects of the three-dimensional shape of proteins.

KADIS, SOLOMON, THOMAS C. MONTIE, and SAMUEL J. AJL, "Plague Toxin," *Sci. American*, 220, No. 3, p. 92 (1969). A protein produced by a bacillus is lethal to mammals.

NOMURA, MASAYASU, "Ribosomes," *Sci. American*, 221, No. 4, p. 28 (1969). These organelles are constructed from proteins and ribonucleic acid.

PERUTZ, M. F., "The Hemoglobin Molecule," *Sci. American*, 211, No. 5, p. 64 (1964). Describes the three-dimensional shape of this transport protein.

PORTER, R. R., "The Structure of Antibodies," *Sci. American*, 217, No. 4, p. 81 (1967). These proteins of our immune system inactivate foreign substances.

SHARON, NATHAN, "Glycoproteins," *Sci. American*, 230, No. 5, p. 78 (1974). Some proteins have sugars linked to them.

See also the references to Chap. 4.

3
Determining Protein Structure

Questions concerning macromolecular structure are clear and straightforward. "What is the three-dimensional structure of chymotrypsin?" "What is the structure of the DNA in the gene for insulin?" "How are the monomers arranged in cellulose?" "What is the relationship between the sequence of amino acids in an enzyme and its activity as a specific catalyst?" But getting the answers is not simple. Even when appropriate analytical techniques are available, many man-years of painstaking research are needed to characterize a single protein, nucleic acid, or polysaccharide.

We cannot describe all the procedures used to analyze macromolecules. To acquaint you with the scope of such studies, the analysis of protein structure will be presented in some detail. Most of the principles and many of the techniques are adaptable to the analysis of other kinds of macromolecules.

3.1 The relationship between protein structure and function

What is to be learned by determining the *structure* of a protein? By the mid-1930s, the features common to all proteins were understood: the structures of the amino acids were known and the role of the peptide bond in joining them together was recognized. However, this information did not explain the differences between proteins or the relationship between these differences and protein function. These questions could be answered only by analyzing and comparing the amino acid sequences and three-dimensional shapes of particular protein molecules.

First, there is the question of how the amino acid sequence of a protein is related to its three-dimensional shape. Although there are many possibilities as to how the amino acid sequences of proteins might influence their overall shapes, only detailed analyses of the shapes of individual proteins can tell us which of the possible interactions between R groups do, in fact, govern three-dimensional shape. The details of protein structure discussed in Chap. 2 were verified only after the three-dimensional configurations of several proteins had been determined.

The second, and perhaps more important, question is how each protein is adapted to serve its particular biological function. How can proteins, which are all chains of amino acids, be different enough to function in such diverse roles as antibodies, hormones, enzymes, or structural proteins? What differences between ribonuclease and hemoglobin give the former its enzymatic activity while making the latter a transport protein? The answers must lie in the differences in amino acid sequences and three-dimensional shapes.

A third reason for the interest in the details of protein structure is to find out why some proteins function while others, possessing small alterations, do not. Many metabolic diseases are a result of the individual producing a nonfunctional protein in place of the normal molecule. For example, people with sickle-cell anemia produce a hemoglobin that is incapable of carrying oxygen under certain conditions, although it differs from normal hemoglobin by only one amino acid. What is "wrong" with sickle-cell hemoglobin? Detailed knowledge of the amino acid sequences and three-dimensional shapes of the protein molecules is necessary to explain such phenomena. In the future, such knowledge may lead to therapeutic advances.

Another reason for determining (at least) the amino acid sequence of a particular protein is that it is the obvious first step toward the chemical synthesis of that molecule. At present, proteins for therapeutic use are obtained by laboriously extracting the proteins from the appropriate organs of cows, pigs, or horses. Chemical synthesis may be one of the ways to obtain large quantities of those proteins which are most difficult to isolate, including many hormones that are found in only minute amounts in any tissue. Back-to-nature enthusiasts notwithstanding, chemically synthesized compounds can often be safer, purer, and cheaper than those obtained from "natural" sources; they are identical to the natural (organic) product.

3.2 Determining the complete structure of a protein involves four basic steps

1. Purification of the protein; that is, the protein must be freed from other macromolecules, particularly other proteins.
2. Determination of the total amino acid composition of the protein; that is, how many molecules of each amino acid does this protein contain?

Determining Protein Structure 47

    3. Determination of the linear sequence of the amino acids in the protein, that is, the order of amino acids.

    4. Determination of the way in which the amino acid chain is folded when the protein is in its preferred three-dimensional shape.

These steps apply to the determination of the structure of any macromolecule. Just substitute the name of the particular macromolecule and its monomer for "protein" and "amino acid."

3.3 To study the structure of a particular protein, that protein must be freed from all the other components of the cell

It is almost impossible to learn anything about the structure of a particular protein when it is part of the normal cellular mixture of lipids, sugars, nucleic acids, other proteins, and so on. One must first obtain a solution containing only the protein to be studied.

    Each step in the purification of a protein essentially involves eliminating a particular set of undesirable, interfering molecules from the total mixture of cellular chemicals. Small molecules, then nonprotein macromolecules, and finally, unwanted proteins are removed. In addition, any procedure used for isolating an active protein must be physically gentle and chemically mild. In Chap. 2 we saw that proteins are easily denatured to inactive forms, which is clearly undesirable if one wishes to study the relationship between protein structure and function.

    For purification to proceed, there must be a method for identifying the desired protein in the presence of many other molecules. After each purification step, it is necessary to determine, rapidly and reliably, that the desired protein is present and active in the solution being saved for use in the succeeding purification step. The test for a particular protein (called an *assay*) usually involves a measurement of the unique biological activity of that protein. A solution can be assayed for a hormone by examining its effects on animals in specific metabolic states; for example, a solution can be assayed for insulin by testing for its ability to relieve diabetes in rats. Antibodies can be detected by their ability to bind specific proteins. Enzyme assays are described in Chap. 4.

    A single cell often contains small amounts of more than 1000 different proteins. An average protein in liver cells makes up less than one-half of 1% of the dry weight of the cells, so purification means getting rid of 99.5% of the cellular material without accidently discarding or destroying the desired protein. Because they are easier to purify, many of the best-studied proteins are those which are made in unusually large amounts by certain very specialized cells. Hemoglobin is a case in point: this macromolecule accounts for 35% of the dry weight of red blood cells, which greatly facilitates its purification. Similarly, chymotrypsin was a logical choice for analysis; since it is secreted from the cells that produce it, its purification required only the elimination of other extracellular material and not the removal of the large array of compounds found inside cells.

The first step in protein purification is to break open the appropriate cells (liver, brain, bacteria, or whatever) by grinding in a mortar and pestle or by homogenizing in a high-speed blender. This produces what is called a *cell-free extract*, a slurry containing the dissolved chemicals of the cytosol and the intact cellular organelles.

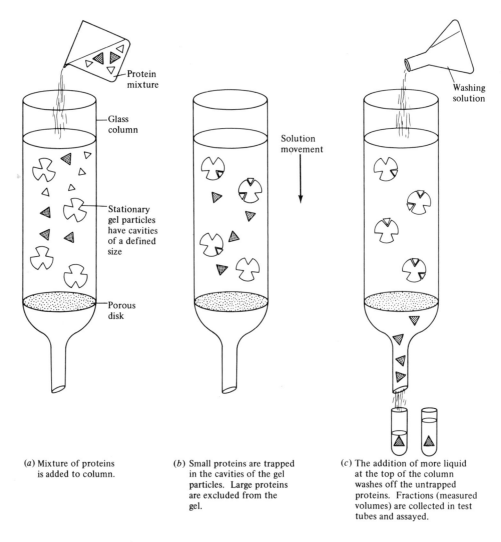

(a) Mixture of proteins is added to column.

(b) Small proteins are trapped in the cavities of the gel particles. Large proteins are excluded from the gel.

(c) The addition of more liquid at the top of the column washes off the untrapped proteins. Fractions (measured volumes) are collected in test tubes and assayed.

Figure 3.1 Molecular-sieve chromatography separates proteins of different sizes.

Determining Protein Structure

(Proteins that are localized within specific cellular organelles can be partially purified by first isolating that organelle.)

Small molecules—such as sugars and free amino acids—can be separated from macromolecules in the cell extract on the basis of the enormous differences in their sizes. The extract is put inside a bag made from a semipermeable membrane which is placed in an aqueous solution; the pores in the membrane are large enough to allow

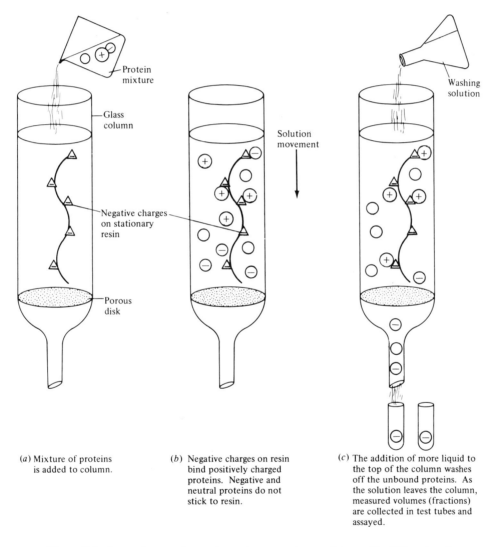

(a) Mixture of proteins is added to column.

(b) Negative charges on resin bind positively charged proteins. Negative and neutral proteins do not stick to resin.

(c) The addition of more liquid to the top of the column washes off the unbound proteins. As the solution leaves the column, measured volumes (fractions) are collected in test tubes and assayed.

Figure 3.2 Ion-exchange chromatography separates proteins with different net charges.

small molecules to pass out of the bag but are too small to allow macromolecules to escape. This procedure is called *dialysis*.

The proteins must now be separated from the other macromolecules in the solution, primarily polysaccharides and nucleic acids. This can be done by adding a chemical which decreases the solubility of proteins, while leaving other macromolecules in solution. A chemical commonly used for this purpose is ammonium sulfate. When ammonium sulfate is added to a solution of macromolecules, the proteins precipitate and are thus separated from the other large molecules. (The proteins are not, however, denatured by this precipitation.) Different proteins precipitate when the solution contains different concentrations of this chemical, so this can sometimes be used to eliminate unwanted proteins.

Further steps in the purification of a protein are designed to separate one particular protein from a mixture of proteins. What are the differences between proteins that might be used to separate them? The most useful ones are size—the total number of amino acids a protein contains—and overall charge, which reflects the amino acid composition of the proteins.

Proteins can be separated according to their size by pouring a solution of the protein mixture over a porous gel that contains cavities of a specified molecular size (Fig. 3.1). High molecular weight proteins are too big to enter these gel cavities; they pass right by them and out of the gel. Small proteins, however, can enter these cavities, tend to linger there, and therefore their progress is retarded. These gels, which are manufactured with a variety of cavity sizes, are called *molecular sieves*. Proteins can also be separated on the basis of the differences in their net charges by chromatographing the mixture of proteins over an insoluble material which carries charged groups (Fig. 3.2). If this material has a positive charge, it will bind proteins with negative charges, but allow positively charged and neutral proteins to pass through (Sec. 3.4).

3.4 The amino acid composition of a protein is determined by breaking its peptide bonds and analyzing the resulting mixture of amino acids

Once a protein is purified, the next step in studying its structure is to determine the number of residues of each amino acid that are present in one molecule of the protein. To release free amino acids, every peptide bond in the protein backbone must be broken; this is done by the chemical addition of a water molecule (hydrolysis), essentially the reverse of peptide bond formation (Sec. 2.3). Because peptide bonds are strong covalent bonds, this breakdown requires drastic treatment, usually heating the protein in an aqueous solution of strong hydrochloric acid for about 20 hours. The result is a solution containing different amounts of each of the amino acids, which must then be separated from each other, identified and the amount of each determined. Amino acids, like peptides and proteins, can be separated from each other on the basis of the differences in their overall charges (Table 2.2), using the techniques of ion-exchange chromatography or electrophoresis.

Ion-exchange chromatography utilizes an ion-exchange resin, an insoluble material that is a support for ionizable groups, often $-SO_3^-\,Na^+$. The negative sulfonic acid groups are covalently linked to the resin; the sodium ions are held onto these negative groups by ionic bonds and can be replaced by other positively charged molecules. The pH of the solution of amino acids to be separated is made acidic so that the amino acid molecules are all positively charged. This solution is passed over the resin and, because they are positively charged, the amino acids displace the Na^+ ions and become bound to the negative SO_3^- groups on the resin. The amino acids with two positive charges, lysine and arginine, are bound most tightly to the charged resin. The resin with the bound amino acids is now washed with a solution at a pH nearer 7, which converts the amino acids to their uncharged or negatively charged forms, in which they are no longer attracted to the stationary $-SO_3^-$ groups. Therefore, as the pH of the washing solution is raised, the (most negative) acidic amino acids are washed off the resin first, followed by the neutral molecules and, finally, the basic amino acids. In practice, the pH of the washing solution is raised gradually to take advantage of very slight differences in the ability of the different neutral amino acids to extricate themselves from the resin; in this way, even very similar molecules can be separated.

Amino acids are colorless compounds, so it is impossible to see exactly when each compound comes off the resin. To locate the amino acids, each fraction which comes through the column is tested for compounds that will react with ninhydrin [Sec. 2.2(b)]; the total amount of color produced indicates the quantity of amino acid in each fraction. Other color-producing reagents can also be used.

Since ninhydrin reacts with all amino acids, how do we determine which ninhydrin-positive fractions correspond to a particular amino acid? Under identical chromatographic conditions, a given amino acid will always come through the column at the same time, and this retention time is unique to each amino acid. The order in which amino acids come off the column can be established by running separate solutions of each amino acid through the column and noting their respective elution times. Then, when a mixture of amino acids is chromatographed, the order (i.e., time) of their appearance is known and one can determine how much, if any, of a particular amino acid is present.

The analysis of amino acid mixtures by ion-exchange chromatography has now been fully automated and commercially available amino acid analyzers are used extensively in research and clinical laboratories. The analysis of a typical amino acid mixture is shown in Fig. 3.3. In medical laboratories, this machine is used to detect amino acids in serum and urine, an important step in the diagnosis of certain diseases.

Electrophoresis is another technique for separating amino acids or other molecules which, like ion-exchange chromatography, takes advantage of the differences in their overall charge at specific pHs. All electrophoretic techniques are based on the principle that unlike charges attract. When placed in a stationary electric field, positively charged amino acids are attracted by and therefore move toward the

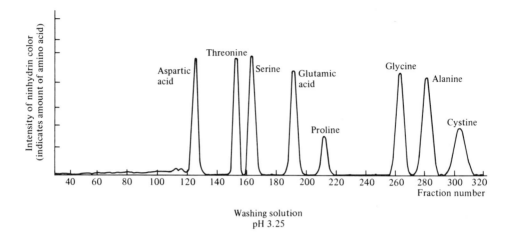
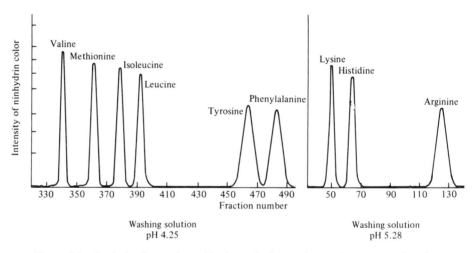

Figure 3.3 Analysis of an amino acid mixture by ion-exchange chromatography. An acidic mixture of amino acids was applied to a negatively charged ion exchange resin in a glass column (see Fig. 3.2). The resin was then washed with solutions of increasing pH, which removed the amino acids one at a time. As the solutions exited at the bottom of the column, 1-milliliter fractions were collected, and each was assayed for amino acids by the ninhydrin reaction (Sec. 2.2). [Reprinted with permission from Spackman et al., *Anal. Chem.*, 30, No. 1, p. 190 (1958). Copyright by the American Chemical Society.]

negative electrode, while negatively charged amino acids move toward the positive electrode. The charged molecules move toward the electrodes at rates that are proportional to their total positive or negative charges.

For most electrophoretic procedures, a drop of the solution to be resolved is placed on a solid support—either paper or gel—which is then submerged in a solution

Determining Protein Structure

at the desired pH; the electrodes are then placed in the solution at opposite ends of the solid support. After electrophoresis, the amino acids can be detected by spraying or soaking the chromatogram in ninhydrin. Analogous to ion-exchange chromatography, these procedures can be calibrated by electrophoresing each amino acid individually.

Which electrode will attract the various amino acids during electrophoresis? At pH 7, the neutral amino acids have no net charge and will not be attracted to either electrode. The positively charged basic amino acids, arginine, histidine and lysine, will move toward the negative electrode, while the negatively charged molecules, aspartic acid and glutamic acid, are attracted by the positive electrode. This is diagrammed in Fig. 3.4.

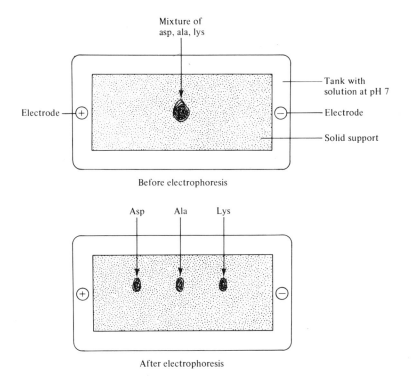

Figure 3.4 Separation of aspartic acid, alanine, and lysine by electrophoresis at pH 7.

Changing the pH of the solution bathing the chromatogram changes the total charge on the amino acids and therefore alters the direction and rate of their migration in the electric field. At acidic pHs, all amino acids are positively charged and are attracted to the negative electrode; the basic amino acids with their +2 charge will move toward the negative electrode faster than the other amino acids. At basic pHs, the acidic amino acids move toward the positive electrode at a faster rate than all the other amino acids. Since it is somewhat difficult to carry out electrophoresis on a large scale, this is used as an analytical rather than a preparative method.

3.5 Determining the linear sequence of the amino acids in a protein

In 1958, F. Sanger was awarded a Nobel prize for having determined the linear sequence of the 51 amino acids of insulin (Fig. 2.3); this was the first protein to be completely sequenced. The techniques Sanger developed have since been refined and extended, although, for large proteins, these procedures remain tedious and time-consuming. Today, the complete amino acid sequences of more than 100 proteins and peptide hormones are known. In spite of such significant achievements, these represent only a small fraction of the proteins found in cells.

How do we approach the project of determining the amino acid sequence of a protein? Although the amino acid composition of a protein is relatively easy to obtain and is vital in the "bookkeeping" sense, such data provide no information about the actual sequence of amino acids in the protein. Remember that there are no rules governing the linear sequence of amino acids in proteins. Any amino acid can be linked to any other one and there is no periodicity or repeating unit in the order of the amino acids in most polypeptide chains. Trying to draw conclusions as to the amino acid sequence of a protein from its amino acid composition is like seeing a hundred cars in a parking lot and then trying to decide in what order they entered the lot—the possibilities are enormous.

For every protein, determining the amino acid sequence must begin with the intact macromolecule. The simplest and most logical method for sequence determination would be to start at one end of the protein, remove amino acids one at a time, and identify each as it is removed. But since all amino acids are linked together by peptide bonds, this approach requires very specialized chemical "scissors" which cut only the terminal peptide bond and leave the remainder of the peptide backbone intact. Two such chemical scissors are available: carboxypeptidase, an enzyme that degrades proteins by cleaving the peptide bond at the carboxy terminal (C-terminal) end of the macromolecule and the Edman degradation technique, a chemical procedure for breaking the peptide bond at the amino terminal (N-terminal) end of the protein. Usually, both techniques are used to close in on the sequence of amino acids in the middle of the polypeptide chain.

The enzyme *carboxypeptidase* cleaves the last peptide bond at the C-terminal end of any protein. If the polypeptide chain shown in Fig. 3.5 were treated with this enzyme for a short period of time, free lysine would be released, leaving behind a chain of 10 amino acids with leucine in the C-terminal position. Lysine could be separated from the polypeptide and identified by electrophoresis or ion-exchange chromatography. The 10-membered polypeptide chain could now be treated with carboxypeptidase, releasing free leucine and leaving a nine-membered chain, and so on. In actual practice, the method does not work this smoothly; the enzyme does not simply remove the C-terminal amino acid and then stop acting while the researcher isolates the shorter polypeptide chain. Rather, carboxypeptidase moves on to the new C-terminal peptide bond as soon as the original one is broken, and, since there is some variation in the speed with which these degradations take place, some

Determining Protein Structure

A TYPICAL PEPTIDE

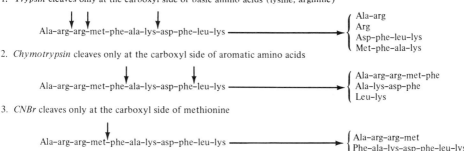

Figure 3.5 Methods for breaking specific peptide bonds.

"second bonds" will be broken before all the "first bonds" are broken. Thus, although a short treatment of this polypeptide with carboxypeptidase would release mostly lysine, some free leucine would invariably also be produced. As the enzyme moves successively farther into the peptide, the mixture of reaction products becomes too complicated to yield useful information.

The *Edman degradation technique* is used to determine the amino acid sequence from the N-terminal end of a protein (Fig. 3.6). The intact protein is first reacted with phenylisothiocyanate (PTH), a chemical reagent that attaches itself preferentially to the free terminal amino group. The remaining unreacted PTH is inactivated and the resulting "tagged" protein is treated with a weak acid solution to release a cyclic derivative of the N-terminal amino acid. The acid treatment is too mild to break other peptide bonds, and so the second product of this reaction is the original protein less one amino acid. Treatment of the peptide in Fig. 3.5 would yield a derivative of alanine and a 10-membered peptide with arginine in the N-terminal position. After the shortened peptide and the derivative are separated, the Edman degradation is applied to the shortened peptide to identify the amino acid that was penultimate to the N-terminus of the original peptide, and so on.

A device known as a "sequenator" uses the Edman degradation method to sequence proteins automatically. In a few years automatic sequencing and synthesis of proteins will probably be commonplace, leaving biologists to tackle more crucial problems, such as how these proteins work.

If we need to determine only the N-terminal amino acid, a protein or peptide can be reacted with fluorodinitrobenzene (Sangers reagent), which reacts with the free amino group of the N-terminal amino acid [Sec. 2.2(b)]. The protein wearing its

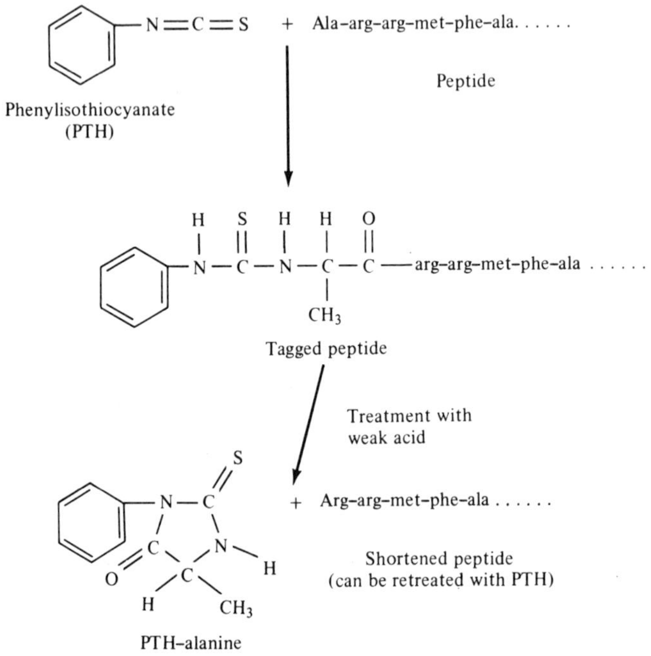

Figure 3.6 Identification of the N-terminal amino acid in a peptide by the Edman degradation.

yellow "tag" is then treated with strong acid, which breaks all the peptide bonds in the molecule, releasing free amino acids plus the yellow DNP derivative of the N-terminal amino acid. Since no shortened peptide is recovered, this technique cannot be used to identify internal amino acid residues. Such treatment of the peptide in Fig. 3.5 would produce DNP-alanine plus one molecule of leucine, two of lysine, one of methionine, and so on.

For technical reasons, the Edman degradation can be used to identify only 40–50 amino acid residues from the N-terminal end of a protein, and carboxypeptidase treatment at best a dozen amino acid residues from the C-terminal end. No combination of these techniques can be used to determine the sequence of the 200 or more amino acids found in an average protein. To determine the sequence of the innermost amino acid residues in such a macromolecule, additional steps are needed. The approach presently used is to break up the protein into short peptides, then separate the peptides and determine the amino acid sequence of each peptide individually. If a protein is broken down into only one set of peptides, the sequence of each peptide can be determined, but it would still not be clear how the peptides fit together, since there would be no overlaps. This problem can be overcome by breaking the protein into two or more different sets of peptides using different cleavage methods, analyzing the peptides in each set separately, and then lining the peptides up by their overlapping and corresponding amino acid sequences.

The sequencing procedures can be illustrated with the peptide in Fig. 3.5. Assume that the amino acid composition is known and that the N- and C-terminal amino acids have been identified as alanine and lysine, respectively. Suppose that the molecule were first broken down into four small peptides which Edman degradation and carboxypeptidase treatment showed to be: ala-arg, arg, asp-phe-leu-lys, and met-phe-ala-lys. From these data, we cannot decide which of the two peptides ending in lysine came from the C-terminal end of the original molecule or where the free arginine fits into the sequence. This dilemma can be resolved by an analysis of a second, different set of small peptides derived from the original molecule. Let us assume that a second fragmentation technique produced three peptides: ala-arg-arg-met-phe, leu-lys, and ala-lys-asp-phe; again, no order can be assigned to these peptides. However, the data from the two sets of peptides can be combined to reveal the complete amino acid sequence of the original peptide. The sequences that occur only once in each set of peptides, such as "asp-phe," "ala-arg," and "met-phe," must have arisen from the same region of the original peptide; in addition, the original molecule contained only one methionine and one aspartic acid residue, so these must also correspond. Using this deductive reasoning, the two sets of fragments can be lined up to reveal the sequence of the internal amino acid residues in the original molecule:

Cleavage 1: ala-arg arg met-phe-ala-lys asp-phe-leu-lys
Cleavage 2: ala-arg–arg–met-phe ala-lys–asp-phe leu-lys
                    Corresponding regions

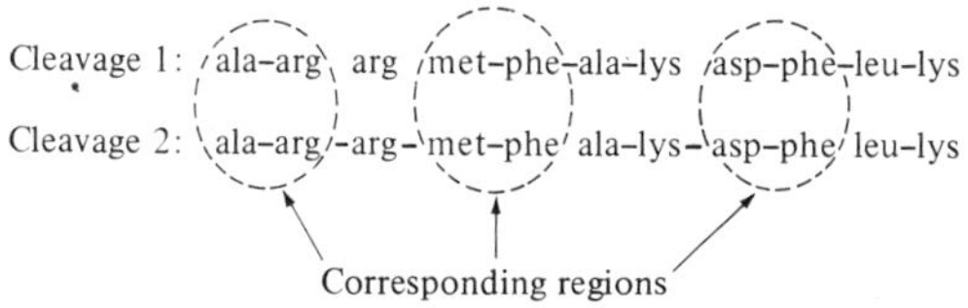

For large proteins, the number of fragments may be large and produce a very complicated jigsaw puzzle, but as long as some overlapping sequences are known, the fragments can eventually be pieced together in the correct order.

Several techniques—some enzymatic, some chemical—can be used to fragment proteins. To yield meaningful data, the methods chosen to cleave a protein must be highly reproducible (they must always yield the same peptides from a given protein), they should produce peptides of a manageable length (5–20 amino acids), and each method used should be guaranteed to yield a unique set of fragments. Three techniques commonly used to cleave proteins are enzymatic treatment with trypsin or chymotrypsin (both pancreatic enzymes) or chemical breakage with cyanogen bromide (CNBr). Each of these methods splits only those peptide bonds next to a few specific amino acids. Trypsin, for example, cleaves peptide bonds which join the carboxyl group of one of the basic amino acids (lysine or arginine) with the amino group of any other amino acid; thus, all the peptides produced by trypsin treatment have a basic amino acid in the C-terminal position, except for the one peptide which came from the C-terminal end of the intact protein. Chymotrypsin breaks peptide bonds involving the carboxyl group of an aromatic amino acid (phenylalanine, tyrosine, or tryptophan). Cyanogen bromide cleaves the peptide backbone on the

C-terminal side of methionine. These fragmentation patterns are illustrated in Fig. 3.5.

Cleavage of an intact protein by any of these techniques produces a mixture of 20 or 30 different peptides which must be separated from each other before their amino acid compositions and sequences can be determined. Because the peptides in the mixture have different amino acid compositions, they differ in overall charge and can therefore be separated by electrophoresis. The total charge on a peptide is the sum of the positive and negative charges on its R groups plus the charges of the terminal carboxyl and amino groups; the total charge is pH-dependent. Since in actual situations the number and composition of the peptides is not known, the mixtures are usually electrophoresed at more than one pH, to find the conditions where the mixture is resolved into the largest number of peptides. For example, the charges on the three products of chymotrypsin cleavage of our sample peptide vary with changes in pH (Table 3.1); only at pH 7 can these peptides be completely separated by electrophoresis. Once separated, the amino acid sequence of each individual

Table 3.1 Charges on Peptides Vary with pH

	Total charge on peptide		
	Acidic pH	pH 7	Basic pH
Ala-arg-arg-met-phe	+3	+2	−1
Ala-lys-asp-phe	+2	0	−2
Leu-lys	+2	+1	−1

peptide can be determined by using a combination of carboxypeptidase treatment, Edman degradation, and/or reaction with DNFB.

To summarize, the determination of the amino acid sequence of a protein proceeds by the following steps:

1. The terminal amino acid residues of the intact protein are identified: the N-terminal residue(s) by Edman degradation or reaction with FDNB, the C-terminal residue(s) by treatment with carboxypeptidase.

2. To obtain the internal amino acid sequence, the protein is broken down into two or more different sets of small peptides by the action of chymotrypsin, trypsin, or cyanogen bromide.

3. Each set of peptides obtained in step 2 is subjected to electrophoresis to separate the individual peptide fragments.

4. The individual peptides in each set are sequenced by the methods used in step 1.

5. By lining up the corresponding and overlapping amino acid sequences found in the peptides produced by different cleavage procedures, the peptide fragments can be ordered to reveal the complete amino acid sequence of the intact protein.

Although different chemical and enzymatic methods are used, the principles of fragmentation, sequencing small fragments, and lining up overlapping segments apply to the sequencing of all macromolecules. These techniques have been very successful in determining the structures of DNA and RNA molecules.

3.6 The three-dimensional shape of a protein can be determined by x-ray crystallography

What amino acid R groups are on the interior of a protein? What proportion of the peptide backbone is wound up in an α-helix? If the protein requires a cofactor, how does the small molecule fit into the protein? Answering these questions requires an analysis of the intimate details of a protein's three-dimensional shape.

Stated simply, analysis of the three-dimensional shape of a protein requires a very powerful microscope—one capable of magnifying a single protein molecule until the details of its structure, including the atoms of each R group, are distinguishable. Such a "microscope" has become available with the development of x-ray crystallography. To understand the principles of this process, consideration of the more familiar light microscopy is instructive. Light microscopy consists of two major steps: first, when visible light is shone on an object, the light waves are bent (diffracted) in a manner dependent upon the size and shape of the internal component parts of the object and, second, the diffracted light waves are collected and recombined by a series of mirrors and lenses to produce the image we see. To produce a clear, well-resolved magnification by light microscopy, the length of the light waves must be about twice that of the distance between the structures being magnified. Protein molecules are very small and the distances between individual atoms within a molecule are incredibly short. The wavelengths of visible light are too long to be useful for magnifying protein molecules; radiation with much shorter wavelengths must be used. The form of radiation with suitably short wavelengths is x-irradiation.

The first step in magnifying a protein using x-rays is similar to the first step of light microscopy: a crystal of the protein is illuminated with a beam of x-rays, and the x-rays are diffracted by the electrons grouped around the centers of the various atoms. Unfortunately, there is no system of mirrors and lenses which can collect the diffracted x-rays and recombine them to produce a meaningful image. Instead, the diffracted x-rays are allowed to hit a photographic film, where they produce a series of spots whose intensities are determined by the arrangement of the atoms in the protein molecule. A large number of different spot patterns are produced by allowing the x-rays to penetrate the protein crystal from different angles. Finally, the data obtained from the spot patterns are analyzed and combined by a complex, usually computerized, mathematical analysis to produce a detailed three-dimensional "picture" of the protein molecule. X-ray diffraction techniques were used to obtain the three-dimensional models of hemoglobin and chymotrypsin shown in Figs. 2.6 and 2.7.

It is easy to see why complete studies (amino acid sequence and three-dimensional structure) have been done on only a small number of proteins. It requires a large commitment of manpower and research resources to do this, but more proteins are completed every year. Only in this way will we understand how proteins work in living cells, with obvious benefits to basic and medical science.

SUMMARY

Determining the complete, detailed structure of a protein, or any other macromolecule, is a difficult task involving complex biochemical techniques. The protein must first be purified from all other cell components. The amino acid composition of the protein is determined and then the linear sequence of these monomers is established using specialized chemical "scissors." Finally, x-ray crystallography is used to visualize the way the chain of amino acids is folded up upon itself.

Such detailed studies of many proteins have revealed how various interactions between amino acid R groups stabilize the three-dimensional shape of proteins and form the active sites for enzymes. We have learned why collagen forms long fibers, ideal for a structural protein, and how the peptide chains and cofactors of hemoglobin are associated in a way that allows the complex to transport oxygen. These structural studies are also advancing our knowledge of metabolic diseases in which defective proteins are made or in which specific proteins are missing.

PRACTICE PROBLEMS

The structural formulas of the amino acids, given in Table 2.2, may be used to answer the following questions.

1. The amino acid sequences of several abnormal hemoglobins differ from that of normal hemoglobin by only one amino acid. In sickle-cell hemoglobin (HbS), for example, a valine residue replaces a glutamic acid residue found in normal hemoglobin (HbA). Some of these amino acid changes give the abnormal hemoglobin an electrophoretic mobility different than HbA. Which of the abnormal hemoglobins listed below could be distinguished from HbA by electrophoresis at neutral pH? At acidic pH? At basic pH?

	Amino acid in abnormal hemoglobin	Amino acid in normal hemoglobin
HbS	Val	Glu
HbI	Glu	Lys
HbM$_{Boston}$	Tyr	His
HbC	Lys	Glu

Determining Protein Structure 61

2. Digestive enzymes are important tools for determining the amino acid sequence of proteins. Show the major product obtained when the peptides listed below are treated with (a) trypsin, (b) chymotrypsin, and (c) carboxypeptidase.

   Arg-arg-arg-arg-arg-arg-arg

   Arg-tyr-arg-phe-lys-tyr

   Leu-met-ala-ser-arg-glu-phe-val

   Gly-ala-his-tyr-thr

3. When the amino acid sequence of insulin was under study, the following peptides were produced from one segment of the protein. Arrange the peptides by their overlapping sequences, which establish the order of these 15 amino acids.

   Phe-val-asp

   Val-glu-ala-leu

   Val-asp-glu

   His-leu-cys-gly-ser-his-leu

   Glu-his-leu-cys

   Ser-his-leu-val-glu

4. You are given two test tubes and told that one tube contains a solution of alanine and the other contains a solution of hemoglobin. Describe two methods you could use to tell which tube contains the alanine and which contains the hemoglobin.

5. A peptide X has the amino acid composition: met_2, lys_2, cys_2, leu_2, val_2. The following information was obtained about its amino acid sequence. What conclusions about X's structure can be drawn from each of these results? What is the structure of peptide X?
   (a) Reaction with DNFB reagent followed by acid hydrolysis gives two molecules of DNP-met.
   (b) Reduction of the peptide with mercaptoethanol gives two equal fragments, each containing one molecule of met, lys, cys, leu, val.
   (c) The fragments, when treated with carboxypeptidase, released mainly leucine.
   (d) The fragments, when treated with trypsin, gave two smaller fragments, one of which contained met, cys, lys.

6. An octapeptide called octase has been isolated from octopus tentacles. Complete acid hydrolysis of octase gave one residue each of leu, arg, gly, met, phe, thr, ala, and ser. When octase's amino acid sequence was studied, the following results were obtained. What conclusions can be drawn from each result? What is the structure of octase?
   (a) Octase was reacted with DNFB and hydrolyzed. The product was DNP-met.
   (b) Octase was digested with carboxypeptidase; free threonine was the major product, but a little free glycine was also obtained.
   (c) When octase was digested with chymotrypsin, two products (A and B) were obtained. Product A contained phe, met, and ala; B contained gly, ser, arg, leu, and thr. B was reacted with DNFB and hydrolyzed to give DNP-leu. (What is the sequence of fragment A?) (What do you now know about B?)

(d) When octase was digested with trypsin, a dipeptide containing thr and gly, and an unknown hexapeptide were obtained. (What is the sequence of fragment B? What is the structure of octase?)

7. Gramicidin S is an antibiotic used in some skin ointments. It is a peptide, containing 10 amino acids and nothing else. Carboxypeptidase treatment of this drug releases no free amino acids. When Gramicidin S is treated with DNFB and the product hydrolyzed, none of the amino acids is tagged with DNP on its α-amino group. Propose a generalized structure for Gramicidin S that would account for these results. (Rather than naming the R groups of specific amino acids, use $R_1, R_2, \ldots, R_{10}$.)

SUGGESTED READING

KADIS, SOLOMON, THOMAS C. MONTIE, and SAMUEL J. AJL, "Plague Toxin," *Sci. American*, 220, Vol. 3, p. 92 (1969). Electrophoresis was used to purify this protein.

PERUTZ, M. F., "The Hemoglobin Molecule," *Sci. American*, 211, No. 5, p. 64 (1964). X-ray crystallography was used to determine the structure of hemoglobin.

STROUD, ROBERT M., "A Family of Protein-Cutting Proteins," *Sci. American*, 231, No. 1, p. 74 (1974). Indicates specificities of proteolytic enzymes used to fragment proteins.

THOMPSON, E. O. P., "The Insulin Molecule," *Sci. American*, 192, No. 5, p. 36 (1955). How the amino acid sequence of insulin was determined.

4 Enzymes

The next time you drink a glass of milk, think for a while about what is happening to this food. The relatively small number of chemicals in milk are converted in your body to the enormous variety of proteins, polysaccharides, nucleic acids, and lipids, which compose muscles, nerves, and body organs; you also obtain energy from the milk. The capacity to convert a small number of compounds, rapidly and efficiently, into thousands of other large and small molecules is unique to living cells; the chemical reactions of metabolism do not occur to any appreciable extent in the absence of cells.

What do living cells contain which promote these reactions? All living cells possess *enzymes, specialized proteins that assist and accelerate* (catalyze) *chemical reactions.* Enzymes are essential to all the chemical reactions which together constitute cell metabolism. Every time a chemical bond is made or broken—when macromolecules are produced or degraded, when one small molecule is converted to another, when energy is obtained from a compound—a protein with enzymatic activity is required to cause the reaction to proceed at a reasonable rate. As an illustration, the conversion of milk sugar, lactose, into a fatty acid involves a series of about 20 chemical reactions; each reaction requires an enzyme specifically designed to promote that step in the series. Without these enzymes, the chances of lactose becoming a fatty acid are nil. Without enzymes, most of the reactions in living organisms would not take place; every facet of cell physiology, growth, and behavior depends on their action.

4.1 A single enzyme molecule can be used over and over again to convert many molecules of substrate into product

The equation for any enzyme promoted chemical reaction can be written in the generalized form

$$A + B \xrightleftharpoons{\text{Enzyme}} C + D$$

The compounds present at the start of the reaction, A and B, are called *substrates* for the enzyme; the molecules formed during the reaction, C and D, are referred to as *products*. Not all enzymatic reactions have two substrates and two products; there are many possibilities—a single substrate can be broken down into two products, or two substrates can combine to form one product, and so on.

Most cellular reactions are reversible; that is, under certain conditions A and B react to form C and D, while under other conditions, C and D react to give A and B. In the equation above, this reversibility is indicated by the double arrows. If one side of a reaction is favored, the arrows are given unequal lengths. For example, a reaction in which the formation of products is favored is written $A + B \rightleftharpoons C + D$.

Enzymes are proteins with catalytic activity; they are organic catalysts. Catalysts are substances that assist or accelerate chemical reactions without themselves being permanently altered during the reaction. (You may be familiar with the catalytic properties of metals such as platinum.) Like their inorganic counterparts, enzymes assist and accelerate chemical reactions, but they themselves are not changed during the reaction. (In the reaction equation shown, this is indicated by writing the name of the enzyme above the reaction arrows, apart from the substrates and products.) Because it is not used up or destroyed in the reaction, a single enzyme molecule can work over and over again, promoting the conversion of thousands of molecules of substrate into product. Each cell requires only a small number of copies of any specific enzyme to process a very large number of particular substrate molecules. Thus, enzymes exert an influence on cellular metabolism which is very large when compared to their quantity in the cell.

4.2 Cells need enzymes to make reactions proceed at a fast rate

All metabolic reactions are energetically and chemically possible even in the absence of enzymes; they are spontaneous reactions. There is nothing mysterious or extraordinary about the reactions that enzymes catalyze, for no enzyme can promote a reaction that is theoretically impossible.

If essential metabolic reactions are chemically and energetically possible in the absence of catalysts, why do cells need enzymes? The critical problem for the cell is *time*. It may take hundreds of years for a reasonable number of substrate molecules

Enzymes 65

to react spontaneously and form products, especially at the neutral pH and relatively low temperatures inside living cells. In order to maintain itself and grow, the cell must take a few low molecular-weight compounds from its environment and convert them into thousands of other compounds. To accumulate sufficient amounts of many chemicals in a reasonable time, the cell needs a faster method than simply waiting passively for hundreds of necessary reactions to occur by themselves. Cells rely on enzymes to overcome this time problem. Enzymes make chemical reactions take place at much faster rates than they occur spontaneously; enzyme-catalyzed reactions go millions of times faster than the corresponding uncatalyzed reactions. Only through the use of enzymes can cells produce and maintain sufficient supplies of metabolically important chemicals to function and grow at a reasonable rate.

The same time problems occur when reactions are carried out in a chemistry laboratory. Most reactions require heat or acidic or basic conditions to make them take place at a reasonable rate. The living cell cannot use heat or pH extremes, so it uses enzymes.

4.3 Each enzyme can catalyze only one type of metabolic reaction

An enzyme can assist and accelerate only one type of chemical reaction. Each enzymatically active protein, with its unique sequence of amino acids, is designed to promote the making or breaking of a single type of chemical bond, and it cannot act upon other chemical bonds. This characteristic is referred to as *enzyme specificity*. (This is discussed further in Sec. 4.6.) Since a given enzyme can accelerate only one of the hundreds of chemical reactions comprising cellular metabolism, each cell must produce hundreds of different, catalytically active proteins.

Current studies of how enzymes act to accelerate chemical reactions are focused on two major areas: (1) how enzymes affect the energy requirements of a chemical reaction, and (2) how enzymes work in the mechanistic sense (i.e., what physical features of substrate and enzyme molecules are important in catalysis). This chapter discusses these aspects of enzyme catalysis in a general way; specific enzymes will be described under the appropriate areas of metabolism.

4.4 Enzyme nomenclature

The current system of enzyme nomenclature is based upon substrate and reaction specificity. In the past, enzymes were not named in this manner and many of the older, trivial names are still in use. To help you in your future reading, the ways in which enzymes are or have been named are listed below.

1. The first enzymes studied were named for their color, their location in the body, or even after the researcher who discovered them. This system became increasingly chaotic as more and more enzymes were discovered. Nonetheless, we regularly use names such as trypsin, chymotrypsin, and catalase.

2. Some enzymes were named by simply adding the suffix "ase" to the name of their substrates. Urease and ATPase are examples. Since most compounds are substrates for several reactions, this system has its limitations.

3. Many enzyme names consist of the major substrate plus a general word (with the suffix "ase") indicating the type of reaction catalyzed. This is the naming system most often found. For example, enzymes catalyzing reactions in which a phosphate group is removed from the substrate (a phosphate monoester) with the addition of water are called phosphatases; alkaline phosphatase has this catalytic ability in solutions with basic pH. Several of the common terms for reaction specificity are defined in Table 4.1.

4. For use in the technical literature, a naming system has been designed to apply to all known, as well as all as yet undiscovered, enzymes. In this scheme, enzyme-catalyzed reactions are divided into six classes. The six classes are explained in Table 4.1. An enzyme is named by listing all of its substrates followed by the reaction class. Thus, alkaline phosphatase becomes orthophosphoric monoester phosphohydrolase. This is a very complex system and you are not expected to remember the names of enzymes in this way.

TABLE 4.1 Enzymes Are Named According to Their Reaction Specificity

Enzymes can be divided into six major classes, depending on the kind of reaction they catalyze (1 through 6 below). Listed under each general class are examples of the frequently encountered common names for groups of enzymes falling into the general class.

	Reaction catalyzed
1. Oxidoreductases	Oxidation and reduction of substrates (usually involve hydrogen transfer)
Dehydrogenases	Transfer of hydrogen atoms from substrate to NAD^+
Reductases	Addition of hydrogen atoms to substrate
Oxidases	Transfer of hydrogen atoms from substrate to oxygen
2. Transferases	Transfer of a chemical group (such as a methyl group, amino group, phosphate group) from one molecule to another
Phosphorylases	Addition of orthophosphate to substrate
Transaminases	Transfer of amino group from one substrate to another
Kinases	Transfer of phosphate from ATP to substrate
3. Hydrolases	Cleavage of bonds by the addition of water
Phosphatases	Removal of phosphate from substrate
Peptidases	Cleavage of peptide bonds
4. Lyases	Addition of groups to double bond (—C=C—, —C=O, —C=N)
Decarboxylases	Removal of carbon dioxide from substrate
5. Isomerases	Rearrangement of atoms of a molecule
6. Ligases	Formation of new bonds using energy from (simultaneous) breakdown of ATP
Synthetases	Joining two molecules together

4.5 Enzymes promote chemical reactions by making them more energetically favorable

All chemical compounds contain chemical energy, including energy in bonds and energy due to favorable or unfavorable interactions between different parts of the molecule. Some compounds have more chemical energy than others, but there is no compound that contains no energy. Thus, when two compounds react to form a product, chemical energy can be given off (an *exergonic* reaction) or taken up (an *endergonic* reaction) in the formation of the reaction products. During an exergonic reaction, the excess chemical energy of the starting compounds does not disappear; it is converted into heat, light, or mechanical energy. Conversely, during an endergonic reaction, nonchemical energy is transformed into the chemical energy contained in the products.

When we describe reactions as endergonic and exergonic, we are referring to the *net* energy changes that accompany particular chemical reactions; these terms do not indicate the fluctuations in the energy content of the compounds while the reaction is actually taking place. In fact, the energy levels of compounds change dramatically as they participate in chemical reactions. To understand how enzymes can make a chemical reaction more favorable in an energetic sense, we must examine the energy fluctuations that occur during uncatalyzed chemical reactions and then ask how this pattern is altered by an enzyme.

Let us look at the overall energy changes during an uncatalyzed exergonic reaction: A + B → C + D (Fig. 4.1). When the starting compounds, A and B, react to form

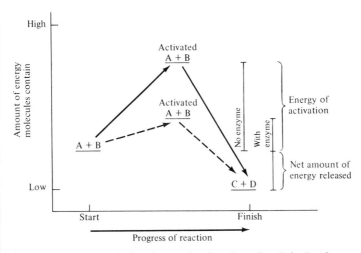

Figure 4.1 Energy changes during the reaction A + B → C + D in the absence (solid line) and presence (dashed line) of an enzyme catalyst. The substrates, A and B, must always be activated before they can form the products, C and D. The energy needed for this activation is lower when the enzyme is present. The *net* amount of energy released is the same with or without the enzyme.

the products, C and D, the net result is a loss of energy, at least in the sense that the sum of the energy in the products is less than the sum of the energy in the starting compounds. Enzymes do not alter the *net* energy loss of a chemical reaction; the same energy change takes place during a reaction whether or not it is catalyzed by an enzyme.

If you know that energy is lost when A and B react to form C and D, you might think that this reaction would proceed spontaneously at a rapid rate, making enzymes unnecessary. Many such exergonic reactions cannot occur spontaneously, however, because in order to react at all the starting compounds must first be activated, that is, A and B must gain an amount of energy considerably greater than that which they ordinarily possess. This increase in energy acts as a "trigger" for the reaction. Once they obtain this extra measure of energy, the starting molecules can react rapidly to form the products C and D.

In a large population of reaction substrates, the energy contents of individual molecules are neither identical nor static; rather, molecules are constantly gaining or losing some energy. As their energy contents fluctuate, molecules of A and B will occasionally obtain the energy required to achieve the activated state. If heat is applied, more of them can attain the condition and reaction will ensue. But under the temperature and pH conditions found in cells, these activations occur only infrequently, and therefore the spontaneous conversion of substrates to products is a rare event. In other words, in terms of energy, the uncatalyzed reaction proceeds at an extremely slow rate because it is only rarely that substrate molecules achieve the activated state.

Enzymes promote chemical reactions by lowering the amount of energy needed to activate the substrate molecules. Enzymes decrease the energy level that A and B must attain before they can react to form C and D. With this lowering of the activation energy barrier, the number of substrate molecules achieving the activated state is greatly increased, and therefore the products are formed at a much faster rate.

4.6 The catalytic ability of an enzyme depends on its three-dimensional shape

The catalysis of a chemical reaction by an enzyme usually consists of three basic steps, in which individual enzyme molecules participate over and over again:

1. The substrate(s) recognizes the enzyme and the two come together in a specific arrangement, the *enzyme–substrate complex.*

2. Through the catalytic activity of certain groups on the enzyme surface, the substrates are converted to products.

3. The products leave the enzyme.

As an example, the mechanism of chymotrypsin action is shown in Fig. 4.2.

Enzymes

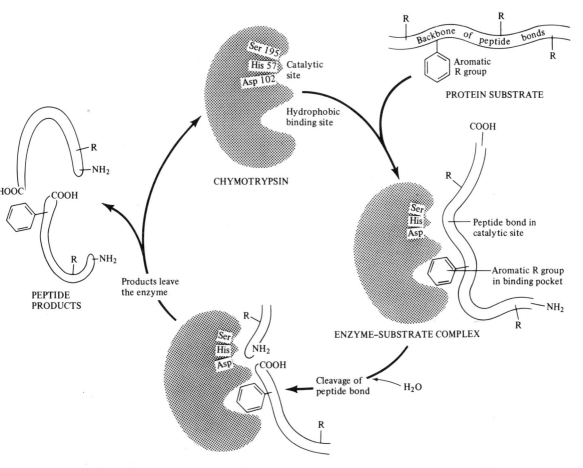

Figure 4.2 Mechanism of chymotrypsin action. The active site of chymotrypsin consists of two parts: a hydrophobic binding site, which holds aromatic R groups, and a catalytic site, where peptide bond cleavage occurs. Three amino acid R groups are responsible for catalysis: the R groups of serine, at position 195 in the amino acid chain, histidine at position 57, and aspartic acid at position 102. These three amino acids can be seen clearly in the three-dimensional models of chymotrypsin in Fig. 2.5.

The study of the mechanism of enzyme catalysis is basically a study of the interactions between the enzyme and its substrates. Since the properties of the substrates are usually known, especially if they are small molecules, the question becomes "What physical features of enzymes give these proteins their ability to catalyze specific chemical reactions?"

What does an enzyme "look like" to a substrate molecule? The substrate can "see" only the surface of the enzyme, that part of the protein which is exposed to their common aqueous environment; in most cases, the substrate can neither recognize nor penetrate the inner, hydrophobic core of the protein. The surface contours of an enzyme are determined by a combination of the folds in the helical backbone of the protein and the positions of hydrophilic, often charged R groups which stick out from the backbone into the solution (Sec. 2.4). Somewhere along the convoluted surface of each enzyme is an indentation, a crevice, whose contours have the precise shape and carry the proper charge to contain and bind the substrate molecule(s). At this place on the enzyme surface—called the *active site*—the substrate(s) sits down on the enzyme; like pieces in a three-dimensional jigsaw puzzle, the substrate and enzyme lock together to form the *enzyme–substrate complex* (Fig. 4.2). The two are held together by weak bonds such as ionic bonds and hydrogen bonds; no covalent bonds are necessary.

When a reaction involves two or more substrates, an enzyme accelerates the reaction simply by bringing the substrates together; the enzyme, in effect, concentrates the substrates. The enzyme's active site serves as a framework that holds the substrates in an orientation in which they can easily react with each other. Such an advantageous juxtaposition of the appropriate chemical bonds of the substrates would occur only rarely when the substrates are floating free in solution.

Enzymatic catalysis of the conversion of substrates to products depends on certain amino acid R groups which protrude into and are an integral part of the active site. When the substrates are in the active site, they are lined up precisely with that part of the enzyme that will perform the catalysis. The susceptible regions of the substrate molecules—where bonds are to be made or broken—are fixed in position next to the catalytic R groups and cofactors. The catalytic R groups exert forces on the substrate bonds to be altered so that the bonds are stretched, distorted, and weakened and the desired reaction becomes more likely. The R groups responsible for catalysis are often charged (e.g., those of the acidic or basic amino acids) or carry a special reactive group (e.g., the hydroxyl group of serine or the imidazole ring of histidine). The particular R groups used for catalysis, of course, depend upon the enzyme in question and upon the nature of the substrate bonds being made or broken.

It is important to realize that those regions of the active site responsible for binding the substrates may not be the ones responsible for catalysis. The R groups lining the active site usually function in only one of these roles. Small organic molecules or metal ions which are necessary for the functioning of many enzymes (Sec. 2.5) are often an essential part of the active site; depending on the enzyme, these cofactors may assist either in the binding of the substrate to the active site or in the catalytic process itself.

Although the active site of an enzyme comprises only a small portion of its surface area, the entire, intact protein is necessary for optimal catalytic activity. In most cases, the cleavage of a few amino acid residues from either end of an enzyme renders the enzyme incapable of exerting its normal catalytic function. In addition, most enzymes can no longer function if a single amino acid within the active site is replaced by a different amino acid, or if other amino acid substitutions change the overall three-dimensional structure of the protein (Sec. 12.10). The explanation of these phenomena lies in the fact that most of the amino acid chain is directly or indirectly needed for the maintenance and activity of the active site. The portions of the protein that line the active site are not one continuous segment of the linear chain of amino acids; rather, when the protein attains its preferred three-dimensional shape, distant parts of the polypeptide chain are brought close together to form the active site. Proper folding of the entire chain of amino acids is necessary to hold the appropriate portions of the molecule together in the active site.

In Sec. 4.3 it was stated that each enzyme catalyzes only one type of metabolic reaction. The reasons for this should now be clear. From this discussion of the mechanism of enzyme action, it is easy to see that the active site of each enzyme is designed to hold a particular substrate and to catalyze one reaction very efficiently; it would not have the correct shape, charge, or catalytic R groups necessary to promote totally different reactions. For example, the active site of chymotrypsin, designed to recognize large proteins and cleave their peptide bonds, could hardly be expected to attach a phosphate group to the simple sugar, glucose. The shape and charge of the active site of each enzyme is designed to bind a particular substrate, and the catalytic groups within each active site can promote only certain reactions; this defines enzyme specificity.

A good example of this specificity is seen among the peptidases, enzymes that catalyze the hydrolysis of peptide bonds in protein (Sec. 3.5). Even though all peptidases catalyze the hydrolysis of peptide bonds, their specificities are different since each can recognize and hydrolyze peptide bonds in different positions in a protein. Chymotrypsin recognizes and hydrolyzes only those peptide bonds next to phenylalanine and tyrosine. Trypsin acts only on peptide bonds next to basic amino acids. The active site of carboxypeptidase can bind and cleave only the last peptide bond at the C-terminal end of a protein.

Much of this description of active sites of enzymes also applies to proteins that function in the transport of small molecules (Sec. 2.1). Transport proteins such as hemoglobin are specific in that they are designed to carry only one type of small molecule. The small molecule recognizes and binds to a particular place on the transport protein in a manner analogous to the binding of a substrate to the active site of an enzyme.

4.7 Enzyme assays

When researchers purify an enzyme, they need to know how much active enzyme is present at each stage in the purification. Every day hospital laboratories examine hundreds of blood and tissue samples to determine, for diagnostic purposes, how much of certain enzymes they contain. Both situations call for accurate means of assaying enzyme quantity. An enzyme assay involves finding the quantity of a particular enzyme in a sample by measuring that catalytic activity which is unique to the enzyme. The amount of catalytic activity is proportional to the number of molecules of enzymatically active protein.

The principles of assay are the same for all enzymes. A measured amount of the material being analyzed (serum, cell extract, etc.) is placed in a solution containing the substrates for the enzyme under study. Then, over a period of time (usually minutes), the solution is monitored either for the appearance of a reaction product or for the disappearance of a substrate. Monitoring is simplest where one of the substrates or products is colored or absorbs light of a specific wavelength. Changes in color and light absorbency of a solution can be measured in a machine called a spectrophotometer. The results of an enzyme assay in which the formation of product is measured are often plotted as follows:

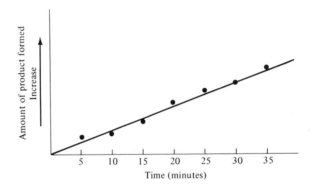

One of the assays for alkaline phosphatase is detailed in Fig. 4.3(a).

The rate of an enzymatic reaction is defined as the amount of product produced per unit of time (seconds or minutes) and is a direct measurement of the catalytic activity in the sample assayed. When tested under standard assay conditions, a given amount of enzymatically active protein always has the same amount of catalytic activity. Thus, the rate of a reaction is proportional to the quantity of enzyme in the sample [Fig. 4.3(b)]. Once an enzyme has been obtained in pure form, the precise amount of catalytic activity per milligram of enzyme protein can be determined and this value is used to calculate the amount of enzyme. However, many enzymes have not been purified completely; in these instances one can only compare the rates of reaction per milligram of *total* protein in different solutions. During the purification of an enzyme from a cell extract, the rate of reaction per milligram of total protein is determined after each purification step; this ratio will increase as the contaminating proteins are removed from the desired enzyme. For clinical purposes, the enzyme activity per milligram of total protein or blood volume in the unknown sample is compared to values previously obtained with healthy individuals.

4.8 External factors affecting enzyme activity

We stated that a given amount of enzymatically active protein always has the same amount of catalytic activity under the same assay conditions. What are "assay con-

Enzymes

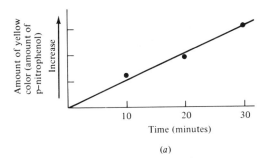

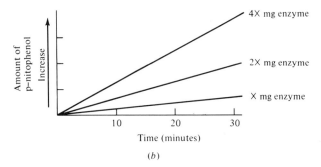

Figure 4.3 An assay for alkaline phosphatase. (*a*) Alkaline phosphatase catalyzes the reaction:

$$p\text{-nitrophenylphosphate} + H_2O \longrightarrow p\text{-nitrophenol} + \text{phosphate}$$
$$\text{(colorless)} \qquad\qquad\qquad\qquad \text{(yellow)} \qquad \text{(colorless)}$$

In the assay, a measured volume of the sample (serum, cell extract, etc.) is mixed with a solution of the phosphate ester, *p*-nitrophenylphosphate; the solution is at pH 10.3, the pH optimum of alkaline phosphatase (Sec. 4.8). As the reaction proceeds, the colorless *p*-nitrophenylphosphate is converted to *p*-nitrophenol, which is yellow. The increase in the yellow color of the solution is measured spectrophotometrically. The results of such an assay can be plotted. (*b*) The rate of the reaction is proportional to the quantity of enzymatically active protein in the assay mixture; doubling the amount of enzyme doubles the rate of the reaction.

ditions"? Four important ones are: (1) substrate concentration, (2) cofactor concentration, (3) temperature, and (4) pH. These factors must be carefully controlled when an enzyme is being assayed, especially when one wants to compare the amount of enzyme in a number of different samples, at different times.

For routine assays, the amount of substrate in the reaction mixture is far in excess of what the enzyme can convert to product in the time period under study. A single molecule of enzyme can act upon hundreds of molecules of substrate in a matter of seconds. If only a small number of substrate molecules are present in the assay mixture, they will be used up very quickly and the reaction will stop. In an extreme case, if 1000 molecules of alkaline phosphatase were incubated with 100 molecules of *p*-nitrophenylphosphate, the reaction curve would become:

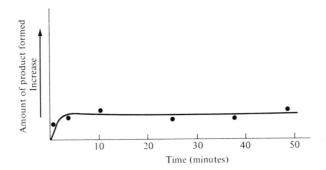

Because the substrate was the limiting factor, this assay does not measure how much enzyme is present. In general, one needs millions of molecules of substrate per molecule of enzyme so that the reaction will continue for a reasonable length of time and the reaction rate will depend on the enzyme concentration.

Although cofactors such as metal ions are not used up during an enzyme reaction, it is important that they be present in the assay mixture in sufficient quantities so that each enzyme molecule can bind its full complement of cofactors. The small molecules are often stripped from proteins during purification and must be replaced before enzyme activity can be measured.

The catalytic activity of an enzyme is dependent upon the temperature and pH at which it is being observed. We saw in Sec. 2.6 that extremes of temperature and pH can denature proteins and destroy their biological activity. Enzymes are no exception. Boiling an enzyme solution usually destroys its catalytic activity, as does exposure to very acidic or very basic solutions.

Enzymes are often sensitive to even less drastic changes in their physical surroundings, and slight changes in temperature may alter enzyme activity. For each enzyme there is a temperature at which it is most active and its activity diminishes when the temperature is changed in either direction away from this optimum. Most enzymes have their greatest activity somewhere between 25 and 37°C, corresponding to the temperature of most cellular environments. (It is interesting that those few bacterial species that live in hot springs or geysers produce proteins that work best at higher temperatures.) Why are enzymes so sensitive to temperature? Temperature changes cause slight changes in the three-dimensional shape of proteins; increases in temperature may cause the protein to loosen up and become less compact. Even small changes in the dimensions of its active site can make an enzyme a less efficient catalyst.

Enzymes are also sensitive to small changes in pH. Many enzymes work most efficiently at a pH near 7, the condition found inside living cells. Their catalytic activity decreases as the pH is shifted away from this optimum. Figure 4.4 shows the

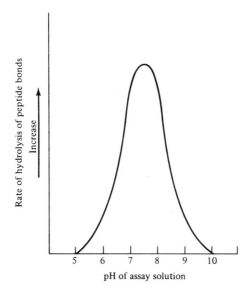

Figure 4.4 The catalytic activity of chymotrypsin is pH-dependent. To construct this curve, chymotrypsin was assayed at several different pHs, and the rate of reaction at each pH was calculated. A plot of the reaction rate versus pH produced this bell-shaped curve (typical of many enzymes) with an optimum at pH 7. (After Finn Wold, *Macromolecules: Structure and Function*, Prentice-Hall, Inc., Englewood Cliffs, N.J., 1971, p. 77.)

pH dependence of the catalytic activity of chymotrypsin, which digests proteins in the intestine. (In contrast, gastric enzymes that digest dietary proteins under acid conditions work optimally at low pH.) Why does pH influence enzyme activity? Slight changes in pH, which do not denature the protein, may change the charges on the R groups within the active site, thereby reducing the enzyme's ability to bind the substrate and catalyze the reaction. Changes in pH can also alter the charges of the substrate molecules, thus changing the character of the enzyme–substrate complex.

4.9 Isoenzymes

Organisms often contain chemically different proteins which catalyze the same metabolic reaction; such chemical variants are called *isoenzymes*. In many animals, proteins with different amino acid compositions perform identical catalytic functions in different body organs such as skeletal muscle, heart muscle, liver, and bone. Isoenzymes can even be found within an individual cell; in this case, the enzymes are usually localized in different cellular compartments (e.g., the nucleus and the cytosol, or the mitochondria and the cytosol).

The human lactate dehydrogenases (LDHs) are a well-documented and medically

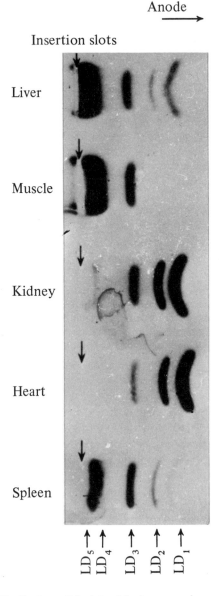

Figure 4.5 Distribution of lactate dehydrogenase isoenzymes in human tissues. (LDH) as shown by gel electrophoresis. Tissue extracts were spotted at the origin (arrow). All five isoenzymes of LDH move toward the positive electrode in the system used; LDH-1 is the most negatively charged and moves the farthest. After electrophoresis, the enzymes were assayed in the gel; the size and intensity of each spot is a direct measurement of the quantity of one isoenzyme in an extract. (After Albert Latner and Andrew Skillen, *Isozymes in Biology and Medicine*, Academic Press Ltd., London, New York, San Francisco. 1968, Fig. 2, p. 7).

important example of isoenzymes. There are five chemically different LDHs in man, all of which catalyze the reversible conversion of lactic acid to pyruvic acid. Each of the five isoenzymes has a characteristic electrophoretic mobility which reflects its particular amino acid composition. Analyses have revealed that several human tissues contain a different combination of the five LDHs: heart muscle contains primarily LDH-1 and LDH-2; liver primarily LDH-5; kidneys contain high levels of LDH-1, LDH-2, and LDH-3; and so on. The distribution of the LDH isoenzymes in human tissues is shown in Fig. 4.5.

4.10 Medical and industrial applications of enzymology

The extensive research on enzyme identification, purification, and mechanism of action is having an impact on our everyday lives. (Consider enzyme detergents!) In both medicine and industry, people are finding more and more uses for enzymes and enzymology. The medical applications of enzymology fall into two categories: first, the use of enzyme levels as diagnostic tools and second, the pharmacologic uses of purified enzymes. Industry is finding that pure preparations of enzymes are efficient catalysts for many chemical reactions, particularly those usually carried out by intact microorganisms.

(a) ENZYMES AND MEDICAL DIAGNOSIS

The medical diagnosis of internal disorders is complicated by the fact that many of these disorders mimic each other; thus, it is often difficult to tell, from symptoms alone, which internal organ is under stress. For example, similar chest pains can result from indigestion or from a serious heart problem. A determination of enzyme levels in the serum is often helpful in pinpointing which, if any, body tissue or organ has been damaged or is malfunctioning. When a tissue is injured or diseased, some cells of that tissue are destroyed and their contents, enzymes included, are released into the bloodstream. Therefore, if an enzyme is normally found predominantly in a tissue other than blood, an increase in its level in the blood indicates that that tissue has been damaged. While total serum levels of specific enzymes are diagnostic in some cases, isoenzymes, such as the organ-specific lactate dehydrogenases (LDHs), are especially important. Normally, serum contains only low levels of the five LDHs. An increase in total serum LDH activity could result from damage to the heart muscle, skeletal muscle, pancreas, or liver. To distinguish these possibilities, one must determine which isoenzyme is responsible for the rise in serum LDH activity. A large increase in the serum level of LDH-1 indicates that heart muscle has been damaged, as in a myocardial infarction. Liver disease is indicated when elevated levels of LDH-5 are found. Characteristic LDH patterns are also diagnostic of such diverse conditions as muscular dystrophy, rheumatoid arthritis, and pancreatic disease. The alkaline phosphatases are another set of isoenzymes used for medical diagnosis. This enzymatic activity is important in several stages of bone formation and is present in high concentration in the liver. Increased levels of specific alkaline phosphatases can be used to diagnose liver disease and bone disorders such as rickets and bone cancer.

Recent advances in enzymology have helped our understanding of the underlying causes of many inherited metabolic diseases, including Tay-Sach's disease, galactosemia, and phenylketonuria (PKU). We now know that many of these rare disorders result because the individual cannot make a specific enzyme; for example, children with Tay-Sach's disease lack one of the lipid-degrading enzymes, which leads to an accumulation of lipid material in their nerve cells. Hopefully, this knowledge will help us find a way of controlling these conditions. Already, assays for the missing enzymes are being used to detect these diseases in fetuses and to identify healthy individuals who are carriers of the diseases (Sec. 12.10).

(b) PHARMACOLOGIC USES OF ENZYMES

In theory, it should be possible to give active enzymes to individuals who cannot make them and to use enzyme therapy in cases where a person produces toxic levels of a chemical that can be enzymatically destroyed. This is not effective in practice because there are few foolproof methods for administering active enzymes to humans. First, there is the problem of getting the enzyme to the tissue where it is needed. Unless properly encapsulated, orally administered enzymes are denatured by the acidic fluid of the stomach. Even if they pass through the stomach, the intact proteins cannot cross the intestinal cell membranes. Enzymes given intravenously are likewise unable to cross cell membranes and enter the deficient organ. (However, in principle, a protein that functions in the blood could be replaced.) An equally serious problem arises because the body views added enzymes as foreign substances and the enzymes trigger the immune response, which leads to the destruction of the enzyme and (often) to allergic reactions. Diabetics who have been treated with bovine insulin for many years require increasingly larger doses, probably owing to the fact that the bovine insulin is inactivated by the body.

There are, however, some special cases in which enzyme treatment is appropriate. A mixture of digestive enzymes (amylase, trypsin, and lipase), in specially coated capsules, is given to patients with problems of duodenal digestion. Enzymes can be applied to the skin and mucous membranes with relatively few problems. Purified preparations of trypsin and chymotrypsin are used to clean the skin after severe burns; these enzymes degrade the proteins in the dead cells, clotted blood, and pus at the burn site. Streptodornase, an enzyme that degrades DNA, is also used to cleanse burned skin. Removal of the debris allows antibiotics to penetrate the area more effectively. Similar degradative enzyme preparations are used to liquefy the thick sputum that collects in the respiratory tract of patients with asthma, emphysema, and cystic fibrosis. In these bronchial and pulmonary disorders, trypsin and a DNA-degrading enzyme are administered by inhalation as an aerosol.

Enzymes

(c) INDUSTRIAL USES OF ENZYMES

The majority of the industrially important enzymes are used to catalyze the breakdown of macromolecules, often carbohydrates or proteins. In many processes, particularly in the food industry, it is necessary to degrade one type of macromolecule (e.g., starch) without damaging other macromolecules (e.g., proteins). The substrate specificity of enzymes makes them ideal for such processes; they are far superior to inorganic acids and bases which break down virtually all macromolecules. Enzymes have the additional advantage that their action can be stopped by heat denaturation once the degradation is completed. Of equal significance to both manufacturers and consumers is the fact that enzymes are essentially nontoxic to humans, although there have been some adverse dermatological side effects with products such as enzyme-containing detergents.

For most industrial applications, relatively impure enzyme preparations are used. The general procedure is to select a microorganism, animal tissue, or plant tissue which is rich in the desired enzyme and then to prepare a water-soluble cell-free extract of that material.

AMYLASES: Amylases are enzymes that degrade starch, a polysaccharide, into a mixture of short chains of glucose plus free glucose. Amylases are used in the preparation of corn syrup, which involves degrading the very large, insoluble cornstarch molecules into smaller, more soluble molecules, mainly maltose (which contains two glucose residues) and glucose, which gives the syrup its sweet taste. The barley extract (mash) used in beer production is important as a source of amylases; the starch molecules in wheat must be broken down to maltose or glucose before they can be fermented by yeast.

In the textile industry, yarns, particularly cotton, are starch-coated prior to weaving to prevent their breakage on mechanical looms. Before dyeing, the woven material is treated with amylases to remove this protective coating.

PECTINASES: Pectinases are analogous in function to amylases, except that they degrade the plant polysaccharide pectin instead of starch. Pectin is a tough structural material found in many fruits and, in the case of apples and grapes, is partially degraded by fungal pectinases to facilitate the industrial extraction of the fruit juice. Fungal pectinases are also used to clarify apple juice, grape juice, and red wine, as consumers will often refuse to buy cloudy beverages.

PEPTIDASES: Peptidases act on proteins to produce short chains of amino acids and individual amino acid molecules. The most commonly encountered peptidase is papain, the active ingredient in meat tenderizers. Papain, which comes from the papaya plant, is also added to bottled beer to destroy the proteins which are a component of chill haze, a fine precipitate that occurs when beer is stored in the cold for long periods of time.

INVERTASE: Did you ever wonder how candies with soft centers are made or how they get the chocolate around the maraschino cherry? The trick involves the use of yeast invertase, an enzyme that cleaves the disaccharide sucrose into its two monosaccharide components, glucose and fructose. The candy center contains a solution saturated with sucrose, which is nearly solid; invertase is added, followed by the chocolate coating. In about 2 weeks, the invertase has split much of the sucrose and the resulting monosaccharides (which are very soluble) go into solution, giving the candy a soft center.

GLUCOSE OXIDASE: Glucose oxidase catalyzes the reaction: glucose + $O_2 \rightarrow$ glucuronic acid + H_2O_2. Powdered eggs, used in cake mixes and other convenience foods, are treated with this enzyme to remove glucose and prevent the interaction between glucose and egg albumin, which leads to bad flavors and a loss of whipping ability. Glucose oxidase is also used to remove molecular oxygen from sugar-containing beverages, such as club soda, in which oxidative changes are detrimental.

4.11 Enzyme activity is destroyed by inhibitors

What do nerve gas and penicillin have in common? The biological effects of these two chemicals, like hundreds of other drugs and poisons, results from their ability to inactivate (inhibit) enzymes. *An enzyme inhibitor is a chemical that combines with an enzyme and reduces or eliminates its catalytic activity.* The decrease in catalytic ability means that the inhibitor has adversely affected the enzyme's active site directly, by damaging or physically clogging up the active site, or, indirectly, by changing the three-dimensional shape of the entire protein, the active site included. Some inhibitors act against only one enzyme, while others inactivate a wide spectrum of enzymes.

Enzyme inhibitors are classified by their mode of action. The three major modes of action are described below. In the vocabulary of enzyme inhibition, *reversibility* refers to whether or not the enzyme–inhibitor complex can dissociate to regenerate the active enzyme. *Competitive* refers to whether or not there is competition between the inhibitor and the natural substrate for a position in the active site.

(a) IRREVERSIBLE INHIBITORS

Irreversible inhibitors combine very tightly (often covalently) with enzymes and cannot be removed without denaturing the protein. Irreversible inhibitors usually attach to, or modify, one particular amino acid R group in the active site, thus both "clogging up" the active site and destroying its ability to bind the substrate and catalyze the reaction. Since the active sites of many enzymes have at least one R group in common, irreversible inhibitors can work against a wide range of enzymes.

(b) REVERSIBLE, NONCOMPETITIVE INHIBITORS

Reversible, noncompetitive inhibitors attach loosely to enzymes, often by ionic bonds to R groups or metal ion cofactors. Because the enzyme–inhibitor association is weak, the protein recovers its catalytic activity when these inhibitors are removed. Noncompetitive inhibitors do not act directly at the enzyme's active site; rather, these inhibitors interact with an R group some distance from the active site. This association alters the three-dimensional shape of the enzyme molecule at the place where the inhibitor lodges, and this change is transmitted throughout the entire protein, including the active site. (This effect is like pushing over a row of dominoes—they all fall if only the first was touched.) Each reversible, noncompetitive inhibitor reacts with specific R groups and will inhibit many enzymes, since the same R groups are found on the exteriors of many different proteins.

(c) REVERSIBLE, COMPETITIVE INHIBITORS

Reversible, competitive inhibitors are fake substrates; they are small molecules which closely resemble the proper substrate for an enzyme, but upon which the enzyme cannot act. Because it resembles the substrate in shape and charge, a competitive inhibitor can fit easily into the active site but, since no reaction takes place, the inhibitor simply remains lodged in the active site, preventing the normal reaction.

Because the enzyme–inhibitor complex is weak, the inhibitor and the natural substrate can compete for the position in the active site, and therefore the inhibition of the enzyme can be relieved by the addition of excess substrate to the enzyme–inhibitor complex. When large quantities of substrate are added to the inhibited enzyme, the substrate molecules (since they "fit" the active site better) displace the inhibitor molecules from the active site and the normal reaction occurs. Thus, the degree of inhibition is a function of the ratio of inhibitor to substrate (rather than the ratio of inhibitor to enzyme as in the other two kinds of inhibition). Because they are substrate analogs, the action of each reversible, competitive inhibitor is specific for one particular enzyme. The existence of reversible, competitive inhibitors provides good evidence for the importance of the three-dimensional (shape) interactions between the substrate and the enzyme.

The control of enzyme activity by both types of reversible inhibitors is the basis of much of the cellular control of metabolism. Metabolic reactions must be balanced with respect to one another so that cell constituents are made in the proper ratios. Since all cellular reactions require enzymes, the rate of these reactions can be kept in balance by increasing or decreasing the activity of certain key enzymes. Intracellular mechanisms for the control of enzyme activity include not only reversible inhibitors, but also reversible activators (Chap. 14).

4.12 Uses and misuses of enzyme inhibitors

Enzyme inhibitors have been used for centuries. From the time of the Roman emperors to the present day, some of the most potent inhibitors of protein function (Table 4.2) have been used to advance man's ambitions and passions. Ancient poisoners recognized the toxicity of metal salts, such as those of arsenic and mercury,

TABLE 4.2 Poisons as Inhibitors of Protein Function

Poison	Mode of action	Comments
Diisopropylfluorophosphate	Inactivates acetylcholinesterase and other "serine" enzymes	Noncompetitive; originally developed as a nerve gas; derivatives used as insect poisons (parathion)
Mercury, arsenic salts	Interact with —SH groups on many proteins	Classic components of poisoners' tool kit; used in medicine (mercurichrome); generally noncompetitive
Diphtheria toxin	Inactivates a protein factor required for protein synthesis in mammals	Produced by certain strains of bacteria
Cholera toxin	Interferes with control of cyclic AMP synthesis	Overproduction of cyclic AMP leads to fluid loss seen in cholera
Cyanide	Combines with cytochrome oxidase and blocks oxidative phosphorylation[a]	—
Malonic acid	Inhibitor of succinic dehydrogenase[a]	Competitive inhibitor
Fluoroacetate	Converted to fluorocitrate; inhibits aconitase[a]	Produced by poisonous plant, *Dichapetalum cymosum*; fatal to animals that eat it
Carbon monoxide	Combines with hemoglobin in blood preventing oxygen transport	Poisonous ingredient in gasoline engine exhaust
Botulism toxin	Prevents release of acetylcholine from nerve terminals	Produced by certain anaerobic bacteria; cause of food poisoning
Antimycin A	Inhibits electron transport	Very toxic to fish; used to kill undesirable fish prior to stocking lakes with commercially valuable species

[a] These reactions are involved in the cells' production of energy from the degradation of small molecules, such as glucose (Chaps. 7 and 8).

that bind to the sulfhydryl groups of cysteine and prevent these groups from forming disulfide bridges, thus destroying the three-dimensional shape of proteins.

More modern poisons include organic phosphate compounds, such as diisopropylfluorophosphate (a deadly nerve gas which is now banned from use in warfare by most nations), which interact specifically with the hydroxyl group of serine. Since the inhibitor forms a covalent bond with the enzyme, inhibition is irreversible. A number of enzymes have serine residues in their active sites, the most important enzyme in this case being acetylcholinesterase, the enzyme that hydrolyzes any excess of the neurotransmitter acetylcholine. Inhibition of acetylcholinesterase causes a buildup of acetylcholine, which deranges nerve function and results in rapid death.

Happily, applications of enzyme inhibitors which are more in the public interest have also been found. Many of the antimicrobial agents used in the treatment of infectious disease are enzyme inhibitors (Table 4.3); these drugs specifically inhibit enzymes found only in bacteria and hence do little harm to the patient.

Table 4.3 Some Antimicrobial Agents and Their Mode of Action Against Bacteria

Agent	Mode of action
Sulfonamide drugs	Compete for p-aminobenzoic acid in enzymatic reaction
Penicillin, ampicillin	Inhibit enzymatic step in the synthesis of cell walls
Tetracycline, erythromycin, chloramphenicol, streptomycin, kanamycin	All inhibit protein synthesis
Rifampicin	New antituberculosis drug, which binds to RNA polymerase and inhibits RNA synthesis

Sulfonamides were among the first and remain among the most popular antimicrobial agents; they are inexpensive and have relatively few and mild side effects. Sulfonamides were discovered by Domagk in the 1930s and it was not until several years later that D. D. Woods found that their antimicrobial activity is reduced in the presence of p-aminobenzoic acid. He correctly reasoned that the chemical similarity between the simple sulfonamide p-aminosulfanilamide and p-aminobenzoic acid (Fig. 4.6) explained how these drugs worked. They inhibit the synthesis of folic acid in bacteria. The sulfa drugs are competitive inhibitors of the enzymatic reaction in the biosynthetic path to folic acid that uses p-aminobenzoic acid as substrate. The structures of two sulfonamide drugs are shown in Fig. 4.6; notice their structural resemblance to p-aminobenzoic acid. Woods' finding also explained why sulfonamide drugs are not effective in open, suppurating wounds; such wounds contain pus and other materials that are a source of p-aminobenzoic acid, which antagonizes the action of the sulfonamide drugs.

Why are sulfonamide drugs effective against bacteria without injuring the human host? The answer is simple: human beings do not have the enzymes for synthesizing folic acid and therefore require this compound as a vitamin. Thus, sulfonamide drugs have nothing to inhibit in human beings.

For cancer chemotherapy, a number of enzyme inhibitors have been tested and found to be effective in eliminating or reducing the spread of malignant cells. Although cancer cells originate from normal mammalian cells, they have rapid and abnormal

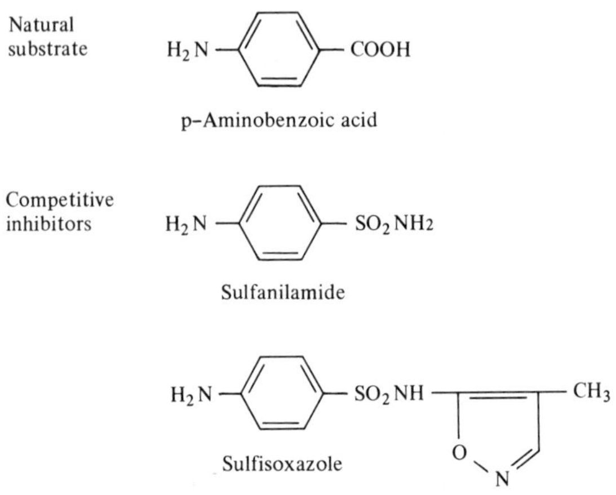

Natural substrate: H₂N—⟨benzene⟩—COOH

p-Aminobenzoic acid

Competitive inhibitors: H₂N—⟨benzene⟩—SO₂NH₂

Sulfanilamide

H₂N—⟨benzene⟩—SO₂NH—⟨isoxazole ring with CH₃⟩

Sulfisoxazole

Figure 4.6 Sulfonamide drugs are competitive enzyme inhibitors. In bacteria, an enzyme needed for folic acid production uses p-aminobenzoic acid as its substrate. The sulfonamide drugs are structurally similar to p-aminobenzoic acid and hence can compete with the natural substrate for a place in the enzyme's active site.

metabolism and growth. The aim of cancer chemotherapy is to inhibit the growth of cancer cells before slower-growing normal cells are damaged. Unfortunately, most anticancer drugs are not specific for cancer cells and do eventually affect normal mammalian cells. One clinically useful anticancer compound is 5-fluorouracil, which inhibits an enzyme involved in the synthesis of one of the monomers of DNA (Sec. 11.8).

SUMMARY

Enzymes are protein molecules that speed the conversion of one organic chemical to another in cells. Enzymes speed up reactions by lowering their activation energy—that extra energy the substrates must acquire before the reaction will occur.

The three-dimensional shape of each enzyme is such that that enzyme can catalyze only one metabolic reaction. The reaction occurs at a specific place on or within the enzyme, called the active site. The active site of each enzyme has the proper shape, size, and charge to bind certain substrates only and to catalyze the conversion of these substrates to specific products.

Enzyme activity is sensitive to outside influences. Because they are proteins, enzymes work best at the neutral pH and moderate temperatures typical of most cells; deviations from these conditions decrease enzyme activity. Enzymes are also inhibited by compounds that damage or block their active sites, or which cause gross alterations in their three-dimensional shapes. Competitive inhibitors—whose structures mimic the enzyme's natural substrate—are of special usefulness because they inhibit specific enzymes; many drugs are competitive inhibitors.

Enzymes

PRACTICE PROBLEMS

1. When an enzyme and its substrate(s) are mixed together, the enzyme catalyzes the conversion of substrate(s) to product(s). For example, mixing trypsin with hemoglobin starts the reaction

$$\text{hemoglobin} \xrightarrow{\text{Trypsin}} \text{small peptides}$$

   Write the equation for the reaction that occurs when the following compounds are mixed at 30°C for 3 hours.
   (a) Lactic acid, lactate dehydrogenase, and NAD^+.
   (b) Alkaline phosphatase and glucose phosphate.
   (c) Chymotrypsin and DNA.
   (d) ATP, glycerol, and glycerol kinase.
   (e) Chymotrypsin and ala-glu-phe-arg-arg.

2. Starch turns purple when exposed to iodine; glucose and maltose do not give this color reaction. Given a potato and an iodine solution, describe how you could assay your saliva for amylase, a major starch-digesting (hydrolyzing) enzyme.

3. Arginase, a liver enzyme, catalyzes the cleavage of the amino acid arginine to ornithine and urea; this enzymatic reaction is critical to nitrogen excretion in man. The data below show the relative amounts of urea produced when arginase acts on arginine at different pHs. From these data, draw the curve of reaction rate versus pH for the activity of arginase. At what pH does arginase have its greatest catalytic activity? Considering the effects of pH on the charges of amino acids, give two possible explanations for the pH dependence of this reaction.

pH	Units urea produced/milligram arginase/minute
5	0
6	5
7	37
8	50
9	80
10	97
11	82
12	60

4. Explain the following enzyme-based phenomena.
   (a) Pencillin does not cure cancer.
   (b) Chymotrypsin inactivates lactate dehydrogenase.
   (c) Meat tenderizers are applied several hours before the meat is cooked rather than during the cooking process.
   (d) Livers from people with niacin deficiencies are expected to have low levels of activity of several dehydrogenases.

5. To assay alkaline phosphatase levels, the enzyme and its substrate, *p*-nitrophenylphosphate (Fig. 4.3), are allowed to react for 30 minutes. The reaction is then quickly

stopped and the amount of colored product measured. Describe two methods you could use to stop the reaction in less than a minute.

6. Enzyme L, isolated from human liver, catalyzes the reaction A + B → C + D. The graph below is the reaction curve when 5 molecules of enzyme L are mixed with 1000 molecules of A and 1000 molecules of B at 25°C. Draw the reaction curves that would be obtained under the conditions described in (a), (b), and (c) below.

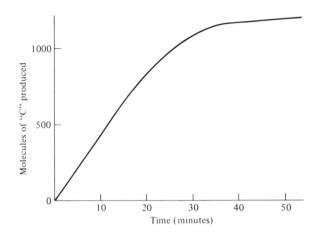

(a) The reaction curves at 90°C and 2°C.
(b) The reaction curve when a large amount of a noncompetitive inhibitor is added to the reaction mixture at the start of the reaction.
(c) The reaction curve when 15 molecules of enzyme L are added to 1000 molecules of A and 1000 molecules of B.

7. Two compounds are known to act as inhibitors of the reaction:

$$\text{isoleucine} + \text{ATP} \xrightarrow{\text{Synthetase}} \text{isoleucyl-AMP} + \text{pyrophosphate}$$

$$\begin{array}{c} H_3C \\ \diagdown \\ CH-CH-COOH \\ \diagup | \\ H_3C NH_2 \end{array} \qquad H_2N-\!\!\bigcirc\!\!-SO_2NH_2$$

(a) Valine (b) PAS

Which of these would you expect to be a competitive inhibitor of the synthetase? Why? What would happen if an excess of isoleucine were added to a solution of the synthetase inhibited by the competitive inhibitor?

8. On the following page are the energy profiles of three different metabolic reactions.
(a) Which of the reaction(s) are exergonic? Which endergonic?
(b) What is the activation energy for each reaction?
(c) Energy given off in cellular reactions is often trapped by the cell to be used later to do cellular work, such as muscle contraction. How much energy would be available to perform cellular work from each of these reactions?

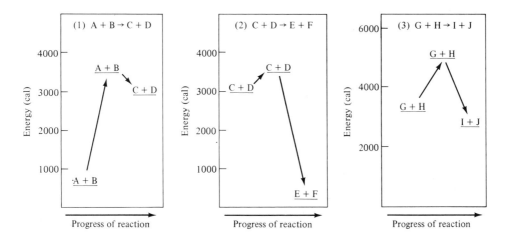

SUGGESTED READING

CHISOLM, J. JULIAN, JR., "Lead Poisoning," *Sci. American*, 224, No. 2, p. 15 (1971). Lead inhibits the enzymes of heme biosynthesis.

DICKERSON, RICHARD E., and IRVING GEIS, *The Structure and Action of Proteins*, Harper & Row, Inc., New York, 1969. See references to Chapter 2.

FRIEDEN, EARL, "The Chemical Elements of Life," *Sci. American*, 227, No. 1, p. 52 (1972). Describes the role of trace elements as enzyme cofactors.

GOLDWATER, LEONARD J., "Mercury in the Environment," *Sci. American*, 224, No. 5, p. 15 (1971). A balanced report emphasizing that mercury-containing compounds range from useful medications to deadly poisons.

KAMATH, S. H., *Clinical Biochemistry for Medical Technologists*, Little, Brown and Company, Boston (1972). Describes spectrophotometers and other machines routinely used in clinical and research laboratories.

PHILLIPS, DAVID C., "The Three-Dimensional Structure of an Enzyme Molecule," *Sci. American*, 215, No. 5, p. 78 (1966). Description of lysozyme.

REED, G., *Enzymes in Food Processing*, Academic Press, Inc., New York (1966). Commercial uses of enzymes are described.

STROUD, ROBERT M., "A Family of Protein-Cutting Proteins," *Sci. American*, 231, No. 1, p. 74 (1974). Gives details of the relationship between the structure and function of chymotrypsin, trypsin, and other enzymes.

5 General Concepts in Metabolism

Good nutrition is based on common sense and means eating a balanced diet of meat, fruit, vegetables, and grains and not too much cake, candy, and fried foods. What do the "proper" foods provide that makes them so important to our bodies? In a general sense, we all know the answers. Meat provides protein to build muscle. Sailors developed scurvy when they had no fresh fruit to provide vitamin C. Candy bars provide quick energy, but eating too many makes us fat. How does this happen? What processes turn food into "us"? Studies of metabolism try to answer these questions from the point of view of the individual cell. *Metabolism is* defined as *the sum of all the chemical reactions that take place in cells*. It includes the processes by which cells use external substances (i.e., food) to build new cells, obtain energy to do work, store calories for future use, and breakdown and excrete unnecessary compounds.

For convenience, metabolism can be seen as having two general purposes: (1) to capture energy from the environment and convert it to a form suitable for use in cellular work, and (2) to transform groups of organic compounds into one another, particularly the transformation of small organic compounds into macromolecules. The two aspects of metabolism are interdependent; neither can occur without the other. On the one hand, cells obtain energy by the exergonic breakdown of organic compounds and, on the other hand, a major use of stored energy is the synthesis of macromolecules.

General Concepts in Metabolism

Cells sequester energy from the environment by two major routes (Chap. 6). Green plants and algae capture light energy through photosynthesis. Animal cells and most microorganisms obtain energy by breaking down organic compounds that are high in chemical energy (particularly sugars and lipids) into products with less chemical energy. During these exergonic degradations, the energy that is released is captured and stored as adenosine triphosphate (ATP). In both processes the stored energy can be used at a later time to perform cellular work. This work includes the synthesis of structural and informational macromolecules, the work of transporting organic molecules and inorganic ions across the cell membrane, the electrical work of nerve cells, and the mechanical work of muscles and flagella.

Only a relatively small number of the organic compounds present in the environment can pass across cell membranes. The second purpose of metabolism is the conversion of this rather small set of compounds into a larger and more diversified set of molecules. Thousands of reactions, each catalyzed by a specific enzyme, are required to split, join together, and rearrange the atoms of these compounds to create the enormous array of carbohydrates, lipids, proteins, and nucleic acids needed for proper cell functioning.

This chapter is intended to give you an overview of metabolism. In later chapters, as you study individual reactions, try to keep in mind how each reaction fits into the larger scheme of things. It is very easy to get involved in studying single reactions or pathways and miss how such reactions fit together in the grand scheme of metabolism. (In short, don't miss the forest for the trees!)

5.1 Every cellular reaction is one step in a metabolic pathway

Metabolic reactions do not occur in a random or haphazard manner, and are not independent of each other. Every metabolic reaction is part of a larger scheme involving many other reactions, and, if traced far enough through the metabolic maze, each reaction is connected to every other. The simplest demonstration of the integration of metabolic reactions is the metabolic pathway. In a metabolic pathway, the product of the first reaction becomes the substrate for the second reaction; the product of the second reaction is the substrate for the third reaction, and so on down the line. Such series of consecutive enzymatic reactions allow the cell to carry out molecular conversions which are impossible to perform by any single reaction.

To illustrate this concept, let us examine the series of reactions whereby the sugar, glucose, is converted to pyruvic acid (pyruvate). This degradation is important because the energy released in several of the chemical steps can be stored in ATP (Chap. 6) and because several of the intermediates formed between glucose and pyruvate are crucial for other metabolic processes (Sec. 5.2). (The complete pathway is shown in Fig. 7.2.) As soon as glucose enters a cell, a phosphate group from ATP is added, forming glucose-6-phosphate. (The "6" designates which of the six carbons of glucose receives the phosphate group.)

Glucose-6-phosphate becomes the substrate for a second reaction; an isomerase catalyzes the rearrangement of glucose-6-phosphate to another phosphorylated sugar, fructose-6-phosphate. Fructose-6-phosphate is, in turn, acted upon by a third enzyme, which adds another phosphate group, forming fructose-1,6-diphosphate. This product becomes the substrate for another enzyme-catalyzed reaction and the cycle continues through five more reactions until glucose is converted into pyruvic acid. Instead of writing separate equations for each reaction in this metabolic pathway, the steps are summarized:

$$\text{glucose} \xrightarrow[\text{ATP} \quad \text{ADP}]{\text{Kinase}} \text{glucose-6-phosphate} \xrightarrow{\text{Isomerase}}$$

$$\text{fructose-6-phosphate} \xrightarrow[\text{ATP} \quad \text{ADP}]{\text{Kinase}} \text{fructose-1,6-diphosphate}$$

Metabolic pathways are classified by whether their "aim" is to build molecules (*synthetic* or *anabolic pathways*) or to break down molecules (*degradative* or *catabolic pathways*). Anabolic pathways are those which start with small, simple compounds—amino acids, sugars, acetyl coenzyme A—and culminate in the production of macromolecules. Most anabolic pathways require substantial energy input to drive them. During glycogen synthesis, for example, two molecules of ATP are used each time a glucose residue is added to the growing glycogen "tree" (Sec. 9.1). In catabolic pathways, macromolecules are broken down into their monomers and the resulting sugars, amino acids, and so forth may be further degraded. Some of the end products of catabolic pathways, such as the carbon dioxide produced during the complete breakdown of glucose, are excreted from the cell. Although they often involve energy-requiring steps, the net result of most catabolic pathways is to produce energy that becomes available for cell work. The metabolic chart presented in Fig. 5.1 shows the interrelationships between some anabolic and catabolic pathways.

The inside of a cell is a busy and crowded place. Every cell contains thousands of molecules of each intermediate in the various pathways, and these molecules are constantly being moved along their respective pathways. Because enzymes are reusable, a relatively small number of enzyme molecules can process the large number of intermediates that flow through a given pathway. Although one pathway may predominate at a particular time in response to certain environmental conditions or in the specialized organs of higher organisms, the basic reactions of metabolism, for example, the conversion of glucose to pyruvate, are carried out by virtually all cells, using the same chemical steps. At this level, there is little difference between bacteria and elephants.

5.2 Metabolic pathways are interconnected by branch-point compounds

When we consider individual metabolic pathways, each seems to be self-contained, but this is mostly a delusion created by our need to simplify and categorize. In reality, *all the pathways of metabolism are interdependent*. The pathways by which

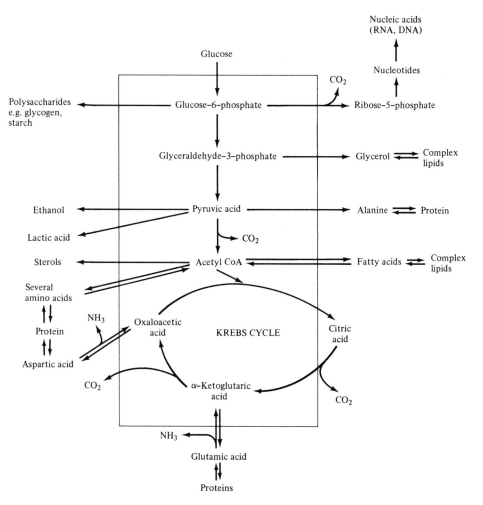

Figure 5.1 Metabolism "in a nutshell." This chart outlines how molecules flow between the pathways of carbohydrate, lipid, nucleic acid, and protein metabolism. There is no real starting point. The reactions going on in a cell at a particular time depend on the external environment, on what the cell has previously stored, and on what biosynthetic work the cell is programmed to do. Many of the arrows represent several enzymatic steps; for example, the conversion of glucose-6-phosphate to glyceraldehyde-3-phosphate requires four reactions.

The major branch-point compounds are indicated within the rectangular area.

Superimposed on these systems are the energy requirements of, and the energy produced by, the various pathways. In general, pathways that build macromolecules, such as protein and fatty acid synthesis, use energy; pathways that degrade molecules, such as glycolysis and the Krebs cycle, produce energy.

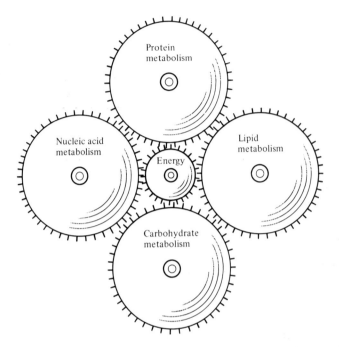

Figure 5.2 Metabolism consists of interlocking pathways. Cellular metabolism consists of five major interdependent parts. No segment of metabolism can function for long unless all the other pathways are working.

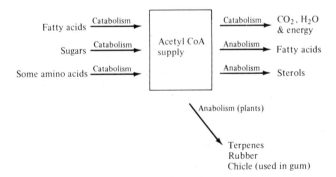

Figure 5.3 Acetyl CoA, a branch-point compound. Acetyl CoA is produced by several catabolic pathways and can either be used to build larger molecules or broken down to provide cellular energy. The fate of any individual acetyl CoA molecule depends on the needs of the cell at the time it is formed.

carbohydrates, proteins, and lipids are produced and degraded are not self-sufficient because compounds produced by one pathway are essential to the functioning of other pathways. In addition, all anabolic and catabolic pathways are linked through the cellular energy supply. The whole of metabolism can be viewed as a set of interlocking gears (Fig. 5.2); either all the central pathways turn or none of them work.

The metabolic intermediates that link pathways together are called "branch-point" compounds, because they occur at a "fork in the road of metabolism." Branch-point compounds are those intermediates produced by one or more pathways that are used as the starting compound for several other pathways. One of the most important branch-point compounds is the activated form of acetic acid, acetyl coenzyme A (acetyl CoA). This intermediate is the end product of several catabolic pathways and serves as the starting material for other anabolic and catabolic pathways (Fig. 5.3). Some other important branch-point compounds are shown in Fig. 5.1. The cellular control mechanisms which operate to determine the fate of any particular molecule of a branch-point compound are discussed in Sec. 5.3 and Chap. 14.

5.3 Cells control metabolic reactions by regulating enzyme activity

How are metabolic pathways controlled? As stated earlier each of the central pathways of metabolism (Fig. 5.1) *can* be carried out by all cells. But, each individual cell may exist in a different and constantly changing environment. To use the available nutrients efficiently and to respond appropriately to external stimuli (temperature, hormones, etc.), a cell must emphasize some metabolic pathways at certain times and then shift to other pathways when environmental conditions change.

Ultimately, of course, the external environment sets the limits on the metabolic pathways that a cell can carry out. No cell can grow in an environment lacking carbon compounds. A human being cannot live without a source of vitamins. There is no way a cell can compensate for these missing compounds.

However, we are more interested in the ways in which a cell can respond to more subtle variations in its environment. There are situations in which, to make economical use of nutrients, a cell must react to the environment by making a choice between metabolic pathways. Imagine that a cell growing in an environment containing glucose as its major carbon source is shifted to an environment containing glucose and alanine. In its new situation, this cell could continue to produce alanine from glucose, but this would be inefficient, since alanine is now available *gratis*. An efficient cell would turn off its own alanine-producing machinery and use alanine from the environment in the synthesis of proteins.

What gives cells this metabolic flexibility? Assuming that the supply of substrate is plentiful, the extent of a metabolic reaction is proportional to the amount or the activity of enzyme available to catalyze the conversion of that substrate to product. Within a cell, enzyme activity can be regulated in two general ways: (1) by increasing or decreasing the catalytic activity of preexisting enzyme molecules, or (2) by increasing

TABLE 5.1 Control of Enzyme Activity

1. The *catalytic activity* of a preexisting enzyme molecule can be altered by:
    (a) The reversible binding of small inhibitor or activator molecules, which induces changes in the three-dimensional structure of the protein and hence of the active site.
    (b) Conversion of inactive to active form (or *vice versa*) by structural alterations in the protein itself, such as removal of amino acids, phosphorylation.
    (c) Changes in the availability of enzyme cofactors such as metal ions.
2. The *total number* of enzyme molecules can be altered by:
    (a) Changes in the rate of enzyme synthesis (by alterations in either transcription or translation).
    (b) Changes in the rate of enzyme degradation.

or decreasing the total number of enzyme molecules. There are any number of ways a cell can accomplish (1) or (2); some are listed in Table 5.1. Chapter 14 describes these mechanisms of regulation in detail.

Any single reaction can be accelerated or decelerated by changes in the activity of the enzyme catalyzing the reaction. To emphasize or deemphasize an entire metabolic pathway, it is not necessary to change the activity of every enzyme in that pathway. Since the extent of a reaction ultimately depends on the availability of its substrate, a significant change in the concentration of any one metabolic intermediate can cause a corresponding change in the amounts of all compounds that are formed after it in the pathway. In other words, "a biochemical chain is as strong as its weakest link." However, the most economical way to regulate a pathway is to change the activity of the first enzyme in that pathway. Inhibition of the first enzyme in a pathway stops the production of end product without wasting any energy or nutrients in the production of unnecessary early intermediates. Since branch-point compounds (Fig. 5.1) are the starting material for many pathways, enzymes that act on these compounds as substrate are prime targets for metabolic control.

5.4 Studying the metabolic maze

How are the details of metabolic pathways elucidated? How do we know that the glucose we eat is converted to pyruvate by a pathway including glucose-6-phosphate and six other intermediates? The first approach to an understanding of metabolism was to determine exactly the nature and amounts of compounds that cells *do* contain. Such analyses produce long lists of compounds found in cells, but do not tell us how or if the compounds are metabolically related.

Biochemists *have* been able to do much more than catalog the chemicals found in cells. As described below, studies using isotopic tracers and studies of cells with blocks in metabolism have been invaluable aids to elucidating the sequences of metabolic events in whole organisms, tissue slices, and preparations of subcellular organelles. In addition, examinations of cell-free systems have revealed many of the fine details of metabolic processes and their regulation.

General Concepts in Metabolism

(a) ISOTOPIC TRACERS

The metabolic fate of a compound can be determined by "tagging" or labeling the molecule in some way and then using the tag to follow the compound along its metabolic journey (vaguely like putting food coloring down the sink and following the color to determine the route the effluent takes on its way to Lake Erie). Early experiments were done by chemically attaching dyes to metabolic intermediates and looking for the appearance of the dye in other compounds. Such chemical tagging is of limited usefulness, however, because the presence of the dye may alter the metabolism of the intermediates under study.

What is needed is a method for labeling the atoms of the compounds involved in cellular reactions such that the scientist can "see" the label but the cell cannot. The use of isotopes has allowed scientists to do exactly this.

Isotopes are alternative forms of an element that occur in nature. The isotopes of a given element all have the same atomic number, but different atomic weights. For instance, all carbon isotopes have the atomic number 6; the most common has an atomic weight of 12, but very small amounts of carbon with atomic weights of 13 or 14 also exist in nature. Because they all have the same atomic number, *all isotopes of a given element have the same chemical properties*, and therefore cells cannot distinguish between isotopes of the same element. All the isotopes of carbon work equally well in metabolic compounds; cells cannot distinguish glucose containing carbon with an atomic weight of 14 (^{14}C-glucose) from the more abundant ^{12}C-glucose. Isotopes used in biochemical research are listed in Table 5.2.

TABLE 5.2 Isotopes Used in Metabolic Studies

The chemical properties of the rare isotopes are identical to those of the more abundant form of the element. Rare isotopes can be detected by their unusual physical properties.

Element	Abundant isotope	Rare isotope	Distinguishing physical property of rare isotope
Nitrogen	^{14}N	^{15}N	Heavy, nonradioactive
Carbon	^{12}C	^{13}C	Heavy, nonradioactive
		^{14}C	Radioactive
Hydrogen	^{1}H	^{3}H	Radioactive
		^{2}H	Heavy, nonradioactive
Phosphorus	^{31}P	^{32}P	Radioactive
Iodine	^{127}I	^{131}I	Radioactive
Sulfur	^{32}S	^{35}S	Radioactive

The advantage of using rare isotopes in metabolic studies is that these isotopes are easily detected by the unusual *physical* properties they give a molecule (Table 5.2). These physical properties are the "tag" which the scientist can follow. Some heavy isotopes, such as ^{2}H and ^{15}N, are stable and, when incorporated into macromolecules, give those macromolecules an unusually high density. A protein formed from

^{15}N-amino acids is denser than its normal ^{14}N counterpart. Other isotopes, such as carbon 14, are unstable (radioactive), and their breakdown produces ionizing radiation which can be detected by special radioactivity counters. A cell grown on ^{14}C-amino acids contains radioactive proteins.

The usual approach to studies of metabolic pathways is to employ specially prepared metabolic intermediates containing high levels of one or the other rare isotope. As an example, suppose that a rat were fed ^{14}C-glucose. Within a short time, its tissues would contain ^{14}C-glucose-6-phosphate and after a few more minutes, ^{14}C-pyruvic acid would be formed and the animal would begin to exhale radioactive CO_2. In addition, label from the ^{14}C-glucose would eventually appear in liver and muscle glycogen. The isolation and identification of such radioactive compounds tells us what intermediates are derived from glucose. The time course of the appearance of labeled carbon in various metabolites indicates the order in which these intermediates are formed. By studying how much of the radioactive label goes into glycogen as opposed to carbon dioxide, one can determine which pathway is quantitatively more important under varying experimental conditions. Even more detailed metabolic analyses can be carried out if a single specific carbon atom in a molecule is radioactively labeled (Fig. 5.4).

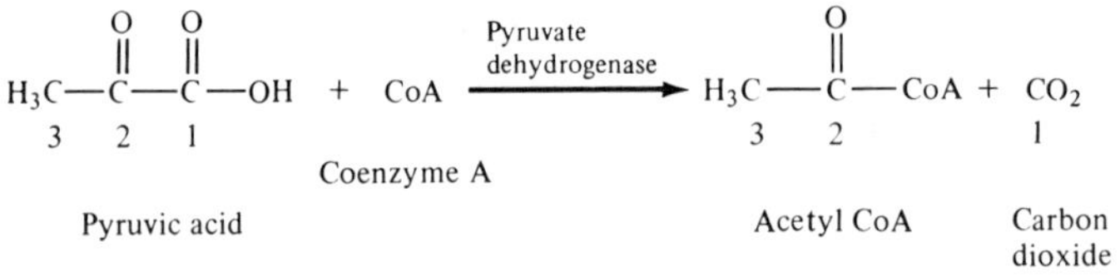

$$H_3C-\underset{3}{\overset{\overset{O}{\|}}{C}}-\underset{2}{\overset{\overset{O}{\|}}{C}}-\underset{1}{OH} + \underset{\text{Coenzyme A}}{CoA} \xrightarrow{\text{Pyruvate dehydrogenase}} H_3C-\underset{3}{\overset{\overset{O}{\|}}{C}}-\underset{2}{CoA} + \underset{1}{CO_2}$$

Pyruvic acid   Acetyl CoA   Carbon dioxide

Figure 5.4 Use of molecules in which a specific atom is isotopically labeled. When pyruvic acid reacts with coenzyme A to form acetyl CoA, one molecule of carbon dioxide is produced (Sec. 7.5). To understand the mechanism of this reaction, one needs to know which of the carbon atoms of pyruvic acid ends up in carbon dioxide. To answer this question, pyruvic acid is prepared in which only carbon 1 is radioactive; when such pyruvate-1-^{14}C is metabolized, radioactive carbon dioxide is produced. If, instead, carbon 2 or carbon 3 were radioactively labeled, no ^{14}C-carbon dioxide would be given off.

(b) METABOLIC BLOCKS

Much of our understanding of metabolic pathways comes from studies of pathways that are not functioning because one of the reactions in the sequence is blocked. When one reaction in a pathway is blocked, the pathway is effectively "dammed up." The substrates of the blocked reaction (and sometimes earlier intermediates) accumulate in the cell as the cell continues trying to produce the end product of the pathway. In extreme cases, the substrate for the blocked reaction is excreted from the cell (Table 5.3).

Blocking a specific step in a metabolic pathway is achieved by inactivating or eliminating the enzyme which catalyzes that step in the sequence. Methods for block-

TABLE 5.3 Effects of Blocking Specific Reactions in a Metabolic Pathway

Hypothetical metabolic pathway:
$$A \longrightarrow B \longrightarrow C \longrightarrow D \longrightarrow E$$
The letters represent intermediates in a hypothetical metabolic pathway. The consequences of blocking any one step in the pathway are shown below.

Step blocked	Intermediates that accumulate	Intermediates not produced
Conversion A → B	A	B, C, D, E
Conversion B → C	B, (A)[a]	C, D, E
Conversion C → D	C, (A, B)[a]	D, E
Conversion D → E	D, (A, B, C)[a]	E

[a] Whether or not the compounds in parentheses accumulate depends on the pathway being studied.

ing specific enzymatic steps fall into two categories: (1) the use of inhibitors of specific enzymes (Chap. 4), and (2) the use of mutants. Mutant cells cannot produce an active enzyme because the genetic information for making that protein is defective (Chap. 12). Mutant strains of microorganisms with defects in specific pathways can be isolated in the laboratory and analyses of the plentiful supply of such strains has been a major source of our knowledge of metabolism.

Regardless of how they are achieved, the consequences of single metabolic blocks (Table 5.3) reveal a great deal about the pattern of cellular metabolism. The accumulation or excretion of unusual compounds as a result of single metabolic blocks has led to the identification of intermediates in specific pathways. The mutants of higher organisms include individuals with inherited metabolic disorders, such as the glycogen storage diseases. These diseases, in which glycogen breakdown is blocked, are good examples of how the characterization of accumulated compounds has helped us to understand normal metabolism. In one such disease, an abnormal type of glycogen molecule accumulates in the patient's liver cells as a result of a block in glycogen utilization, and its presence indicates that the abnormal molecule is an intermediate usually found in only trace amounts. Analyses of products that accumulate in different glycogen storage diseases have allowed the identification of several intermediates on the route of glycogen catabolism.

Metabolic blocks also tell us about the sequence of events (i.e., the order in which intermediates are formed). Let us assume that we know a pathway consists of intermediates A, B, C, D, and E, but we are unsure of the order of their conversion along the pathway. Suppose that a metabolic block allows the cells to produce A and B, but not C, D, or E. This result leads to the conclusion that A and B are formed earlier on the pathway than C, D, and E. If a different blocking agent is available which allows the cells to produce A, B, C, and D, but not E, one can conclude that E is the final product of the metabolic sequence. By summing up the results of several different blocks in a pathway, it is often possible to place all the intermediates in an unequivocal order. The use of mutants in determining metabolic pathways in this way was first developed by Beadle and Tatum, and they were awarded a Nobel prize for this accomplishment (Sec. 12.9).

(c) EXPERIMENTS WITH CELL-FREE EXTRACTS (*in vitro* EXPERIMENTS) REVEAL THE FINE POINTS OF METABOLISM

Once a metabolic pathway has been characterized by studies of the whole cell or organism, using the techniques described above, the final proof is to reconstruct the reaction or pathway in a cell-free system (*in vitro*). All the components thought to be necessary for a metabolic pathway (enzymes, cofactors, etc.) are purified and mixed together with the first substrate in the pathway; if the predicted intermediates and end products of the pathway are produced, one can be certain that the components of and conditions for the operation of the pathway have been correctly identified.

Only by studying reactions or pathways in the test tube, separate from other related pathways, is it possible to understand the details of enzyme action and hence the chemistry of each step in a given pathway. In addition to changing the amounts of substrates and end products of a pathway, it is possible to add suspected control compounds (hormones, inhibitors, or activators) to a test-tube reaction. Hence, *in vitro* systems allow us to elucidate aspects of the regulation of metabolic reactions.

SUMMARY

Metabolism is the sum of all the chemical reactions that take place in cells; each of these reactions is catalyzed by a specific enzyme. Several enzymes working in sequence to perform a specific chemical conversion constitute a metabolic pathway. Metabolic pathways are interrelated, providing substrates for, and using products from each other. Cells can control the rate of a given reaction, and hence the rate of the entire pathway of which it is a part, by changing the amount or the activity of the enzyme which catalyzes that reaction.

Many techniques are used to probe the metabolic "maze." Two important ones are (1) the use of isotopic labels to follow a particular compound along its metabolic route, and (2) studying the consequences of blocking a single metabolic reaction with enzyme-specific inhibitors or by mutation. The most detailed metabolic analyses are done in cell-free systems.

PRACTICE PROBLEMS

Use the metabolic pathways shown in Fig. 5.1 to answer the following questions. You may also want to review Chap. 1.

   1. The anabolism and catabolism of different macromolecules are connected by branch-point compounds. For example,

$$\text{proteins} \xrightarrow{\text{Catabolism}} \text{acetyl CoA} \xrightarrow{\text{Anabolism}} \text{fatty acids}$$

General Concepts in Metabolism 99

Fill in the blanks in the following schemes with the appropriate branch-point compound.

(a) Glucose $\xrightarrow{\text{Catabolism}}$ _____ $\xrightarrow{\text{Anabolism}}$ complex lipids.

(b) Protein $\xrightarrow{\text{Catabolism}}$ _____ $\xrightarrow{\text{Anabolism}}$ complex lipids.

(c) Polysaccharides $\xrightarrow{\text{Catabolism}}$ _____ $\xrightarrow{\text{Anabolism}}$ nucleic acids.

(d) Glucose $\xrightarrow{\text{Catabolism}}$ _____ $\xrightarrow{\text{Anabolism}}$ sterols.

(e) Fatty acids $\xrightarrow{\text{Catabolism}}$ _____ $\xrightarrow{\text{Anabolism}}$ proteins.

2. Anabolic pathways use energy, while catabolic pathways generate energy in the form of ATP. Which of the following pathways would you expect would produce energy? Which would require energy?
   (a) Acetyl CoA to complex lipids.
   (b) Glucose to acetyl CoA.
   (c) Acetyl CoA to sterols.
   (d) Glucose to glycogen.
   (e) Acetyl CoA to carbon dioxide and water.
   (f) Acetyl CoA to rubber.

3. A eucaryotic cell contains many specialized organelles. Some metabolic pathways are restricted to specific organelles. Where in the cell would you expect most DNA synthesis to occur? What organelle is needed for the incorporation of amino acids into proteins? Where do the reactions of the Krebs cycle occur?

4. Complete the following chart by indicating which of the isotopes can be used to label the molecules listed.

	^{14}C	^{15}N	^{32}P	^{3}H
Polysaccharide	Yes	No		
DNA and RNA				
Protein				
Fatty acids				
ATP				
Pyruvic acid				

5. Although their caloric intake may be adequate, human beings cannot survive very long on a diet of starch, a nitrogen source, and the required vitamins and minerals. What essential compounds are missing from this diet? What class of macromolecules cannot be made? What could be added to this diet to make it nutritionally complete?

6. A bacterial culture is growing in a liquid medium containing ^{14}C-glucose (all carbons radioactive), a nitrogen source, and sufficient supplies of vitamins and minerals. After

10 generations, the bacteria are removed from the medium and their components examined. Which of the following compounds would you expect to be radioactive?

Acetyl CoA
Glycerol
Aspartic acid
Phosphofructokinase
Water
RNA

If the growth medium had also contained large amounts of nonradioactive oxaloacetic acid, which compound(s) would contain greatly reduced levels of radioactivity?

7. The reactions shown in Fig. 5.1 are not balanced; that is, they do not show how many molecules of each substrate are needed to produce a molecule of product. Yet, it is obvious that every atom in a molecule entering a pathway must be accounted for in the end products. For example, glucose contains six carbon atoms, but pyruvic acid only three. Therefore, each glucose molecule can give two molecules of pyruvic acid. Balance the following equations so that the number of carbon atoms in the substrate equals that in the product. (Glycerol is a three-carbon compound, ribose has five carbons, and, for the purpose of these equations, acetyl CoA has two carbons.)
   (a) 1 Glucose → __ glycerol.
   (b) __ Pyruvic acid → 1 acetyl CoA + __ CO_2.
   (c) __ Acetyl CoA → 1 16-carbon fatty acid.
   (d) 1 Glucose → __ ribose + __ CO_2.
   (e) __ Glucose → 1 16-carbon fatty acid + __ CO_2.

SUGGESTED READING

Large (2 feet × 3 feet) metabolic "maps" are available from P·L Biochemicals, 1037 West McKinley Ave., Milwaukee, WI 53205.

DAVIDSON, ERIC H., "Hormones and Genes," *Sci. American*, 212, No. 6, p. 36 (1965). Summarizes the metabolic effects of many animal hormones.

LIEBER, CHARLES S., "The Metabolism of Alcohol," *Sci. American*, 234, No. 3, p. 25 (1976). Alcohol disrupts many metabolic pathways in liver cells.

MACALPINE, IDA, and RICHARD HUNTER, "Porphyria and King George III," *Sci. American*, 221, No. 1, p. 38 (1969). A lively account of the far-reaching consequences of a single, inherited defect in metabolism.

PITTS, FERRIS N., JR., "The Biochemistry of Anxiety," *Sci. American*, 220, No. 2, p. 69 (1969). Excessive lactic acid, a normal metabolite, causes anxiety neurosis in susceptible individuals.

SCRIMSHAW, NEVIN S., and VERNON R. YOUNG, "The Requirements of Human Nutrition," *Sci. American*, 235, No. 3, p. 50 (1976). Summarizes the relationships between anabolic and catabolic pathways and the cellular energy cycle.

YOUNG, VERNON R., and NEVIN S. SCRIMSHAW, "The Physiology of Starvation," *Sci. American*, 225, No. 5, p. 14 (1971). Humans respond to fasting by adaptive changes in metabolism.

6

Energy for Cell Work: the Role of ATP

Like building a house, running a mile, or mowing the lawn, *all work done by cells requires energy.* Cells need energy to stockpile small molecules from the environment and to use these small molecules to build large macromolecules, such as proteins, nucleic acids, and polysaccharides. The work done by specialized cells is obvious: muscle cells contract, nerve cells transmit stimuli, and luminescent cells give off light.

This chapter presents a general view of how cells obtain, store, and use energy. The following three chapters describe the chemistry of energy capture and storage in greater detail.

6.1 ATP: the "middleman" between energy-releasing and energy-requiring processes

Most degradative pathways in the cell, including the catabolism of sugars, lipids, and amino acids, release energy as heat or chemical-bond energy; biosynthetic pathways and all other kinds of cellular work require energy. These energy-releasing and energy-requiring processes must be coupled so that energy is available when and where work needs to be done.

In principle, energy-releasing and energy-requiring processes could be directly coupled: each time work is to be done, the cell could take in a small molecule, degrade it, and immediately use the energy released for the desired purpose. But this simple mechanism presents problems. The availability of external supplies of degradable molecules fluctuates and the large number and diversity of both energy-releasing and energy-requiring processes makes such direct supply-and-demand coupling cumbersome and complex.

To avoid these complications, the cell uses a "middleman," a molecule that picks up and holds the energy from any and all energy-releasing processes and then donates this energy to any and all energy-requiring processes when needed. *The middleman in the cellular energy cycle is adenosine triphosphate (ATP).* The use of ATP as energy currency (Fig. 6.1) provides the cell with the means to take the energy from whatever

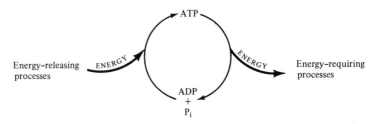

Figure 6.1 Cellular energy cycle.

catabolic pathways are operating at a given time and conserve this energy to do work at a later time.

What is special about the chemistry of ATP which makes it an ideal energy intermediate? ATP is an *energy-rich* or *high-energy* compound because of the nature of the bonds between its terminal phosphate groups (Fig. 6.2); when these bonds are broken by the addition of water (hydrolysis), energy is released. The bonds between the terminal phosphate groups are called *high-energy bonds* (designated by a squiggle, $\sim$), but remember, it is the hydrolysis of the bonds, rather than the bonds per se, which releases energy. In addition to energy, the hydrolysis of ATP produces adenosine diphosphate (ADP, Fig. 6.2) and inorganic phosphate ions (PO_4^{2-} or P_i):

$$ATP + H_2O \longrightarrow ADP + P_i + \text{energy for cell work}$$

In the cell, the reactions that break phosphate bonds are catalyzed by specific enzymes (kinases, Table 4.1) and the phosphate group may be transferred to another molecule rather than being released as P_i.

The fact that the hydrolysis of its phosphate bonds releases rather large amounts of energy makes ATP a "high-energy compound," but does not, in and of itself, explain why ATP is a middleman in the intracellular energy cycle. To understand how ATP "fits" in this cycle, one must examine the energetic relationships between ATP and other metabolic intermediates, particularly other phosphorylated compounds. As shown in Fig. 6.3, the amount of energy released when ATP is hydrolyzed (or, conversely, the amount of energy needed to make ATP from ADP and P_i) lies *between*

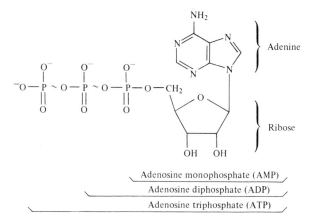

Figure 6.2 Chemical structure of ATP. ATP consists of three parts: (*a*) the purine base, adenine, which is linked to (*b*) the five-carbon sugar, ribose, which is attached to (*c*) three phosphate groups in a linear arrangement. The high-energy bonds between the terminal phosphate groups are indicated by a squiggle (∼).

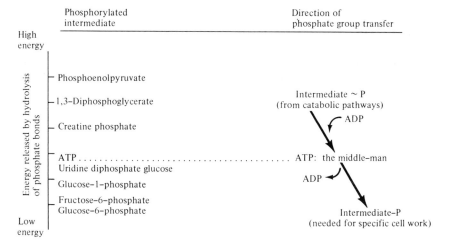

Figure 6.3 Phosphate groups are transferred through compounds of decreasing energy. Phosphorylated compounds (intermediate∼P) whose phosphate bonds have a very high energy of hydrolysis are intermediates in degradative pathways. These molecules donate their phosphate groups to ADP, forming ATP. Compounds required for particular kinds of cell work (intermediate-P) are activated when they accept a phosphate group from ATP. Each of these phosphate-group transfers is an exergonic reaction.

the energy of hydrolysis of several catabolic intermediates and what are generally considered "low-energy" compounds. This means that *the cellular production and cellular utilization of ATP are both exergonic processes.* The breakdown of "very high energy" molecules, such as phosphoenolpyruvate, releases more than enough energy for ATP production.* At the other side of the energy cycle, the breakdown of ATP provides more than enough energy for the synthesis of compounds needed for biosynthetic pathways, such as glucose-1-phosphate. In other words, ATP is an energy middleman, not only in the sense of being a physical shuttle, but also in terms of its energy content.

For certain kinds of cell work, the energy stored in ATP must be transferred to a second intermediate that serves as the energy source during the actual work process. The second high-energy compound is often another nucleoside triphosphate [guanosine triphosphate (GTP), uridine triphosphate (UTP), or cytidine triphosphate (CTP)]. For example, we will see that both ATP and GTP are needed to supply the energy for protein synthesis (Chap. 13).

In summary, ATP is the critical energy carrier in the cell's energy cycle. During energy-releasing processes, energy is stored in high-energy phosphate bonds made when ADP is converted to ATP. The high-energy bonds are later cleaved to supply energy for cell work. The utilization of ATP as the common intermediate among these diverse processes provides great flexibility; the cell can degrade whatever molecules are available and, when necessary, use the stored energy for biosynthesis, active transport, or mechanical work.

6.2 Energy-requiring processes: kinds of cell work

Cell work includes the many diverse metabolic processes on which the cell "spends" energy, that is, the processes that use up ATP. Cell work falls into three major categories: biosynthesis, active transport, and mechanical work.

Biosynthesis is the production of molecules, particularly macromolecules, from their smaller precursors.

Active transport is the movement ("pumping") of ions or small molecules against a charge or concentration gradient, usually across a membrane.

Mechanical work involves the interaction between specific cell components leading to contraction or motion.

Specific examples of each type of work are given in Table 6.1.

---

* In the case of photosynthesis, the chemical energy of ATP is derived from the light energy of the sun (Sec. 9.6).

Energy for Cell Work—The Role of ATP 105

TABLE 6.1 Cell Work

Class	Example
ACTIVE TRANSPORT	
Ions or molecules moved between cell and environment (homocellular transport)	Most cells maintain constant intracellular pH, accumulate sugars, amino acids, other small organic molecules
	In nerve cells, rapid redistribution of Na^+ and K^+ across membranes causes transmission of electrical waves
Ions or molecules moved across cell or layer of cells (transcellular)	Excretion of waste products into the urine by epithelial cells lining kidney tubules
	Excretion of HCl into gastric juice by epithelial cells of gastric mucosa
Ions or molecules moved between organelle and cytoplasm (intracellular)	Accumulation of Ca^{2+}, Mn^{2+}, and other ions by mitochondria
BIOSYNTHESIS	
Production of small metabolic intermediates from ions, elements, or simple organic compounds	Microorganisms and plants make many of their own amino acids, nucleotides, and other macromolecular precursors
	Symbiotic nitrogen fixation: bacteria in root nodules of (leguminous) plants reduce atmospheric N_2 to make ammonia
Production of macromolecules from their monomers	All cells synthesize proteins from amino acids, polysaccharides from simple sugars, nucleic acids from nucleotides, and lipids from acetyl CoA
MECHANICAL WORK	
Unidirectional motion or contraction using actomyosin systems	Skeletal muscles of animals; "catch" muscle of clams and other mollusks
Two- or three-dimensional ("whiplash") motion of cilia and flagella	Flagella of many microorganisms and spermatozoa are used for propulsion
	Cilia of microorganisms, cells of human lungs, trachea, and female reproductive tract are used to move substances past the cells
Intracellular movement of cytoplasm and organelles by contractile filaments (microtubules)	During division of eucaryotic cells, microtubules aid in chromosome separation and rearrangement of nuclear material and cytoplasm

In metabolically active and/or growing cells, biosynthesis is, quantitatively, the most important type of work; it includes the energy-requiring pathways by which all cells produce proteins (Chap. 13), polysaccharides (Chap. 9), nucleic acids (Chaps. 12 and 13), and lipids (Chap. 10). All cells are capable of active transport; it is the predominant form of work in nerve cells. Extensive mechanical work is primarily limited to muscle cells, sperm, and motile single-celled organisms.

There exist a few highly specialized cells that do work that does not fall into these categories. An example is the emission of light by luminescent organisms, such as fireflies. Here, too, ATP energy is used; chemical energy (ATP) is converted into light energy.

6.3 Muscle cells use ATP energy for mechanical work

We know that ATP carries energy between energy-releasing and energy-requiring metabolic processes. How is this done? To answer this question we will look closely at one specific example: the way in which skeletal muscle cells use the chemical energy in the phosphate bonds of ATP to do the mechanical work of contraction. The major questions to be asked are: (1) What metabolic pathways supply muscles with ATP? and (2) On a molecular level, how does ATP interact with the major muscle proteins, actin and myosin, during contraction?

As is true for most cell work, the ATP used to drive muscle contraction comes from the degradation of glucose by glycolysis (Chap. 7) and the Krebs cycle/electron transport system (Chap. 8). [The glucose comes either from the blood or from the breakdown of muscle glycogen (Chap. 9).] In muscles working at a submaximal rate, the more efficient Krebs cycle/electron transport system supplies most of the ATP. When muscles are contracting rapidly, the rate of oxygen influx from the blood is too slow to allow the (aerobic) Krebs cycle/electron transport system to keep up with the ATP demand and most of the ATP is supplied by the less efficient anaerobic process of glycolysis.* This situation is described more fully in Chap. 7.

There is an additional source of ATP which is unique to muscle cells: the phosphorylation of ADP by phosphocreatine, a high-energy compound (Fig. 6.3). The only metabolic reaction in which phosphocreatine is known to participate is the enzyme-catalyzed exchange reaction:

$$\text{phosphocreatine} + \text{ADP} \xrightleftharpoons{\text{Creatine kinase}} \text{creatine} + \text{ATP}$$

This reaction produces no net increase in high-energy phosphate bonds. Phosphocreatine serves only as a reservoir in which energy is deposited when ATP levels rise and from which energy is withdrawn when ATP levels fall. Because of the presence of the phosphocreatine storehouse, the ATP concentration in muscle cells before and after a single contraction is nearly identical; ATP levels do not fall until the phosphocreatine supply is depleted.

---

* Several proteins involved in the electron transport system contain iron, and it is these molecules that give red muscle its distinctive color. Certain less-active muscles have lower amounts of these iron-containing proteins and hence are not colored. This difference is clear in domestic fowl; we obtain white meat from the less active breast and red meat from the legs and thighs, which contain more active muscles.

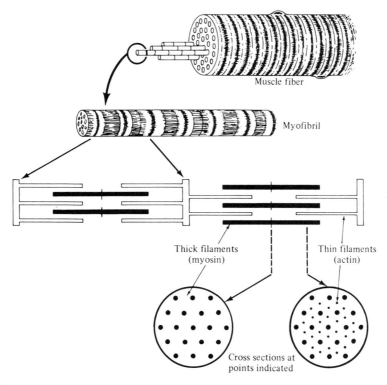

Figure 6.4 Structure of a skeletal muscle fiber. Skeletal muscle fibers are multinucleate cells, each containing many parallel bundles of filaments called myofibrils. Microscopy shows that each myofibril contains parallel thick and thin filaments interspersed in an orderly array. The major protein of the thin filaments is actin; the major protein of the thick filaments is myosin.

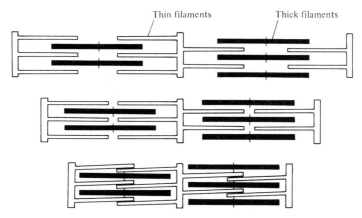

Figure 6.5 Sliding filament model of muscle contraction. The thin filaments move relative to the thick filaments.

The contractile unit of the muscle cell is the myofibril, which contains the water-insoluble filamentous proteins, actin and myosin (Fig. 6.4). Actin, the protein of the thin filaments, is a globular protein (G-actin) which can polymerize to form the fibrous protein (F-actin).

$$n(\text{G-actin-ATP}) \longrightarrow (\text{F-actin-ADP})_n + n\text{P}_i$$

It is believed that F-actin consists of two strands of G-actin "beads" coiled around each other. Myosin, found in the thick filaments, consists of two identical polypeptide chains; it is a huge, rod-shaped molecule with two projecting heads. In the myofibril,

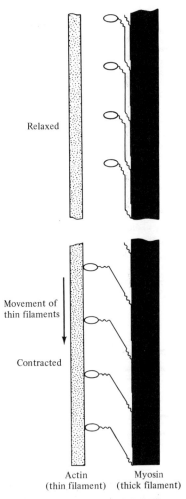

Figure 6.6 Formation of cross-bridges between actin and myosin. Alternate making and breaking of cross-bridges "pulls" the actin along the myosin. The cross-bridges are stable in the absence of ATP. (After Albert L. Lehninger, *Biochemistry*, Worth Publishers, Inc., New York, 1970, p. 593.)

the projecting heads of each myosin molecule make bridges with the surrounding actin molecules. The myosin "heads" resemble the tiny hooks of Velcro, the non-sticky "adhesive" which can replace zippers. Myosin has enzymatic activity; in the presence of calcium ions (Ca^{2+}), it can catalyze the hydrolysis of ATP to ADP and P_i.

Contraction of the myofibril occurs by movement of the thin, actin-containing filaments relative to the thick, myosin-containing filaments (Fig. 6.5). This sliding mechanism results from the alternate making and breaking of cross-bridges between the two proteins (Fig. 6.6). The cross-bridges appear to be *stable* in the *absence* of ATP, and are *dissociated* in the *presence* of ATP. When a nerve impulse reaches a muscle fiber, it causes the release of Ca^{2+}, which stimulates the ATPase activity of myosin. This reduces the ATP concentration around the filaments and allows the cross-bridges to form. When ATP levels rise again, the bridges are broken. As ATP levels continue to fluctuate, new cross-bridges are made and then broken, pulling the actin farther along the myosin. The fall in Ca^{2+} concentration after the nerve impulse has passed stops this cyclic cross-bridge formation, and the muscle relaxes.

6.4 Active transport: the Na^+–K^+ pump

The cell membrane* is a very effective permeability barrier. While some ions and small molecules can move freely across the membrane, such simple diffusion can, at best, only equilibrate concentrations of components of the internal and external environments. This would not suffice when a cell has to survive under conditions when only a low concentration of an essential nutrient is present. To sustain the complex processes of metabolism, cells need a mechanism for concentrating certain substances from their surroundings and excreting others from their cytoplasm. This pumping work, called *active transport*, requires energy from ATP.

Several features are common to all active transport systems. The substances being moved are carried by proteins located in the cell membrane. Transport proteins (permeases) are highly specific for the ions or molecules they carry. Analogous to the active site of an enzyme, the binding site on a transport protein must have the proper size, shape, and charge to contain its particular "substrate." Membrane-bound transport systems also have "sidedness"; the protein must have a different configuration when facing inside the cell as compared to outside the cell to ensure that ions or molecules are always moved in the correct direction.

It is critical for most cells to regulate their internal concentrations of sodium (Na^+) and potassium (K^+) ions. Relative to the environment, aerobic plant and animal cells establish high internal K^+ and low internal Na^+ levels (Fig. 6.7). The high K^+ is necessary for the catalytic activity of many enzymes and for ribosome function during protein synthesis (Chap. 13); high concentrations of Na^+ ions can inhibit some

---

* The macromolecular organization of cell membranes is discussed in Chap. 10.

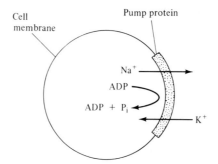

Figure 6.7 Cells regulate their internal concentrations of Na$^+$ and K$^+$. Using energy from ATP, the Na$^+$–K$^+$ pump works to extrude Na$^+$ ions from the cell and bring in K$^+$ ions.

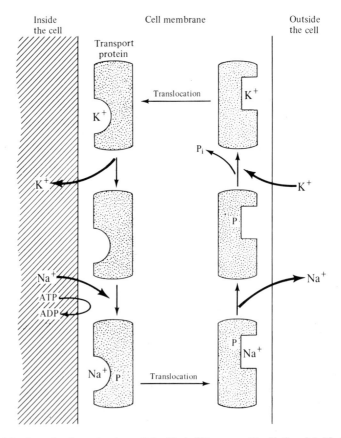

Figure 6.8 Postulated mechanism of the Na$^+$–K$^+$ pump. To distinguish Na$^+$ ions from K$^+$ ions, the transport protein must have different configurations on the two sides of the membrane; the nature of this conformational change is unknown.

Energy for Cell Work—The Role of ATP

of these processes. The transmission of impulses by nerve cells depends on changes in these two ion gradients.

Both the uptake of K^+ ions and the extrusion of Na^+ ions are the work of the Na^+–K^+ pump. This is no minor cellular activity! Each of the many pump proteins extrudes about 20 Na^+ ions per second and it has been estimated that one-third of the ATP consumed by a resting cell is used for this work.

Years of experimentation, primarily with erythrocytes (immature red blood cells), have revealed the molecular details of the mechanics of the Na^+–K^+ pump. First, the movements of Na^+ and K^+ are interdependent; the pump cannot transfer either ion alone, but must alternately excrete Na^+ ions and bring in K^+ ions. Second, the pump protein has ATPase activity; that is, as ions are moved, ATP is cleaved to ADP and P_i. The ADP and P_i remain within the cell. Using ATP labeled with radioactive phosphate, it was found that the pump protein is phosphorylated in the presence of Na^+, while the addition of K^+ causes its dephosphorylation. These experimental results have been incorporated with the features common to all transport systems to produce the molecular model of the Na^+–K^+ pump show in Fig. 6.8.

The active transport of mineral ions by plant roots further illustrates the importance and magnitude of this kind of cell work. Trace minerals (Table 1.5) are required by both plant and animal cells. But only plants can extract these inorganic ions from the soil; animals obtain essential minerals by eating plants. The highly branched root systems of plants are specially adapted to concentrate minerals from soil water where the ions are present at very low concentrations. The active transport systems in root cells share the general features of other cell pumps. Ion-specific carrier proteins transport the ions into the plant cells from the soil water; ATP supplies the energy for this process.

In terms of amount, the major ions accumulated by plants are potassium, calcium, magnesium, nitrogen, sulfur, potassium, chlorine, and silicon. Land plants "mine" about 5 billion metric tons of these trace minerals each year. Compare this to the 1 billion metric tons of metals* mined annually by human technology. Although they grow in "mineral-poor" soil, plants are "outmining" man by 5 to 1!

6.5 The magnitude of the cellular energy cycle

Cells need large amounts of energy for biosynthesis, active transport, and mechanical work. The magnitude of this energy requirement becomes clearer when we examine the quantitative aspects of ATP production and utilization.

The sum of the concentrations of the three adenine nucleotides (ATP, ADP, and AMP) in a cell is between 2 mM and 15 mM.† Most of this is ATP; obviously, from

---

* Primarily iron, copper, zinc, and lead.

† A mole is 6.23×10^{23} molecules. A 1 molar concentration (1 M) is 6.23×10^{23} molecules/liter of solution. A 1 millimolar (1 mM) concentration is 6.23×10^{17} molecules/milliliter of solution.

what we have said, the ratios of the three do fluctuate, but the ATP concentration usually exceeds that of ADP while AMP levels remain relatively low. In addition, the intracellular concentration of ATP is high in comparison with other small molecules involved in energy production and storage (Table 6.2). Knowledge of the intracellular level of ATP does not, however, tell us how fast it is produced and utilized.

TABLE 6.2 Concentrations of Some Molecules Involved in Energy Release and Storage in Human Erythrocytes

	Concentration (mM)
Glucose	5
Phosphoenolpyruvate	0.023
Pyruvate	0.051
Lactate	2.9
ATP	1.85
ADP	0.138
P_i	1.0

The rate at which cells spend ATP is impressive. An active cell requires roughly 2 million (2×10^6) ATPs per second. In dividing *E. coli* (bacterial) cells, it is estimated that half the ATP in the cell is broken down to ADP and P_i every second. Even in (metabolically) less active liver cells, more than half the available ATP is used up every minute. Of course, the ADP formed is quickly rephosphorylated to maintain the supply of ATP.

Supplying the energy to make this amount of ATP is an awesome process. Let us consider an aerobic cell growing on glucose as its energy source. Each glucose molecule, when catabolized to CO_2 and water, yields 36 molecules of ATP (Chaps. 7 and 8). Therefore, each cell needs to degrade about 55,000 molecules of glucose* per second (4.8 billion molecules per day!) to supply energy for cell work.

Although large amounts of energy are needed for cell work, there is a limit to the quantity of ATP that a cell can store economically. For example, 36 ATPs are a much larger mass of material than one glucose molecule. So, for storing energy over a long period of time, it is more economical for the cell to store glucose or one of its polymers (glycogen, starch, etc.) than it is to store an energetically equivalent amount of ATP. Energy can be stored even more compactly as fat; the energy yields from equal weights of carbohydrates and fat are in a ratio of about 4:9.

The compactness and efficiency of fat as a form of stored energy is clearly illustrated in the case of the ruby-throated hummingbird. At the beginning of its annual migration across the Gulf of Mexico, this tiny bird weighs 7 grams, of which 2 grams are fat. This 2 grams of fat supplies all the energy for its long flight. If the hummingbird had to carry all its energy as ATP, it would need 700 grams of ATP!

---

* A gram of glucose contains 3.5×10^{21} molecules.

Energy for Cell Work—The Role of ATP

SUMMARY

Cells need energy for biosynthesis, active transport, and mechanical work; this energy is obtained from the breakdown of simple sugars, lipids, and amino acids, and, in the case of green plants, from sunlight.

Energy-requiring and energy-releasing metabolic processes are coupled by the "high-energy" middleman, ATP. The energy released by catabolic pathways is stored in the terminal phosphate bonds of ATP; these bonds are later broken to provide energy for cell work. Thus, ATP acts as an "energy shuttle" to assure that the cell will have energy available at the right place and at the right time.

PRACTICE PROBLEMS

1. Three classes of cell work are biosynthesis, active transport, and mechanical work. For each class of work, give (a) an example carried out by all eucaryotic cells and (b) two examples carried out only by certain highly specialized eucaryotic cells.

2. Cells are highly organized to carry out different kinds of work. Work is often performed only at a specific site within a cell, and/or it may require the direct participation of a particular organelle. What kinds of work are associated with the following cell structures?
    (a) Cell membrane.
    (b) Nucleus (give two examples).
    (c) Ribosomes.
    (d) Flagella.
    (e) Microtubules.

3. Theology aside, why is it that "Man cannot live by bread alone"?

4. One advantage in using ATP as a middleman between energy-releasing and energy-requiring processes is that of timing—it allows the cell to store energy when it can and to use energy when it must. For an algal cell in one of the Great Lakes: (a) list two external variables that determine when the cell can store energy, (b) list two factors that determine when the cell uses energy, and (c) list two kinds of work that are performed only during certain stages of the cell cycle.

5. When studying enzymes, we saw that their three-dimensional structure was critical for their enzymatic activity. The same rule holds for proteins with nonenzymatic functions. (a) What features of the three-dimensional structure of actin and myosin are necessary for their contractile activity? (b) What features of the three-dimensional structure of the Na^+–K^+ pump protein are essential for its transport function?

6. The model for the Na^+–K^+ pump given in Fig. 6.8 incorporates the features common to all active transport systems. Construct an analogous model for the secretion of hydrogen ions (H^+) by kidney tubule cells assuming that one Na^+ is taken up for each H^+ secreted and that one ATP is required for each cycle of the pump.

7. Cells store energy in the phosphate bonds of ATP. Why do cells not store energy as heat?

SUGGESTED READING

COHEN, CAROLYN, "The Protein Switch of Muscle Contraction," *Sci. American*, 233, No. 5, p. 36 (1975). A sequel to the Murray and Weber paper referenced below.

ECCLES, SIR JOHN, "The Synapse," *Sci. American*, 212, No. 1, p. 56 (1965). How nerve cell activity depends on a Na^+–K^+ pump.

EPSTEIN, EMANUEL, "Roots," *Sci. American*, 228, No. 5, p. 48 (1973). An excellent article about the active transport of trace minerals by plant roots.

HAYASHI, TERU, "How Cells Move," *Sci. American*, 205, No. 3, p. 184 (1961). Compares three kinds of mechanical work done by cells: amoeboid movement, the motion of cilia and flagella, and muscle contraction.

LEHNINGER, ALBERT L., "How Cells Transform Energy," *Sci. American*, 205, No. 3, p. 62 (1961). A good introduction to the mechanics of the cellular energy cycle. Describes how specialized organelles use light or chemical energy to make ATP.

LEHNINGER, ALBERT L., *Bioenergetics*, W. A. Benjamin, Inc., New York, 1965. Chapter Four is a very readable description of high-energy phosphate bonds and the role of the ATP/ADP system in the cellular energy cycle.

McELROY, WILLIAM D., and HOWARD H. SELIGER, "Biological Luminescence," *Sci. American*, 207, No. 6, p. 76 (1962). Beautiful pictures of organisms converting chemical energy to light.

MURRAY, JOHN M., and ANNEMARIE WEBER, "The Cooperative Action of Muscle Proteins," *Sci. American*, 230, No. 2, p. 58 (1974). Good diagrams of muscle structure and how it changes in the presence of calcium ions.

SATIR, PETER, "How Cilia Move," *Sci. American*, 231, No. 4, p. 44 (1974). The motion of these hairlike appendages is powered by ATP.

SCRIMSHAW, NEVIN S., and VERNON R. YOUNG, "The Requirements of Human Nutrition," *Sci. American*, 235, No. 3, p. 50 (1976). Our diet must include enough degradable nutrients (i.e., sufficient calories) to provide energy for cellular work.

WOODWELL, GEORGE, "The Energy Cycle of the Biosphere," *Sci. American*, 223, No. 3, p. 64 (1970). Describes the long-range influences of human technology on the delicately balanced energy cycles of ecosystems.

7 The Production of Energy & Intermediates from Glucose

In spite of the fact that sugar, as a dietary constituent, is continually maligned as the source of weight, dental, and other problems, few living organisms could survive without sugars. Carbohydrates* are essential as a source of energy and in the formation of many intermediate metabolites that can be used to form the carbon skeletons for practically all constituents of living cells. Thus, molecules of carbohydrate are broken down and, subsequently, other molecules are synthesized from the breakdown products. In addition, many of the steps in the catabolism of carbohydrates result in the harnessing of energy in the form of ATP. The cellular degradation of the carbohydrate, glucose (Fig. 7.1), provides the best example of nature's ability to extract every ounce of biochemical potential and every bit of energy (calories) from a typical foodstuff.

The source of this essential sugar depends on the type of cell. Animals and most microorganisms require simple sugars or polysaccharides in their diets. Polysaccharides cannot cross cell membranes; (extracellular) digestive enzymes degrade polysaccharides to their monomers (monosaccharides), which cells can then assimilate and

---

* The words "sugar" and "carbohydrate" are often used interchangeably. In general, carbohydrates are divided into simple monosaccharides (such as glucose, galactose, and ribose) and more complex polysaccharides which contain chains of monosaccharide molecules (e.g., starch is a polymer of glucose).

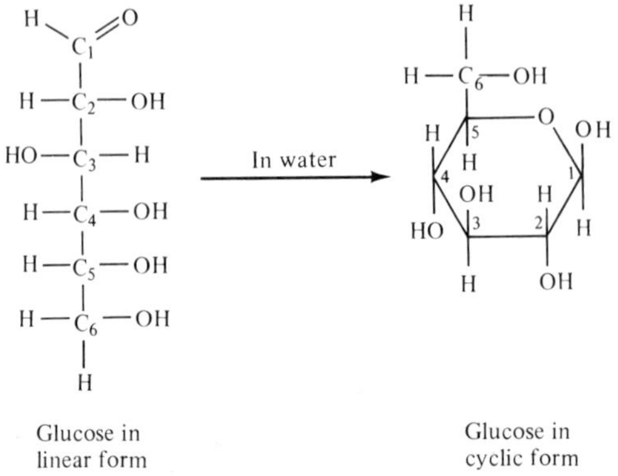

Figure 7.1 Structure of glucose. In solution, the open-chain form of glucose cyclizes to form a ring structure when the aldehyde at carbon 1 reacts with the alcohol at carbon 6 to form a hemiacetal. Analogous cyclic hemiacetals are formed by other six-carbon and five-carbon sugars.

metabolize. Plants make monosaccharides from carbon dioxide and water during photosynthesis (Sec. 9.6). However, regardless of the source of the sugars, the primary degradative pathway is essentially the same in all cells.

It may seem a chore to have to follow the degradation of glucose through the many steps of glycolysis, the Krebs cycle, and the electron transport system. However, this series of reactions allows living systems to use glucose to its full potential, and it is only by knowing these steps that we can appreciate the importance of sugar metabolism. An understanding of these steps also illustrates how delicate control systems balance metabolic pathways so that the cell does not degrade more glucose than necessary, even when it is presented with a large excess of this nutrient, and how different environmental conditions, such as a lack of oxygen, can affect the way in which glucose is metabolized. It is to understand such processes that we study biochemistry.

The end result of aerobic* glucose degradation in cells is exactly the same as it is when this sugar is thrown into a fire:

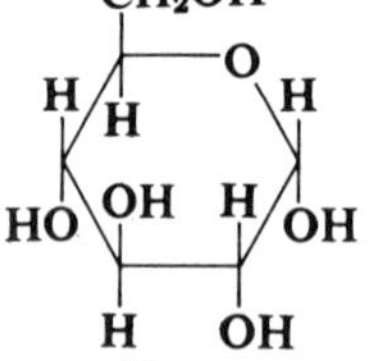

$$\text{Glucose} + O_2 \longrightarrow CO_2 + H_2O + \text{energy}$$

But what goes on in this reaction is totally different in the two cases. When sugar is burned in a fire, it is rapidly converted to carbon dioxide and water, and a single burst

---

* Aerobic conditions imply the presence of a source of oxygen (O_2), such as air. Anaerobic conditions are those in which no oxygen (air) is present.

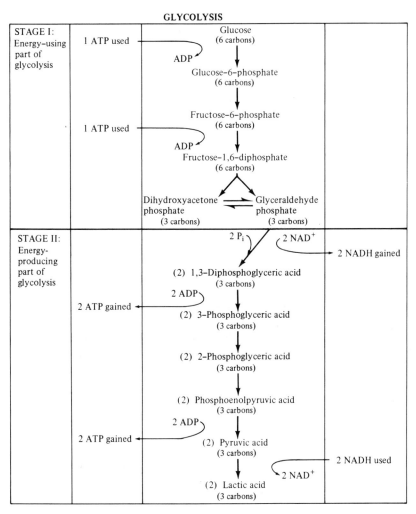

Figure 7.2 Glycolysis. During glycolysis, each glucose molecule is broken down into two molecules of pyruvic acid. From this sequence of reactions, the cell gains two ATPs and uses up two NAD^+s. Under anaerobic conditions, pyruvic acid may be converted to lactic acid, as shown here, or to ethanol and CO_2, depending on the type of cell. Under aerobic conditions, pyruvic acid is converted to acetyl CoA, which is further degraded in the Krebs cycle (Chap. 8).

The arrows in this diagram show the direction of the reactions when glucose is being degraded. Many of these reactions can be reversed when the cell needs to synthesize glucose. The number of carbon atoms in each intermediate is indicated; complete chemical structures are given in the text.

of heat energy is released. Cells cannot afford to release all the energy obtainable by the oxidation of glucose in a single burst of heat, since such a large release of heat would denature and inactivate proteins. To avoid this problem, the cellular oxidation of glucose takes place in a series of steps with the release of small packets of chemical energy at each step.

When glucose is burned in oxygen, the theoretical energy release is 686,000 calories per mole (about 4000 calories per gram, one-fifth of a teaspoon). In the cell, the amount of energy captured during the oxidative metabolism of glucose is not as much as this, but as we will see, there is efficient production and storage of energy in the form most useful for cell work, such as biosynthesis, muscle action, and active transport (Chap. 6). Most of the energy produced is stored in the high-energy bonds of ATP, so that the cell captures the glucose energy, not as heat, but as chemical energy, and there is no drastic increase in the temperature of the cell.

We will now discuss the pathways of glucose degradation in detail. The first series of glucose-degrading reactions is the glycolytic pathway, shown in Fig. 7.2. This figure no doubt boggles the mind, but it is not necessary to memorize the steps in detail. However, a good understanding is worth striving for. This chapter deals with the early events in glucose degradation, which do not require oxygen. The next chapter explains the reactions of the Krebs cycle in which acetyl CoA is degraded to carbon dioxide and water in the presence of oxygen.

It is probably correct to say that it was man's interest in alcoholic beverages that motivated him to unravel the steps of glycolysis! Yeasts were the first organisms to be studied in detail, since they were the source of products both useful and pleasant for man. Early experiments showed that alcohol production required the presence of carbohydrate and inorganic phosphate; subsequent studies led to the isolation and characterization of intermediates and enzymes. The process of working out the pathway of glycolysis took many years, and it still continues in its more subtle and detailed biochemical aspects. Yeasts are no longer the sole subjects for these investigations as mammalian cells are now amenable to study in tissue culture; the details of glycolysis and its regulation have been confirmed in many different organisms.

7.1 Glycolysis. I. The energy-using steps of glucose breakdown

In the first stage of glycolysis, the six-carbon sugar, glucose, enters the cell and is activated prior to cleavage into two halves.

Most cell membranes allow free passage (diffusion) of small, uncharged molecules like glucose into the cell until the concentrations on both sides of the membrane become identical. Once it enters the cell, glucose is kept inside by conversion to a charged form that cannot diffuse back out across the cell membrane. This is accomplished by phosphorylation, the product being glucose-6-phosphate:

The Production of Energy and Intermediates from Glucose

[Chemical structure of Glucose] + ATP —Hexokinase→ [Chemical structure of Glucose-6-phosphate] + ADP

There are several things of interest in this first reaction. To begin with, it uses up a molecule of ATP, which is what the cell is supposed to be making! As discussed in Chap. 5.3, regulation of the first enzyme in a pathway is an efficient method of metabolic control; in mammals, the phosphorylation of glucose is believed to be stimulated by the pancreatic hormone, insulin (Chap. 9). [There are a large number of regulatory hormones in mammals, and many of them exert their control by acting directly or indirectly on reactions involved in the metabolism of sugars (Chap. 9).] In addition, glucose-6-phosphate is the starting material for another metabolic pathway, the pentose phosphate pathway, which produces ribose (5-carbon) sugars necessary for the synthesis of nucleic acids and NADPH, the hydrogen donor in many anabolic pathways.

In the second step of glycolysis, glucose-6-phosphate undergoes a rearrangement, or isomerization, catalyzed by the enzyme phosphoglucose isomerase. The product of this reaction is the six-carbon sugar, fructose-6-phosphate.

Subsequently, yet another molecule of ATP is used to form fructose-1,6-diphosphate from fructose-6-phosphate. The reaction is catalyzed by phosphofructokinase, which is probably one of the most important and also most studied enzymes in this pathway.

[Chemical structure of Fructose-6-phosphate] + ATP —Phosphofructokinase→ [Chemical structure of Fructose-1,6-diphosphate] + ADP

This enzyme is important in the regulation of glycolysis. As we have described earlier (Chap. 5), the activity of enzymes can be influenced by reversible inhibitors and activators. The catalytic activity of phosphofructokinase is stimulated by fructose-6-phosphate, AMP and ADP, but is inhibited by ATP and citrate (Fig. 7.3). All of this makes very good sense. Since one object of glycolysis is to produce ATP, and since the presence of excess AMP, ADP, or fructose-6-phosphate means that the cell is

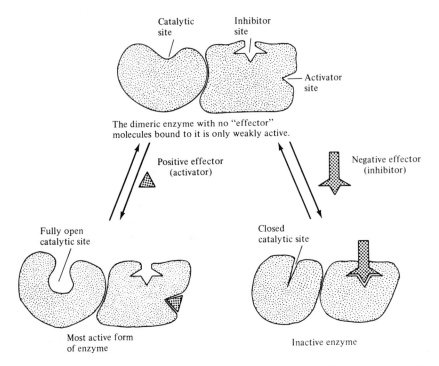

Figure 7.3 Small effector molecules control enzyme activity. The regulation of phosphofructokinase is typical of enzymes that are controlled by effectors (Sec. 5.3). This glycolytic enzyme is activated by fructose-6-phosphate, AMP, and ADP; it is inhibited by ATP and citric acid.

There is good biochemical evidence that activators and inhibitors (which are reversible) bind to a site on the enzyme, the regulatory site, that is separate and distinct from the active or catalytic site of the enzyme. When an inhibitor binds, it causes subtle conformational changes in the shape of the enzyme in such a way that activity at the catalytic site is changed. The catalytic and regulatory sites are often on different subunits of multimeric enzymes.

deficient in ATP, these molecules are activators of the enzyme, stimulating it to get on with the business of degrading more glucose, thereby making more ATP. Conversely, an excess of ATP means that the cell is catabolizing more glucose than necessary; excess ATP inhibits phosphofructokinase and prevents wasteful overmetabolism of sugars. Excess citrate inhibits the enzyme for reasons that will be described later (Chap. 8), but in essence, the control of phosphofructokinase activity serves to balance the rates of functioning of glycolysis and the Krebs cycle. This is just one example of the many delicate controls that operate in cell metabolism.

Fructose-1,6-diphosphate is a six-carbon sugar and the fourth step in glycolysis is the splitting of this sugar into two three-carbon halves.

$$\underset{\text{Fructose-1,6-diphosphate}}{{}^{2-}O_3POCH_2 \begin{array}{c} O \\ H \: HO \\ H \quad \quad OH \\ OH \: H \end{array} CH_2OPO_3^{2-}} \xrightleftharpoons{\text{Aldolase}} \underset{\text{Dihydroxyacetone phosphate}}{\begin{array}{c} H \\ | \\ H-C-O-PO_3^{2-} \\ | \\ C=O \\ | \\ H-C-OH \\ | \\ H \end{array}} + \underset{\text{Glyceraldehyde-3-phosphate}}{\begin{array}{c} H-C=O \\ | \\ H-C-OH \\ | \\ H-C-O-PO_3^{2-} \\ | \\ H \end{array}}$$

The products of this cleavage arise from a chemical reaction known as a reverse aldol condensation—hence the name aldolase for the enzyme; the exact distribution of the carbon atoms in a reaction such as this can be determined by the use of radioactively labeled precursors (Sec. 5.4). The three-carbon products, dihydroxyacetone phosphate and glyceraldehyde-3-phosphate, are interconvertible, and thus they are essentially equivalent in metabolic terms. Dihydroxyacetone phosphate is converted to glyceraldehyde-3-phosphate and it is glyceraldehyde-3-phosphate that participates in the next step of glycolysis.

Dihydroxyacetone phosphate is not, however, useless in itself, for it is the branch-point compound that links the catabolism of sugars to the synthesis of complex lipids (Sec. 10.2). Thus, glycolysis becomes linked to other biochemical pathways. There are very few "exclusive" pathways in metabolism; all biochemical pathways are linked to other pathways, and they can be considered as one interconnected network. For this reason, inhibitors that cause derangement at one step in metabolism can often affect many pathways. It is little wonder that administration of drugs (e.g., birth control pills, tranquilizers, etc.) or blocks in the metabolism of key amino acids (such as phenylalanine in the genetic defect PKU) or hormones (diabetes) often have undesirable and far-reaching effects on metabolism.

In this first stage of glycolysis, no energy has been produced and two molecules of ATP have been used for each molecule of glucose. Although this may appear to be the opposite of what the cell set out to do, the glucose molecule is now "primed" for the reactions that involve energy production.

7.2 Glycolysis. II. The release of useful energy

Glyceraldehyde-3-phosphate is converted to 1,3-diphosphoglycerate (an oxidation reaction) by the action of a dehydrogenase that removes hydrogen atoms* (electrons) from the substrate and transfers them to the coenzyme nicotinamide adenine dinucleotide (NAD^+).

---

* For our purposes, we can discuss hydrogen atoms and electrons in (biochemically) equal terms. An oxidation reaction involves either the addition of oxygen to a molecule, the removal of hydrogen atoms (proton + electron) or the removal of an electron. Reduction reactions are the converse, involving the addition of electrons or hydrogen atoms, or the removal of oxygen from a molecule.

$$\begin{array}{c}
\text{H—C=O} \\
| \\
\text{H—C—OH} \\
| \\
\text{H—C—O—PO}_3^{2-} \\
| \\
\text{H}
\end{array} + \text{NAD}^+ + \text{P}_i \xrightarrow{\text{Glyceraldehyde phosphate dehydrogenase}} \begin{array}{c}
\text{O} \\
\parallel \\
\text{C—O} \sim \text{PO}_3^{2-} \\
| \\
\text{H—C—OH} \\
| \\
\text{H—C—O—PO}_3^{2-} \\
| \\
\text{H}
\end{array} + \text{NADH} + \text{H}^+$$

Glyceraldehyde 3-phosphate

1,3-Diphosphoglyceric acid

This is a key reaction in the glycolytic sequence. The electrons and hydrogen ion* stored in NADH can be used in other anabolic pathways or, under aerobic conditions, they can provide energy for ATP formation (Chap. 8). Note that in this reaction, inorganic phosphate (P_i) is added directly to a substrate molecule without the mediation of ATP.

The product, 1,3-diphosphoglycerate, is a high-energy phosphate intermediate that can be used to make ATP from ADP exergonically (Fig. 6.3). This happens in the next step, catalyzed by phosphoglycerate kinase. Reactions catalyzed by kinases can work in both directions but are not always reversible; it is interesting to note that this kinase favors the dephosphorylation of the substrate to form ATP, whereas the other kinases in this pathway favor reactions in which ATP is used up.

1,3-Diphosphoglyceric acid is important for another reason: it can be converted to 2,3-diphosphoglyceric acid by the action of a specific isomerase. 2,3-Diphosphoglyceric acid is found in red blood cells and is used to modulate the oxygen-carrying power of hemoglobin. When 2,3-diphosphoglyceric acid binds to hemoglobin, the ability of this molecule to transport oxygen is much reduced. Fetal hemoglobin can compete favorably with adult hemoglobin for oxygen because it binds 2,3-diphosphoglyceric acid less well than adult hemoglobin.

$$\begin{array}{c}
\text{O} \\
\parallel \\
\text{C—O} \sim \text{PO}_3^{2-} \\
| \\
\text{H—C—OH} \\
| \\
\text{H—C—O—PO}_3^{2-} \\
| \\
\text{H}
\end{array} + \text{ADP} \underset{}{\overset{\text{Phosphoglycerate kinase}}{\rightleftharpoons}} \begin{array}{c}
\text{O} \\
\parallel \\
\text{C—OH} \\
| \\
\text{H—C—OH} \\
| \\
\text{H—C—O—PO}_3^{2-} \\
| \\
\text{H}
\end{array} + \text{ATP}$$

1,3-Diphosphoglyceric acid

3-Phosphoglyceric acid

---

*In this and subsequent dehydrogenase reactions, two H^+'s and two electrons are removed from the substrate. NAD^+ accepts both electrons and one H^+.

The product of this reaction, 3-phosphoglycerate, is not able to transfer its remaining phosphate to ADP to form ATP. For this to occur, rearrangement of the molecule must take place to provide a suitably reactive compound. This is brought about by a pair of reactions that produce a three-carbon compound, phosphoenolpyruvate, containing a high-energy phosphate bond (Fig. 6.3).

$$\begin{array}{c}\text{O}\\\diagdown\\\text{C—OH}\\|\\\text{H—C—OH}\\|\\\text{H—C—O—PO}_3^{2-}\\|\\\text{H}\end{array} \xrightleftharpoons{\text{Isomerase}} \begin{array}{c}\text{O}\\\diagdown\\\text{C—OH}\\|\\\text{H—C—O—PO}_3^{2-}\\|\\\text{H—C—OH}\\|\\\text{H}\end{array}$$

3-Phosphoglyceric acid   2-Phosphoglyceric acid

$$\begin{array}{c}\text{O}\\\diagdown\\\text{C—OH}\\|\\\text{H—C—O—PO}_3^{2-}\\|\\\text{H—C—OH}\\|\\\text{H}\end{array} \xrightleftharpoons{\text{Enolase}} \begin{array}{c}\text{O}\\\diagdown\\\text{C—OH}\\|\\\text{C—O}\sim\text{PO}_3^{2-}\\||\\\text{H—C—H}\end{array} + \text{H}_2\text{O}$$

2-Phosphoglyceric acid   Phosphoenolpyruvic acid

In the latter reaction the long arrow indicates that the forward reaction is strongly energetically favored, although, like most enzymatic reactions, it is reversible.

The enolase that catalyzes the formation of phosphoenolpyruvate is very sensitive to inhibition by fluoride ion. The use of inhibitors to study the steps of biochemical pathways has been mentioned (Chap. 5); in studies of glycolysis, fluoride can be used to block the formation of phosphoenolpyruvate. Similarly, iodoacetic acid inhibits glyceraldehyde-3-phosphate dehydrogenase and can be employed to interfere with an earlier step in glycolysis.

In the next step of glycolysis, phosphoenolpyruvic acid is dephosphorylated to form ATP from ADP, and also pyruvic acid.

$$\begin{array}{c}\text{H—C—H}\\||\\\text{C—O}\sim\text{PO}_3^{2-}\\|\\\text{C—OH}\\\diagup\\\text{O}\end{array} + \text{ADP} \xrightarrow{\text{Pyruvate kinase}} \begin{array}{c}\text{H}\\|\\\text{H—C—H}\\|\\\text{C=O}\\|\\\text{C—OH}\\\diagup\\\text{O}\end{array} + \text{ATP}$$

Phosphoenolpyruvic acid   Pyruvic acid

This reaction is not reversible under normal intracellular conditions. Subsequently, pyruvic acid can be degraded in several different ways, depending on the type of cell and whether or not oxygen is available.

7.3 How much energy can we get from glycolysis?

The formation of ATP when phosphoenolpyruvate is converted to pyruvic acid is the last energy-producing step in glycolysis. It is appropriate to ask (from an energy point of view) whether this series of steps was worth all the trouble. We see (Fig. 7.2) that for each molecule of glucose catabolized, two ATP molecules are used up in the first stage (glucose to glyceraldehyde phosphate), but for each half-molecule of glucose (glyceraldehyde phosphate), two molecules of ATP are made in the second stage. Thus in the degradation of one glucose molecule to two pyruvic acid molecules, the cell realizes a net gain of two molecules of ATP. This seems a poor yield in terms of energy, when we realize that, from a theoretical viewpoint, glucose contains enough chemical energy to make about 60 molecules of ATP when it is oxidized completely to carbon dioxide and water. However, the most efficient biochemical production of energy from sugars is yet to come: the oxygen-requiring reactions of the Krebs cycle and electron transport system.

Nonetheless, the low yield of ATP energy from anaerobic glycolysis is quite sufficient to support muscle function (for a 4-minute miler this is substantial). Many bacteria and yeast can grow in the absence of oxygen and rely on anaerobic glycolysis as the major pathway of energy production. Remember, also, that the production of energy is not the only aim of sugar catabolism, since a number of important and useful intermediates such as glucose-6-phosphate and 2,3-diphosphoglyceric acid are also formed.

7.4 The fate of pyruvic acid under anaerobic conditions

Depending on the type of cell, pyruvic acid can be converted to lactic acid or to ethanol. Both conversions require NADH and regenerate NAD^+ for reuse in the earlier steps of glycolysis.

(a) LACTIC ACID PRODUCTION

In certain cells, the product of anaerobic glycolysis is lactic acid, formed by the reduction of pyruvic acid.

$$\underset{\text{Pyruvic acid}}{\begin{array}{c} H \\ | \\ H-C-H \\ | \\ C=O \\ | \\ \underset{O}{C-OH} \end{array}} + NADH + H^+ \xrightleftharpoons{\text{Lactate dehydrogenase}} \underset{\text{Lactic acid}}{\begin{array}{c} H \\ | \\ H-C-H \\ | \\ H-C-OH \\ | \\ \underset{O}{C-OH} \end{array}} + NAD^+$$

Note that the hydrogen atoms are provided by the reduced coenzyme NADH and that this reaction regenerates NAD^+, which is again capable of accepting hydrogen atoms. This is a very important feature of the formation of lactic acid under anaerobic conditions. Without the formation of lactic acid, the cell's supply of NAD^+ would dwindle to nothing (it would all be in the form of NADH) and glycolysis would stop completely. The formation of lactic acid from pyruvate is solely for the regeneration of NAD^+; lactic acid is a "dead-end" product.

In muscle cells, lactic acid is formed when energy demands are great and when the oxygen supply is limited. Athletes often produce large amounts of lactic acid during heavy exercise. The buildup of lactic acid in working muscles and in the bloodstream is one manifestation of what is known as an *oxygen debt*, acquired when oxygen cannot enter muscle cells at a fast-enough rate to meet energy requirements. Under such circumstances, the less-efficient (anaerobic) energy-producing pathways are used to power muscle. The "panting" of a person after strenuous activity is a result of the body's desire to pay off this oxygen debt and convert lactic acid back into glycogen (Fig. 9.3).

Lactate dehydrogenase is an interesting enzyme because it exists in multiple forms (isoenzymes), and these forms are characteristic of different organs and tissues of the human body. As has been described previously (Sec. 4.9), determination of the amounts of various organ-specific isoenzymes in blood provides a convenient diagnostic test for identifying malfunctioning organs.

Many bacteria convert pyruvic acid to lactic acid, a conversion that is important in the production and preservation of food. When shredded cabbage or immature cucumbers are soaked in a salt brine, bacteria already present on these vegetables

TABLE 7.1 Production of Alcoholic Beverages

The production of different alcoholic beverages depends on the source of carbohydrate and on the particular yeast strain used to carry out the glycolytic reactions. A yeast that makes good wine will not make good beer, since some yeasts can make and tolerate more alcohol than others. However, no yeast can tolerate more than 15% alcohol; high-proof beverages such as whiskey and gin are obtained by concentrating the alcohol by distilling the ferment.

Source of carbohydrate	Product
Grapes	Wine
Rice	Sake
Barley, rye, or corn	Whiskey
Honey	Mead
Potato	Vodka
Barley malt, grain	Beer

convert the plant sugars to lactic acid. As the pickling solution becomes more acid, other species of microorganisms can no longer grow, hence the preservative value of pickling. The preservation of olives is a similar process, as is the silage method for the storage of animal feeds. Anaerobic fermentation lowers the pH of the mash, which protects it from further bacterial spoilage.

Lactic acid-producing bacteria are also used to prepare a variety of soured milk products, such as yogurt, sour cream, and buttermilk. Cultures of lactic acid-producing bacteria are added to the appropriate milk product and, as the solution becomes acidic, the milk proteins precipitate to give a semi-solid product. A quick way to make a form of sour cream for cooking (if you forgot to buy some) is to add a little vinegar (acetic acid) to milk or cream.

(b) ETHANOL PRODUCTION

Certain microorganisms, notably yeast, regenerate NAD^+ from NADH under anaerobic conditions by converting pyruvic acid to ethanol (ethyl alcohol).

$$\text{Pyruvic acid} \xrightarrow{\text{Pyruvate decarboxylase}} \text{Acetaldehyde} + CO_2$$

$$\text{Acetaldehyde} + NADH + H^+ \xrightleftharpoons{\text{Alcohol dehydrogenase}} \text{Ethanol} + NAD^+$$

It is these two reactions, carried out by certain strains of yeast, that provide us with alcoholic beverages (Table 7.1). The distinctive flavors of different drinks are due to components of the particular mash that is fermented and also to metabolic reactions that produce other alcohols, ketones, aldehydes, and similar flavoring components. The formation of vinegar is the result of the oxidation of ethanol to acetic acid by bacteria of the genus *Acetobacter*. These bacteria are present in the air, so if you do not want your wine to turn to vinegar, you should keep it corked (or drink it quickly).

The ability of yeasts to convert pyruvate to carbon dioxide and ethanol also explains the use of this organism in bread making. As the yeast degrades the sugars in the bread dough, the carbon dioxide formed makes the bread "rise." Fortunately or unfortunately, baking kills the yeast and drives off the ethanol.

Many organic compounds, such as acetone, trimethylene glycol, and butyric acid, are made from carbohydrates using different organisms. Such bacterial fermentations are important industrial processes.

7.5 The fate of pyruvic acid under aerobic conditions

The degradation of glucose under anaerobic conditions does not release very much of the potential energy of glucose. To obtain more efficient energy conversion, cells use molecular oxygen (O_2) to complete the oxidation of glucose. Under aerobic conditions, glucose molecules are degraded completely to CO_2 and water, providing additional intermediates important to cell metabolism and capturing a much larger proportion of the available energy of glucose in the form of ATP. The degradation of glucose beyond the pyruvic acid stage involves many reactions that will be discussed in Chap. 8. For the moment it will suffice to discuss the first reaction in the sequence: the production of acetyl coenzyme A (acetyl CoA). This molecule is the branch-point compound which links the metabolism of amino acids, lipids, and sugars (Fig. 5.1).

Acetyl CoA can be considered simply as an activated form of acetate:

$$CH_3-\overset{O}{\underset{}{C}}-S-CH_2-CH_2-\underset{H}{N}-\overset{O}{\underset{}{C}}-CH_2-CH_2-\underset{H}{N}-\overset{O}{\underset{}{C}}-\underset{HO}{\overset{H}{C}}-\underset{CH_3}{\overset{CH_3}{C}}-CH_2-O-\overset{O}{\underset{-O}{P}}-O-\overset{O}{\underset{-O}{P}}OCH_2 \cdots$$

←Acetate→ ←Cysteine unit→ ←Pantothenate unit→ ←ATP unit→

ACETYL COENZYME A
(acetyl CoA)

The acetate residue is linked to the coenzyme by a thioester bond. Coenzyme A is a complex organic molecule made from cysteine, ATP, and a vitamin known as pantothenic acid. This explains why pantothenic acid is an essential dietary factor; without pantothenic acid, coenzyme A could not be made, and therefore a key branch-point compound would be missing.

The formation of acetyl CoA from pyruvic acid represents the first step in the degradation of glucose in which a carbon atom is removed as carbon dioxide. This reaction requires several enzymes and cofactors and has a complex mechanism.

$$\text{coenzyme A} + \text{pyruvic acid} + \text{NAD}^+ \xrightarrow[\substack{\text{Pyruvate} \\ \text{dehydrogenase complex}}]{\substack{\text{Thiamine pyrophosphate} \\ \text{Lipoic acid} \\ \text{FAD}}} \text{acetyl CoA} + \text{NADH} + CO_2$$

For economy and convenience, most cells keep all the enzymes needed for this reaction in one "package," known as the pyruvate dehydrogenase complex. Thus, as intermediates are formed, they are presented directly to the next enzyme for the next step. Other enzyme complexes that come in convenient packages are known to occur.

This reaction involves five different vitamins or vitamin derivatives. Lipoic acid, thiamine pyrophosphate (vitamin B_1), and FAD (riboflavin, vitamin B_2) are cofactors for the enzymes in the pyruvate dehydrogenase complex. NAD^+ (niacin) and coenzyme A (pantothenic acid) are substrates for the reaction. This illustrates the role of vitamins in normal cellular metabolism. But, before you rush to your local supermarket to buy expensive vitamin pills, you should realize that these, like all other vitamins, are present in ample quantities in normal, sensible diets.

However, there *are* deficiency states in which reactions requiring cofactors derived from vitamins can be severely reduced. The role of vitamin B_1 in the formation of acetyl CoA from pyruvate was first demonstrated when a vitamin B_1-deficient diet was fed to animals. Under these circumstances the concentration of pyruvic acid in blood increased, indicating that the dietary deficiency decreased the metabolism of pyruvic acid. A deficiency of vitamin B_1 in human beings leads to a disease known as "beriberi," which is characterized by neuritis. The symptoms of this disease cannot be ascribed to any one reaction, since vitamin B_1 is a cofactor in many different metabolic reactions.

SUMMARY

In most cells, the major source of energy is catabolism of the simple sugar, glucose. The degradation of glucose begins with the glycolytic pathway, in which this six-carbon sugar is broken down into two molecules of pyruvic acid, a three-carbon compound. For each glucose molecule entering glycolysis, the cell gains two ATPs which are eventually used for cell work. Glycolysis also produces NADH and important intermediates for various anabolic pathways. An important feature of the glycolytic pathway is that it can operate under anaerobic conditions.

The fate of the pyruvic acid produced by glycolysis is determined by the type of cell and whether or not oxygen is available. Under anaerobic conditions, pyruvic acid is converted to excretable products, such as lactic acid or ethanol. Both anaerobic conversions regenerate NAD^+ needed in the earlier steps of glycolysis. Under aerobic conditions, pyruvic acid is degraded to CO_2 and acetyl CoA; the latter compound is further degraded in the energy-producing reactions of the Krebs cycle and electron transport system.

PRACTICE PROBLEMS

Consult Figs. 5.1 and 7.1 in answering the following questions.

1. (a) A man decided to make some wine. He filled a large jar with crushed grapes (a source of sugar and phosphate), water, and yeast, closed the jar tightly, and set it in a kitchen corner. One week later, the jar exploded. Why?
   (b) Undaunted, he made a new mixture and placed the container in the refrigerator. This time it did not explode, but even after several months the mixture had not turned to wine. Why not?

2. Under *anaerobic* conditions, what is the cell's maximal net gain of ATPs from one molecule of the following compounds? (Assume that all the compounds are inside the cell.)
   (a) Alanine.
   (b) Glucose.
   (c) Fructose-1,6-diphosphate.
   (d) Glucose in the presence of fluoride ions.
   (e) Glucose-6-phosphate at 85°C.

3. A child with a rare disease developed symptoms of drunkenness following vigorous exercise, although he did not seem to tire easily. The pH of his blood did not fall (i.e., $[H^+]$ did not increase) during such activity. Suggest a possible biochemical mechanism for this disease.

4. A contracting muscle in a frog's leg was injected with disulfiram, which prevents the binding of NAD^+ to dehydrogenases. Give one possible biochemical explanation why the contractions quickly stopped.

5. Why can't yeast produce alcohol from pure sugar unless phosphate ions are added to the fermentation mixture?

6. The metabolic fate of dietary glucose depends on the overall biosynthetic and energy requirements of the body. In the following situations, what would you predict would happen to the bulk of the glucose? Would it be:
   (a) not metabolized?
   (b) used as a source of energy? Would it be catabolized to acetyl CoA or only to the three-carbon stage?
   (c) diverted away from glycolysis and into a biosynthetic pathway? Which pathway(s)?
   The situations are:
   (1) the glucose in the twentieth mint you ate while watching the late, late show.
   (2) the glucose in the cookie eaten by a basketball player at halftime.
   (3) the glucose eaten by a diabetic who has been without insulin for 40 hours.
   (4) the glucose consumed by a thiamine-deficient individual.
   (5) the glucose eaten by a person with a hypothetical disease in which large amounts of ribose are excreted in the urine.

7. If your goal were to produce the maximal amount of radioactive CO_2 from ^{14}C-glucose (all carbons labeled) enzymatically, would you prefer an *in vitro* system that converts glucose to ribose and CO_2 or one that converts glucose to acetyl CoA and CO_2?

SUGGESTED READING

LEHNINGER, ALBERT L., "How Cells Transform Energy," *Sci. American*, 205, No. 3, p. 62 (1961). Excellent overview of glycolysis, the Krebs cycle, and photosynthesis.

LEHNINGER, ALBERT L., *Bioenergetics*, W. A. Benjamin, Inc., New York, 1965. Highly readable description of the generation of ATP in anaerobic cells is found in Chapter 5.

MARGARIA, RODOLFO, "The Sources of Muscular Energy," *Sci. American*, 226, No. 3, p. 84 (1972). Good description of the relationships among energy-storing and energy-releasing reactions in muscle.

PITTS, FERRIS N., JR., "The Biochemistry of Anxiety," *Sci. American*, 220, No. 2, p. 69 (1969). Lactic acid, usually a harmless product of anaerobic glycolysis, may trigger anxiety attacks in susceptible individuals.

8 Degradation of Carbohydrates: Energy & Intermediates from Krebs Cycle & Electron Transport

The Krebs cycle and the electron transport chain are the cell's primary metabolic furnace. These two pathways, acting together, "burn" acetyl CoA in oxygen to produce carbon dioxide, water, and energy. While the useful energy released by a conventional furnace is in the form of heat, that from the Krebs cycle and electron transport chain is captured as chemical bond energy in ATP. The enzymes and carrier proteins which constitute the Krebs cycle and electron transport system act in concert and for this reason are localized within the same organelle, the mitochondrion.

8.1 Acetyl CoA—Molecule at the crossroads

The formation of acetyl coenzyme A (acetyl CoA) from pyruvic acid is an irreversible step in which a molecule of carbon dioxide is released (Sec. 7.5). The details of the subsequent degradation of acetyl CoA were worked out over many years by many investigators, but since the outline of this pathway was first suggested by H. A. Krebs in 1937 (he was awarded a Nobel prize for this feat), the pathway has been named the Krebs cycle. It is also called the citric acid cycle or the tricarboxylic acid cycle (TCA cycle). The steps are known in some detail, although many fine points of the system still escape the modern biochemist, and the enzymology and regulation of the Krebs cycle remain subjects for intensive study.

Many different catabolic pathways can provide acetyl CoA for the Krebs cycle. In addition to glycolysis, the degradation of certain amino acids and lipids also produces acetyl CoA (Fig. 8.1). Thus, the cell can obtain energy from a number of

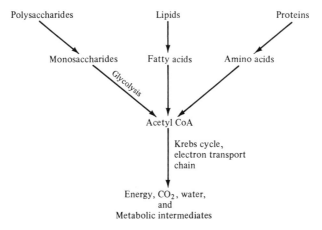

Figure 8.1 Acetyl CoA is the end product of many degradative pathways.

chemically different molecules by converting them to acetyl CoA and processing this molecule in the Krebs cycle and electron transport chain.

It is common for biochemists to have metabolic charts hung on the walls of their offices or laboratories, and it should come as no surprise to learn that the Krebs cycle is in the center of these charts. It is the "heart" of metabolism.

It should be becoming clear that carbohydrates, lipids, and proteins are not only interrelated, but also *interconvertible*. The reactions of the Krebs cycle allow cells to convert carbohydrates to proteins, and vice versa. We shall see later how carbohydrates and proteins can be converted to lipids* through the common intermediate, acetyl CoA. Plant seeds and certain bacteria can convert lipids to carbohydrates; however, animal cells lack the enzymes for this interconversion.

8.2 Reactions of the Krebs cycle

In outline, the acetate (two-carbon) moiety of acetyl CoA is transferred to a carrier molecule that enters into a series of reactions that convert the acetate to carbon dioxide and hydrogen, eventually to regenerate the original carrier molecule. The hydrogen atoms are picked up by special carrier molecules which introduce them into

---

* Proteins are converted into lipids less efficiently than are carbohydrates, a fact that provides the biochemical basis for some slimming regimes.

the electron transport chain, where their available chemical energy is used to make ATP. In simplified form, the Krebs cycle looks like this:

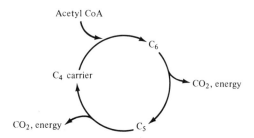

The details of the Krebs cycle are shown in Fig. 8.2. Although all the steps in the cycle have key roles in metabolism, we will limit our discussion to those conversions which are important in regulation and those which involve hydrogen transfer.

Some organisms, such as plants and certain bacteria, possess a modified version of the Krebs cycle, the glyoxylate cycle, in which acetyl CoA is added to oxaloacetic acid (as in the Krebs cycle) and converted back to this compound by a different cycle of reactions. The glyoxylate cycle (Sec. 8.4) is found most often in organisms that use acetate, rather than glucose, as a carbon source.

(a) ACETATE ENTERS THE CYCLE

The first step in the Krebs cycle is the condensation of acetyl CoA with the "carrier" molecule of the cycle, oxaloacetic acid. This produces the six-carbon tricarboxylic acid (three "—COOH" groups) citric acid. The two carbon atoms from the acetate have now been introduced into the cycle and can promote one complete round of reactions. Since each molecule of glucose gives rise to two molecules of acetyl CoA, the degradation of one molecule of glucose promotes two rounds of the Krebs cycle. The coenzyme A released as acetate is introduced into the cycle is reused as a carrier.

An interesting inhibitor affects this first step. Fluoroacetic acid is the product of a poisonous African plant, and when a cell absorbs fluoroacetic acid, it mistakes it for acetic acid and converts it to fluoroacetyl CoA, which then donates its two carbon moiety to oxaloacetic acid, forming fluorocitric acid. Up to this point the enzymes involved have not detected the presence of the foreign fluoride atom and are "fooled" into using the poison as acetate. Why, then, is fluoroacetic acid a poison? The reason is that fluorocitric acid is a competitive inhibitor of the isomerase that converts citric acid to isocitric acid. Fluorocitric acid *looks like* citric acid, but it is not a substrate for the enzyme.

In conjunction with its role as an intermediate in the Krebs cycle, citric acid has an important function in cell regulation. Citric acid is an inhibitor of phosphofructokinase, the enzyme that catalyzes an early step of glycolysis (Sec. 7.1). Under normal conditions, all the available citric acid is catabolized in the Krebs cycle to provide energy. However, when an excess of acetyl CoA leads to an overproduction of citric

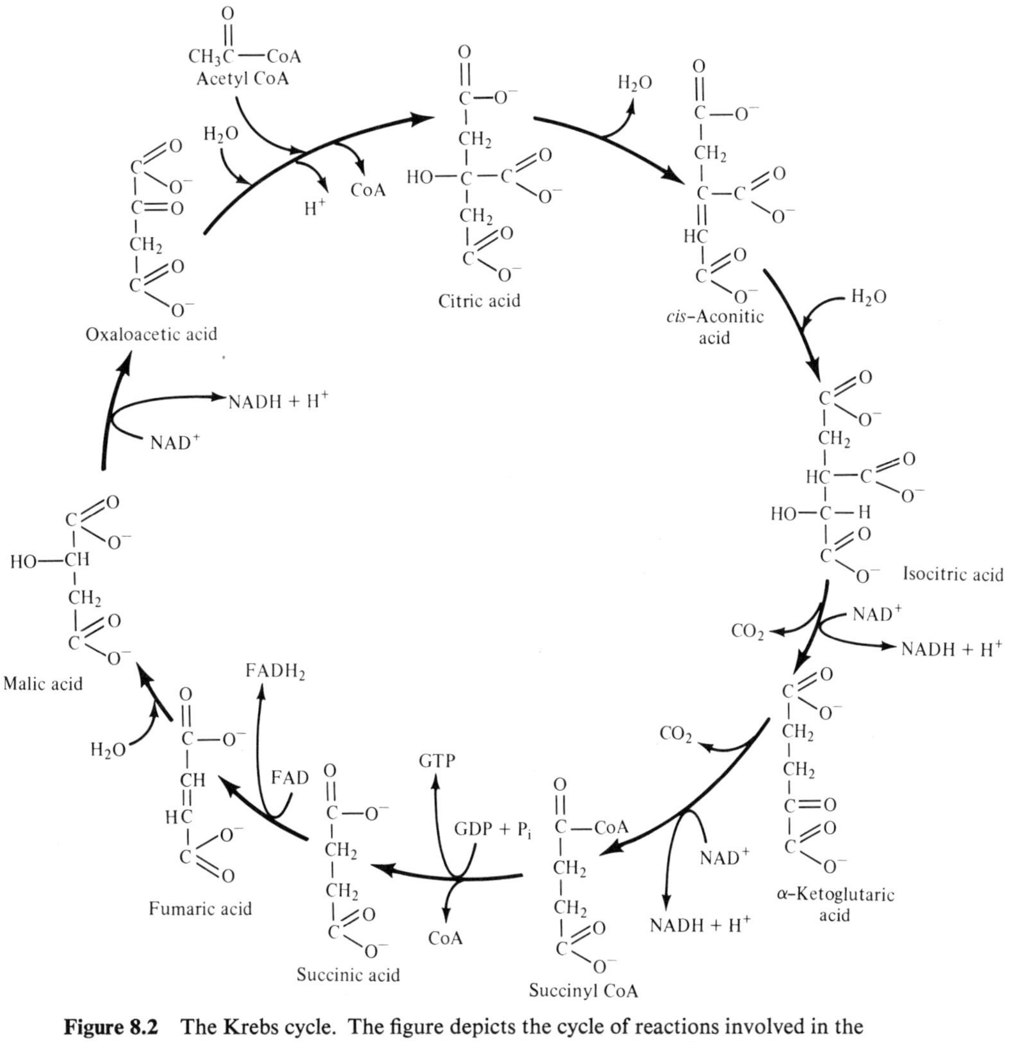

Figure 8.2 The Krebs cycle. The figure depicts the cycle of reactions involved in the aerobic degradation of acetyl CoA to CO_2 and hydrogen within the mitochondria of eucaryotic cells. During the cycle, hydrogen atoms are removed from certain carboxylic acids, transferred to NAD^+ or FAD, and oxidized to water in the electron transport system. Carbon atoms are removed as CO_2 to complete the degradation of acetyl CoA.

acid, the excess citric acid inhibits phosphofructokinase, thus preventing nonproductive continuation of glycolysis and wastage of raw materials. When citric acid levels fall, the inhibition of phosphofructokinase is relaxed and glycolysis recommences to provide more starting materials for the Krebs cycle (Figs. 8.3 and 7.2). This is an example of the delicate control systems that keep a balance between various metabolic pathways in the cell. In this particular case, the cell balances the production and

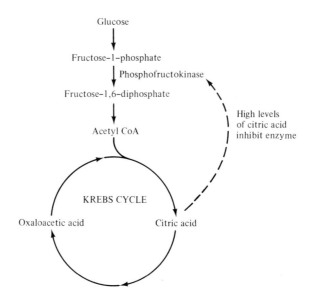

Figure 8.3 Control of glycolysis by inhibition of the enzyme phosphofructokinase by citric acid. In this way, cells can control the rate of glycolysis so that no more acetyl CoA is made than can be utilized by the Krebs cycle.

utilization of acetyl CoA. Thus, when a cell is metabolizing carbohydrates under aerobic conditions, the glycolytic pathway is neatly maintained at a rate sufficient to fuel the Krebs cycle.

(b) SUBSTRATE-LEVEL PHOSPHORYLATION IN THE FORMATION OF SUCCINIC ACID

As will be described in Sec. 8.4, the formation of ATP by the electron transport system provides the bulk of the useful energy that may be extracted from reactions taking place in the Krebs cycle. However, in the conversion of α-ketoglutaric acid to succinic acid,* the nucleotide diphosphate known as guanosine diphosphate (GDP) reacts with inorganic phosphate to form guanosine triphosphate (GTP). GTP, like ATP, is a high-energy compound, and high-energy phosphate groups can easily be donated by this molecule, for example, to ADP. Thus, the formation of GTP is formally equivalent to the production of ATP; the energy of GTP is useful energy obtained from the degradation of acetyl CoA.

The formation of GTP from guanosine diphosphate and P_i is an example of a *substrate-level phosphorylation*; by this we mean that the formation of the nucleotide

---

* This conversion involves succinyl CoA as an intermediate and exemplifies the way in which coenzyme A acts as a biological carrier molecule.

triphosphate (GTP or ATP) is a direct result of the reaction between the various substrates and the nucleotide diphosphate, rather than the result of additional steps such as the electron transport system. ATP is formed by similar substrate-level phosphorylations during glycolysis (Sec. 7.2).

(c) NAD⁺ AND FAD: THE HYDROGEN CARRIERS

When glucose is burned in the presence of oxygen, the hydrogen atoms of the sugar are removed and combined with oxygen to form water. Likewise, in four different steps of the Krebs cycle, for example, the conversion of succinic acid to fumaric acid, oxidation occurs by the removal of hydrogen atoms from the substrate. How is this done? The hydrogen atoms are not released as free hydrogen gas, neither are they combined directly with oxygen; instead, they are transferred from the substrate to a carrier that shuttles the hydrogen atoms into the electron transport system, where their energy is used for ATP production.

In most cases, the carrier that accepts or donates hydrogen atoms or electrons in such oxidation–reduction reactions is nicotinamide adenine dinucleotide (NAD^+), a complex organic molecule that contains the B vitamin, nicotinamide, as an integral part of its structure. A second hydrogen carrier is flavin adenine dinucleotide (FAD) with a structure based on riboflavin, another member of the B vitamin complex.

NAD^+ and FAD are unsaturated, electron-deficient molecules which are able to accept hydrogen atoms. (A hydrogen atom consists of an electron and a proton [H^+].) Each NAD^+ can accept two electrons and one proton; each FAD accepts two electrons and two protons. When NADH or $FADH_2$ donate a pair of electrons to the electron transport chain, the protons that accompanied these electrons are released into the cell cytoplasm (Fig. 8.6).

NAD^+ and FAD are coenzymes that function in collaboration with enzymes named dehydrogenases; they participate in many diverse oxidation–reduction reactions. In addition to feeding electrons into the electron transport system, the reduced (hydrogen-rich) carriers can transfer their hydrogen atoms to unsaturated substrates and reduce them; both processes lead to the regeneration of NAD^+ or FAD. We are already familiar with the role of dehydrogenases in the production of lactic acid and ethanol (Sec. 7.4), and with succinic dehydrogenase, the Krebs cycle enzyme that catalyzes the oxidation of succinic acid to fumaric acid; the latter is competitively inhibited by the "look-alike" malonic acid (Table 4.2).

In spite of the fact that NADH and $FADH_2$ receive hydrogen from substrates, they are not normally used to provide hydrogen for reduction in biosynthetic reactions. For biosynthetic purposes, NADPH, a derivative of NADH carrying an extra phosphate group, is used. Hydrogen atoms can be transferred between these carriers:

$$NADH + NADP^+ \rightleftharpoons NAD^+ + NADPH$$

The direction of this exchange reaction is controlled by the requirement for various substrates, such as NAD^+ or NADPH, and provides cells with a convenient method for balancing the rates of biosynthetic and degradative pathways. When NADH is in excess, NADPH is produced for use in biosynthetic reactions with the concomitant regeneration of NAD^+.

(d) THE REGENERATION OF OXALOACETIC ACID

Naturally, a cycle must end up where it began. Oxaloacetic acid, the acceptor for the acetate residue from acetyl CoA, must be regenerated to allow subsequent rounds of the Krebs cycle. The "final" reaction of the cycle is the conversion of malic acid to oxaloacetic acid; this reaction also produces NADH.

8.3 The flow of intermediates into and out of the Krebs cycle

The Krebs cycle does not function solely for the degradation and production of energy from acetyl CoA; it also serves to provide important intermediates for other metabolic pathways (Fig. 8.4).

The keto acids, oxaloacetic acid and α-ketoglutaric acid, are examples of intermediates that link the Krebs cycle to other pathways. Both of these dicarboxylic acids can be converted to amino acids by transamination. The reverse (amino acid $\rightarrow$ keto acid) can also occur. Transamination involves the exchange of an amino group between two molecules. For example:

$$\text{HOOC}-\underset{\underset{\text{O}}{\|}}{\text{C}}-\text{CH}_2\text{CH}_2\text{COOH} + \text{CH}_3-\underset{\underset{\text{NH}_2}{|}}{\text{CH}}-\text{COOH} \xrightleftharpoons{\text{Transaminase}}$$

α-Ketoglutaric acid   Alanine

$$\text{HOOC}-\underset{\underset{\text{NH}_2}{|}}{\overset{\overset{\text{H}}{|}}{\text{C}}}-\text{CH}_2\text{CH}_2\text{COOH} + \text{CH}_3-\underset{\underset{\text{O}}{\|}}{\text{C}}-\text{COOH}$$

Glutamic acid   Pyruvic acid

In this particular transamination reaction, the amino group of alanine is transferred to α-ketoglutaric acid, forming glutamic acid and pyruvic acid. Glutamic acid is then used in the synthesis of other amino acids, incorporated into protein, or used in nitrogen excretion, while pyruvic acid is an important product of glycolysis and can, of course, be converted to acetyl CoA for the Krebs cycle. By the reverse reaction, the carbon skeleton of glutamic acid can be fed into the Krebs cycle and degraded to provide energy and other intermediates. Thus, we can see how reversible transamination reactions can link the Krebs cycle, amino acid metabolism, and glycolysis.

Transamination, although simple in principle, is actually a very complex reaction. The enzymes catalyzing such reactions require a coenzyme, pyridoxal phosphate, that is made from vitamin B_6. In transamination, the amino group of the substrate amino acid is first transferred to pyridoxal phosphate and then to the keto acid that is the

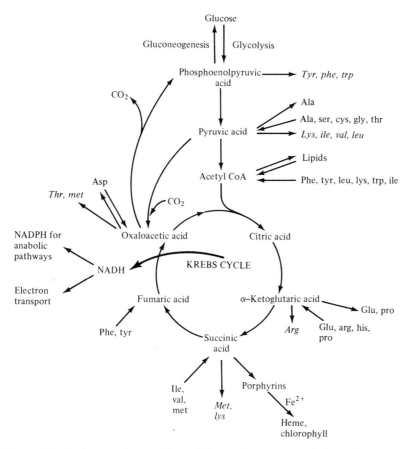

Figure 8.4 Krebs cycle intermediates link together many anabolic and catabolic pathways. The degradation of the carbon skeletons of most amino acids yields Krebs cycle intermediates. Hence, amino acids can be catabolized to provide energy or can serve as the starting material for pathways using the intermediates of the cycle, for example, gluconeogenesis. (The three-letter symbols for the amino acids are defined in Table 2.2.)

The supply of Krebs cycle intermediates is depleted by the synthesis of amino acids and porphyrins, as well as gluconeogenesis. Higher animals, including man, synthesize only about half their necessary amino acids; plants and most microorganisms make all the amino acids they require, which places a greater drain on supplies of Krebs cycle intermediates. (Pathways unique to plant cells and microorganisms are shown in italics.)

The release of hydrogen atoms in the Krebs cycle is as important to overall cellular metabolism as the rearrangement and oxidation of carbon atoms. The hydrogen atoms are either donated to the electron transport chain for the production of energy or used for reduction reactions in biosynthetic pathways.

other substrate of the reaction; the vitamin acts as a carrier for the amino group. Pyridoxal phosphate is an important coenzyme and also plays a role in the cellular metabolism of ammonia.

Another useful intermediate provided by the Krebs cycle is succinic acid, used by bacteria and plants in the biosynthesis of the amino acids methionine and lysine. Succinic acid is also used in the biosynthesis of a large class of complex compounds known as porphyrins (Sec. 11.5). Some typical porphyrins are chlorophyll, the heme of hemoglobin and the cytochromes, and vitamin B_{12}.

The multiple uses of Krebs cycle intermediates could lead to difficulties, and it is possible to envisage situations in which the cell's demand for intermediates from the Krebs cycle could lead to a shortage of oxaloacetic acid, a condition that would be lethal, since the "carrier" would be depleted. To avoid this problem, other ways of forming oxaloacetic acid are available to provide backup supplies to ensure that the Krebs cycle never runs down because of lack of this essential carrier.

The most important method for replenishing the supply of oxaloacetic acid is the enzymatic carboxylation of pyruvic acid*:

$$\text{pyruvic acid} + CO_2 \xrightarrow{\text{Pyruvate carboxylase}} \text{oxaloacetic acid}$$

The extent of this reaction is controlled by the intracellular level of acetyl CoA. When oxaloacetic acid is in short supply, the levels of acetyl CoA build up and activate pyruvate carboxylase, thus stimulating the formation of more oxaloacetic acid. When sufficient oxaloacetic acid is available, there is little free acetyl CoA and hence no stimulation of the reaction.

A second source of oxaloacetic acid is provided by the removal of the amino group of aspartic acid by deamination or transamination.

The glyoxylate cycle, found only in plants and bacteria, is a unique route for the production of Krebs cycle intermediates. Certain plant seeds (peanuts, cottonseeds) contain stored lipids, rather than starch, which must provide both energy and intermediates during germination and seedling development. Clearly, lipids can be used to provide ATP energy, but, in addition, developing plants must synthesize large amounts of the polysaccharide cellulose, and hence the stored lipids must be converted to glucose.

The glyoxylate cycle begins with the production of malic acid from acetyl CoA and glyoxylate:

Acetyl CoA + Glyoxylate ⟶ Malic acid

---

* In addition, the pyruvate carboxylase reaction is an important step in the biosynthesis of carbohydrates; it enables the glycolytic pathway to be reversed so that glucose is formed in a process known as *gluconeogenesis* (Sec. 9.4).

Degradation of Carbohydrates

Malic acid is oxidized to oxaloacetic acid, a reaction that produces two NADHs which fuel oxidative phosphorylation. Oxaloacetic acid can then be converted to glucose by means of gluconeogenesis, or to isocitric acid, the substrate for the second reaction unique to the glyoxylate cycle:

$$\text{isocitric acid} \longrightarrow \text{succinic acid} + \text{glyoxylate}$$

By superimposing these reactions on those of the Krebs cycle and gluconeogenesis, germinating plants are able to convert fatty acids to glucose. As the plant develops, photosynthesis takes over as the producer of glucose for energy and polysaccharide production, and the enzymes of the glyoxylate cycle disappear.

8.4 Energy from the Krebs cycle: ATP is formed by oxidative phosphorylation

We have been discussing various functions of the Krebs cycle and must now consider how it produces useful energy for cellular work. Substrate-level phosphorylation occurs during the formation of succinic acid, but this is not the major source of ATP in the aerobic cell. The bulk of the ATP is produced by *oxidative phosphorylation* in the electron transport system of mitochondria. In principle we can consider electron transport and oxidative phosphorylation as the progressive release of energy from electrons (obtained in certain reactions of the Krebs cycle) as these electrons are passed along a series of compounds of decreasing energy; the "driving force" for this process is the terminal reaction in which oxygen is the final electron acceptor. In this final step, molecular oxygen combines with two electrons and two protons to form water.

The concept of a waterfall provides a simple analogy. The water flowing down a waterfall possesses potential energy that can be harvested and then converted to mechanical or electrical energy; this energy can be withdrawn in one large step by having a big waterwheel at the bottom of the waterfall, or in several small steps by using several waterwheels (Fig. 8.5). In principle, the electron transport system (Fig. 8.6) is no different from a descending series of waterwheels.

What reactions provide the electrons for the electron transport system? Any cellular oxidation reaction in which hydrogen atoms are released can provide input for the electron transport system. Most important for the production of ATP in the degradation of glucose, however, are the oxidation reactions in the Krebs cycle. During the Krebs cycle there are four reactions in which hydrogen atoms are removed from substrates (Fig. 8.2) to NAD^+ or FAD. These acceptors then introduce their electrons into the electron transport system. As the electrons are passed from molecule to molecule through compounds of lower and lower energy, the balance in energy between the sequential intermediates is released and used to generate ATP from ADP and inorganic phosphate. For every pair of hydrogen atoms removed by NAD^+ in the Krebs cycle, a pair of electrons passes into the electron transport system, yielding

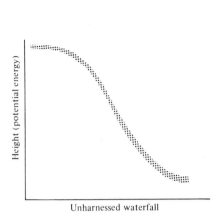
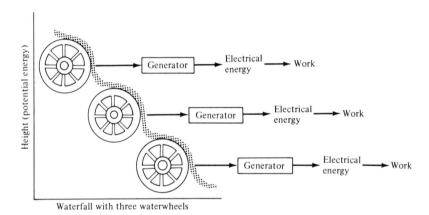

Figure 8.5 Harnessing the potential energy of a waterfall. The potential energy of falling water can be harnessed if the movement of the water (potential energy) is used to drive a turbine wheel, which in turn can be used to turn a generator to produce electricity. Several waterwheels are spaced at intervals in the fall to tap off energy in three small portions rather than one large portion. By analogy, the electron transport/oxidative phosphorylation system removes energy from electrons in small usable packages (ATP) rather than one large amount.

three molecules of ATP. For every pair of hydrogen atoms removed by FAD, two molecules of ATP are produced, because $FADH_2$ introduces its pair of electrons into a lower-energy intermediate than NADH (Fig. 8.6).

What is the nature of the molecules involved in the electron transport system? We have seen that the electrons removed by the oxidation of substrates in the Krebs cycle are accepted by one of two electron acceptors, NAD^+ or FAD. These electrons are transferred to the cytochromes in the electron transport system. The cytochromes are complex molecules containing a heme–metal cofactor complexed to a protein. The metal ion (iron) is the "business end" of the cytochromes and plays an important role

Degradation of Carbohydrates

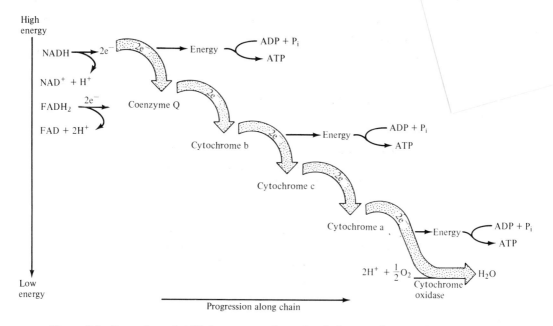

Figure 8.6 Formation of ATP by passage of a pair of electrons from carrier to carrier in the electron transport chain. As each pair of electrons ($2e^-$) is passed from beginning to end in this system, three ATP molecules are formed when ADP combines with inorganic phosphate (P_i). The thick lines show the path of the electrons; the cytochromes and coenzyme Q are electron carrier molecules of decreasing energy capacity in the order indicated in the figure. They are all parts of the same electron "waterfall," with energy generators at the points indicated. Note that electrons carried by $FADH_2$ do not enter the system at the start of the chain and produce only two molecules of ATP by oxidative phosphorylation. These reactions take place in the mitochondria of cells and require oxygen, which is used in the last step in which the enzyme cytochrome oxidase catalyzes the formation of water.

in the acceptance and transfer of the electrons. In certain steps where the electrons are passed from a molecule with high energy to one with lower energy, the energy released is used to phosphorylate ADP to give ATP.

The process just described is known as *oxidative phosphorylation* and is tightly coupled to the flow of electrons through the electron transport chain. Electrons will not flow through the system unless ADP is simultaneously phosphorylated. Hence, both the regeneration of NAD^+ and FAD and the production of energy are dependent on a continued supply of phosphate and adenosine diphosphate.

The electron transport system also requires a continuous supply of molecular oxygen. The reason for the oxygen dependence is the terminal reaction of the sequence, in which water is formed from oxygen and protons and electrons. In the absence of oxygen, this step cannot occur, which results in blockage of the entire pathway. Thus, since the Krebs cycle is dependent on a constant supply of NAD^+ regenerated in electron transport, the Krebs cycle is indirectly an aerobic system. In any series of reactions, one only needs one aerobic step to make the whole series oxygen-dependent.

The reaction between molecular oxygen and hydrogen is catalyzed by cytochrome oxidase. This reaction "pulls" the whole energy-producing system and, if it is inhibited, the effects on metabolism are cataclysmic. This unhappy situation occurs either in the presence of cyanide or carbon monoxide, both potent inhibitors of cytochrome oxidase and, of course, favorite tools of poisoners. Living organisms are essentially asphyxiated by these poisons. Certain antibiotics block specific steps in the electron transport/oxidative phosphorylation system and have been used as inhibitors to elucidate the nature of some of the reactions of the electron transport chain.

Perhaps one of the most important, and most puzzling, aspects of electron transport is the fact that this series of reactions takes place within the mitochondrial membrane. Membranes are complex matrices of lipid and protein (Sec. 10.4), and most of the proteins needed for electron transport and oxidative phosphorylation appear to be an integral part of this matrix. The relationship between the topology of the membrane, the enzymes, and the reactions of electron transport is of great interest.

Not all living systems use oxygen as the final hydrogen acceptor in the electron transport/oxidative phosphorylation system. For example, the sulfur bacteria have a modified electron transport system and produce H_2S, not H_2O, as the final product.

8.5 Energy from glucose: the balance sheet

In looking back over the pathway for the catabolism of glucose, it is perhaps difficult to discern that, in this lengthy series of steps, the products of the oxidation of glucose and other sugars really are the same as those obtained if the sugars were burned in a fire:

$$\text{glucose} + 6\,O_2 \longrightarrow 6\,CO_2 + 6\,H_2O + \text{energy}$$

Although the end products are the same, the benefits of gradual, stepwise enzymatic degradation to the cell are clearly manyfold. The route to carbon dioxide and water provides many important intermediates and useful energy in the form of ATP. Because of the sequential nature of the degradation, cells can capture a high percentage of the energy released by the oxidation of glucose without the self-destructive production of heat. The percentage of energy captured is higher, in fact, than could be obtained with the best available engineering techniques.

Table 8.1 summarizes those steps in glycolysis and the Krebs cycle that provide energy in the form of ATP. Under aerobic conditions, approximately 40% of the energy theoretically obtainable from glucose is captured in ATP. Under anaerobic conditions there is only about $\frac{1}{18}$ of this yield.

8.6 Cytochrome c and evolution

The cytochrome proteins are of tremendous biological interest—not only because of their unique function, but because analysis of the structures of cytochrome proteins from various sources has provided an insight into the ways in which proteins and

TABLE 8.1 Production of Energy During the Cellular Oxidation of Glucose

	Reaction	ATP used (−) or formed (+) per molecule of glucose		Comments
		Anaerobic	Aerobic	
GLYCOLYSIS	Glucose → glucose-6-P	−1	−1	
	Fructose-6-P → fructose-1,6-di-P	−1	−1	
	Glyceralde-3-P → 1,3-diphosphoglyceric acid	0	+4	2 molecules of NADH from each glucose, because C_6 skeleton is split in half; each NADH gives 2 ATP[a]
	1,3-Diphosphoglyceric acid → 3-phosphoglyceric acid	+2	+2	Substrate-level phosphorylation
	Phosphoenolpyruvic acid → pyruvic acid	+2	+2	Substrate-level phosphorylation
	Pyruvic acid → acetyl CoA	0	+6	As above, each glucose molecule promotes two rounds of Krebs cycle; 2 NADH = 6 ATP from electron transport
KREBS CYCLE	Isocitric acid → α-ketoglutaric acid	0	+6	2 NADH = 6 ATP
	α-Ketoglutaric acid → succinic acid	0	+8	2 NADH = 6 ATP; 2 GTP = 2 ATP by substrate-level phosphorylation
	Succinic acid → fumaric acid	0	+4	2 $FADH_2$ = 4 ATP; $FADH_2$ enters electron transport later than NADH
	Malic acid → oxaloacetic acid	0	+6	2 NADH = 6 ATP
	Totals	+2	+36	

[a] Unlike the NADHs formed in subsequent (mitochondrial) reactions (6–10), these NADHs are made in the cytoplasm. Each NADH in the mitochondria yields 3 ATPs. However, 1 ATP must be used to transport each molecule of *cytoplasmic* NADH into the mitochondria, and hence the *net* gain per NADH formed in the cytoplasm is only 2 ATPs.

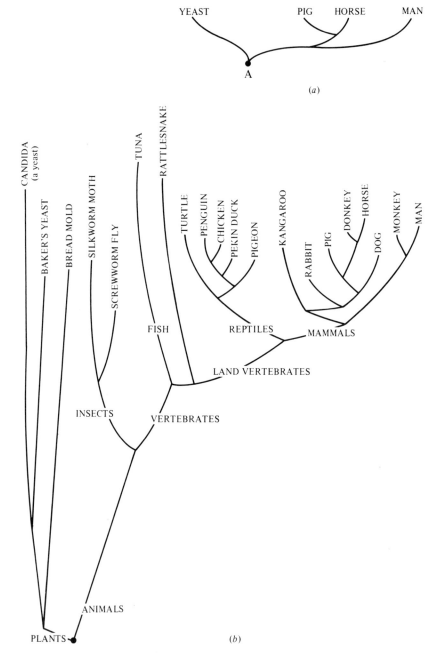

Figure 8.7 Amino acid sequence relationships between the cytochrome c molecules of various species indicate evolutionary relationships. The starting point A represents the primordial form of cytochrome c. The distance along the lines between the species is proportional to the number of amino acid differences in their cytochrome c's; the shorter the line, the more closely related are the organisms in an evolutionary sense (a) Simple evolutionary tree. (b) More complex evolutionary tree. [After Walter M. Fitch and Emanuel Margoliash, *Science* 155: 282 (1967).]

living species have evolved during the last several hundred million years. Because of the constant possibility of spontaneous mutation (Sec. 12.9), the amino acid sequences of proteins have changed during the evolution of living forms. Only those changes which leave the protein functional (active) are passed on from one generation to the next. A "bad" mutation would not be transmitted.

Because cytochrome c is a component of the electron transport system, it is essential to all forms of aerobic life; without cytochrome c the electron transport system could not provide energy. In general, cytochrome c has evolved very little. Those peptide sequences that are critical for the function of the protein in electron transport must be conserved from species to species; too many changes in these sequences would be lethal. However, alterations in other, less critical sequences can be tolerated and by comparing the amino acids of these sequences between different species, we can obtain the details necessary for construction of an evolutionary "tree."

The amino acid sequences of cytochrome c proteins from more than 40 species have been determined. The amino acid sequence of yeast cytochrome c differs from the sequences of both horse and man at 45 different sites, horse cytochrome c differs from human cytochrome c at 12 different sites, and pig cytochrome c differs from human cytochrome c at 10 different sites. If we assume that cytochromes closely related in structure are closely related in an evolutionary sense, we can begin to construct an evolutionary "tree," as shown in Fig. 8.7. From the information given above, we can construct a simple tree; it is clear that yeast must belong to a separate evolutionary branch from horse, pig, or man. The evolutionary trees are nothing more than family trees of the type constructed for the relationships (often illicit) among royal families in Europe. As more cytochrome c sequences are determined, we can add additional branches to the tree, as shown in the lower part of Fig. 8.7. As might be expected, close relationships are seen among such mammals as man, monkeys, and rabbits. Reptiles, fish, insects, and yeast show good relationships among members of their own species but are quite distinct from members of other species. Protein phylogenies of this type are in excellent agreement with evolutionary schemes based on anatomical comparisons.

SUMMARY

Acetyl CoA, produced by the catabolism of sugars, lipids, and amino acids, is degraded to carbon dioxide and water by the coordinated activities of the Krebs cycle and electron transport system. These reactions take place in mitochondria.

The Krebs cycle is at the center of both energy and carbon compound metabolism; it consists of a series of reactions that convert the acetate portion of acetyl CoA to CO_2 and hydrogen. Krebs cycle intermediates can be diverted to provide the carbon skeletons for amino acids, porphyrins, and other metabolic intermediates. Conversely, the supply of Krebs cycle intermediates can be replenished when these compounds are degraded.

Two hydrogen carriers, NAD^+ and FAD, accept hydrogen atoms (protons plus electrons) from reactions in the Krebs cycle and deposit the electrons in the electron transport system, where their energy is used for ATP production. The electron transport system consists of a series of linked reactions in which energy is released in small packets as the electrons are passed through compounds of decreasing energy. The energy released during electron transfer is used to phosphorylate ADP (oxidative phosphorylation), thus supplying the ATP needed for cell work. The major electron carriers are the iron-containing proteins, the cytochromes; the final step is the reaction of hydrogen and oxygen to produce water.

PRACTICE PROBLEMS

Refer to Table 8.1 and Fig. 8.4 when answering the following questions.

1. Cells can obtain energy from the aerobic catabolism of many compounds. Arrange the list below in decreasing order of the maximum energy (number of ATPs) obtainable from one molecule of each compound.

    Acetyl CoA
    Carbon dioxide
    Glutamic acid
    Ammonia
    Glucose
    Alanine

2. Beer is made by growing yeast in large vats without agitation, using grain as the carbohydrate source. Would you expect cyanide to inhibit this process?

3. Several small molecules contain chemical energy which can either be used directly for doing cellular work or can be converted to the energy of ATP. In this sense, NADH is energetically equivalent to three ATPs. How many molecules of ATP can be derived from one molecule of each of the following compounds?

    NADPH
    GTP
    Creatine phosphate
    FAD

4. Would you expect to find mitochondria in a strictly anaerobic organism?

5. Aerobic cells make ATP by both substrate-level phosphorylation and oxidative phosphorylation. Describe the differences between these two processes. What fraction of the ATPs made during the aerobic degradation of glucose are produced by each of these methods? What fraction of the ATPs made during the degradation of acetyl CoA are produced by each of these processes?

6. Some cells are strictly aerobic; they cannot obtain energy from glucose unless glycolysis, the Krebs cycle, and the electron transport chain are all functioning. If any of these pathways is inhibited, energy production stops. What product(s) of the electron transport system are needed to keep the Krebs cycle going? What product(s) of the electron transport chain are needed to allow glycolysis to continue?

7. A bacterial cell is growing in a nutritionally rich environment and catabolizing glucose aerobically. Which of the following inhibitory treatments would act directly on the Krebs cycle? Which would act directly on the electron transport system?

Degradation of Carbohydrates 147

    (a) Deficiency of B vitamins.
    (b) Addition of penicillin.
    (c) Addition of carbon monoxide.
    (d) Heating to 100°C for 30 minutes.
    (e) Removal of inorganic phosphate from the cell's environment.
    (f) Addition of citric acid.
    If the cell were functioning anaerobically, which treatments would be effective in stopping glucose catabolism?

8. An unenergetic and undernourished-looking child was admitted to the hospital. Consultation with the mother revealed that for the past 2 months, the youngster's diet consisted mainly of glucose (in the form of soda pop) and a daily vitamin pill.
   (a) By what general pathways were the child's muscle cells synthesizing the compounds necessary to make actin and myosin?
   (b) The child's diet was supplemented with meat. After this supplementation, it was found that the amount of glucose converted to pyruvate per minute was reduced, although the child was clearly healthier and had more energy. How do you explain this?

9. Enzyme Y is essential to all cells. It has been isolated from many organisms and its amino acid sequence found to differ slightly between species. Which segments of its amino acid sequence would you predict would be identical regardless of the source of the enzyme?

    Some representative data on enzyme Y are given below. In terms of evolution, which organism is most closely related to man? Which is most distantly related to man?

	Number of amino acid differences
Man and bread mold	31
Man and bumblebee	23
Man and orangutan	3
Man and dog	7

SUGGESTED READING

DICKERSON, RICHARD C., "The Structure and History of an Ancient Protein," *Sci. American*, 226, No. 4, p. 58 (1972). Describes the evolution of cytochrome *c*.

GREEN, DAVID E., "The Mitochondrion," *Sci. American*, 210, No. 1, p. 63 (1964). Electron micrographs and schematic drawings illustrate the highly organized structure of the aerobic cell's "metabolic furnace."

LEHNINGER, ALBERT L., *Bioenergetics*, W. A. Benjamin, Inc., New York, 1965. Chapter 6 describes mitochondrial ATP formation.

YOUNG, VERNON R., and NEVIN S. SCRIMSHAW, "The Physiology of Starvation," *Sci. American*, 225, No. 4, p. 14 (1971). Describes both the long- and short-term effects of calorie (i.e., glucose) deprivation on the human body, particularly how metabolism changes in response to this deficiency.

9 Carbohydrate Formation & Storage: How Cells Control Energy Supplies

Glucose is the major energy source for all cell work. We have seen how glucose is broken down by glycolysis and the Krebs cycle, and how the energy released during its degradation is captured in the form of ATP. Because it is their major fuel, cells have developed alternative pathways for making glucose from other nutrients and also for storing glucose when an excess is present; the existence of these pathways (Fig. 9.1) ensures that this all-important sugar is available at all times.

Since glucose is so important, it is essential for animals to maintain relatively constant blood levels of this sugar. But the dietary intake of glucose varies; one day we gorge ourselves on cake and ice cream and the next morning we skip breakfast. Our bodies must be able to accommodate to our fickle behavior! Thus, it is worth considering two situations:

1. What happens when excess glucose is consumed?
2. What happens when an external supply of glucose is not immediately available at the onset of a period of intense (energy-requiring) activity?

The answer to the first question is that part of the glucose is used to provide for immediate needs, part is catabolized to acetyl CoA and stored as lipid (Chap. 10), and part is stored in liver and muscle in a polymer of glucose known as glycogen. In the second situation, the organism must use its stored glycogen, which is broken down to glucose and then catabolized in the usual way.

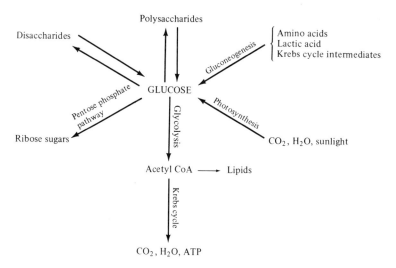

Figure 9.1 Glucose is a branch-point compound linking several anabolic and catabolic pathways.

Under some circumstances, yet another situation can arise. Certain body organs and tissues have an absolute dependence on a continuous supply of glucose provided through the bloodstream. Organs such as the brain can use no other compound as an energy source and have only a limited capacity to store glucose. When an external source of glucose is not available, the body responds by making glucose and glycogen from small molecules such as lactic acid and amino acids usually in the liver. This process (essentially the reverse of glycolysis) is known as *gluconeogenesis* and it maintains appropriate levels of blood glucose under conditions of starvation.

This leads us to the problem of how a living organism decides among these various possibilities. When to break down glycogen? When to make it? We will see that a complex but interesting group of hormone-controlled reactions ensures that the right balance between glycogen synthesis and breakdown is achieved. We must also consider the pathways controlled in this manner.

9.1 Glucose is stored in polysaccharides

In animals, glucose is stored as glycogen, a branched polymer made by linking together several hundred glucose molecules. The glycogen molecule can be thought of as tree-shaped, containing short linear chains of glucose which are connected at branch points (Fig. 9.2). Two types of glycosidic linkages hold the glucose molecules together; the predominant type links glucoses "end-to-end" (1,4-glycosidic bonds), but at each branch point there is a linkage between the "end" of one glucose and the "side" of another glucose (1,6-glycosidic bonds). The synthesis of glycogen from glucose monomers and its degradation are outlined in Fig. 9.2.

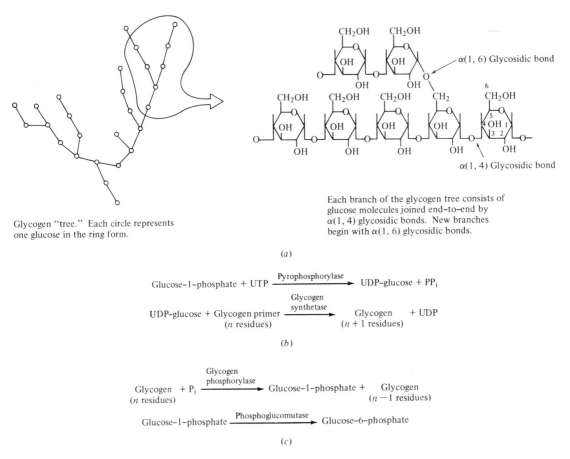

Figure 9.2 Glycogen and its metabolism. (*a*) Glycogen is a nonlinear polymer of glucose. (*b*) Glycogen synthesis: glucose molecules are added one at a time to the growing glycogen tree. One molecule of uridine triphosphate (UTP, a high-energy compound similar to ATP) is needed to activate each glucose molecule prior to its addition to the growing polysaccharide. Glycogen synthetase catalyzes the formation of $\alpha(1,4)$-glycosidic bonds; a "branching enzyme" forms the $\alpha(1,6)$-glycosidic linkages. (*c*) Glycogen breakdown: glycogen is degraded by the sequential removal of individual glucose molecules. Glycogen phosphorylase cleaves only $\alpha(1,4)$-glycosidic bonds; a "debranching enzyme" breaks $\alpha(1,6)$-glycosidic linkages. Glucose-6-phosphate can be used directly in glycolysis.

Although glycogen accumulation is important in energy storage, several rare human diseases are characterized by an excessive accumulation of glycogen in muscle, liver, and/or the renal tubules. Interestingly, most of these "glycogen storage diseases" result from the absence (or low level) of one of the enzymes of glycogen breakdown (i.e., they do not result from overproduction of glycogen). One example is McArdle's syndrome, in which muscle cells lack glycogen phosphorylase and cannot degrade glycogen to glucose; thus, glucose is not always immediately available to provide energy, and people suffering from this disease have a decreased tolerance for physical activity and demonstrate little increase in blood lactic acid after exercise. In Cori's disease, the sufferer lacks the debranching enzyme and cannot degrade glycogen completely, leading to inefficient use of this storage polysaccharide.

Starch, the storage polysaccharide of plants, is also a polymer of glucose. The structural differences between starch and glycogen lie in the shape of the final macromolecules. The biosynthesis of starch is similar to that of glycogen, with the exception that it is ADP-glucose and not UDP-glucose which donates the sugar to the growing macromolecule.

In one respect, the cellular biosynthesis of polysaccharides is unlike the synthesis of other macromolecules such as proteins and nucleic acids; polysaccharides are not synthesized to fit a predetermined pattern. Thus, their size and shape may vary considerably from molecule to molecule. There is no single type of glycogen, no single type of starch; although the monomers are always the same and each polysaccharide contains characteristic monomer–monomer bonds, there are few limits on the number of monomers per branch or on the overall size of the macromolecules.

Not all polysaccharides function as storage molecules; some play a structural role, particularly in the rigid walls of plants, the shells of soft-shelled animals, and the connective tissues of vertebrates. Cellulose, another polymer of glucose and the most prevalent carbohydrate on earth, forms the sturdy material which gives plants their rigidity. Agar, a polymer of galactose (Sec. 9.2), is a structural component of seaweeds. Chitin, a polymer of *N*-acetylglucosamine, is a major structural component of the hard exoskeletons of insects and crustaceans, e.g., lobsters. The mucopolysaccharides, polymers of N-acetylglucosamine and other glucose derivatives, are found in the extracellular spaces of vertebrates. They include hyaluronic acid, which acts as a lubricant for flexible joints; chondroitin sulfate, a major structural component of cartilage, bone, and the cornea; and heparin, which is found in the extracellular spaces of liver, lung, and some other tissues. Dextran is a polysaccharide of glucose produced by the bacterium *Leuconostoc mesenteroides* when grown on the disaccharide sucrose as carbon source. Dextran is used medically as a plasma extender in blood transfusions.

9.2 Monosaccharides and disaccharides

There are many 6-carbon monosaccharides in nature in addition to glucose; we have already encountered fructose (in glycolysis). Six-carbon sugars often occur in the form of disaccharides: molecules containing two monosaccharides joined by a glycosidic bond. The most common disaccharides are lactose, sucrose, and maltose (Table 9.1); you will recognize these sugars as our major nutrient forms of carbohydrate.

For cells to obtain energy from disaccharides, they must first be cleaved enzymatically into their two components; if necessary, the monosaccharides are converted to glucose and metabolized by the glycolytic and Krebs cycle routes. For example, in human beings, lactose is broken down to glucose and galactose by the action of lactase

TABLE 9.1 Disaccharides

Disaccharides are the major carbohydrates used as nutrients by man; their breakdown produces monosaccharides, which are converted to glucose and catabolized by the energy-yielding pathways of glycolysis and the Krebs cycle. Disaccharides are synthesized (by cows or plants) from activated monosaccharides.

Disaccharide	Dietary source for humans	Structure	Biosynthetic precursors
Sucrose	Sugar beet, sugar cane	Glucose–Fructose	UDP-glucose and fructose-6-phosphate
Lactose	Milk	Galactose–Glucose	UDP-galactose and glucose
Maltose	Digestion of starch by amylase	Glucose–Glucose	Starch

in the small intestine; galactose is enzymatically converted to glucose, which then enters glycolysis. A block in the conversion of galactose to glucose causes galactosemia, a recessively inherited disease which is particularly serious during infancy, when milk is the only nourishment. Infants with galactosemia vomit or have diarrhea following milk consumption; they often become mentally retarded. Fortunately, when galactosemia is diagnosed, further brain damage can be prevented by the strict avoidance of foods containing galactose.

An important class of monosaccharides includes the five-carbon sugars, for example ribose and deoxyribose, that are important components of nucleic acids (Chaps. 12 and 13). The ribose sugars can also be used as energy sources by cells. The pathways of ribose synthesis and degradation are linked to the pathways for six-carbon monosaccharides through a series of reactions known as the *pentose phosphate pathway*. This pathway begins with the conversion of glucose-6-phosphate into ribose-5-phosphate; one carbon atom is lost as carbon dioxide during this conversion. Additional transformations in this pathway provide a variety of other five- and six-carbon sugars. The pentose phosphate pathway has additional significance in that it is the source of NADPH for biosynthetic reduction reactions.

9.3 Gluconeogenesis: how glucose is made from noncarbohydrate precursors

Under certain circumstances it is necessary for cells to *make* glucose from noncarbohydrate precursors by reactions that essentially reverse glycolysis. This process, called *gluconeogenesis*, can start from any one of a variety of compounds, including lactic acid, Krebs cycle intermediates, or amino acids.

In animals, these biosynthetic reactions, which take place primarily in the liver, are usually a response to the energy needs of other tissues and/or to blood levels of various metabolites. For example, glucose is often produced through gluconeogenesis to satisfy the demands of certain cell types, such as nerves. Gluconeogenesis may also be used to clear the blood of metabolites excreted by organs other than the liver, for example the lactic acid produced by working muscles.

The production of glucose from noncarbohydrate precursors occurs by the following pathways:

amino acids $\longrightarrow$ Krebs-cycle intermediates $\longrightarrow$ oxaloacetic acid $\longrightarrow$ phosphoenolpyruvic acid $\longrightarrow$ glucose

Amino acids which can be used for the synthesis of glucose or glycogen are called *glycogenic amino acids*; all the amino acids except leucine are glycogenic. Some, such as aspartic acid and glutamic acid, can be converted directly into Krebs cycle intermediates by transamination. However, most amino acids must go through multistep catabolic pathways before their carbon skeletons are converted to Krebs cycle intermediates and ultimately to carbohydrates.

Because certain reactions of the Krebs cycle and glycolysis are irreversible, gluconeogenesis cannot occur by a direct reversal of these two pathways. The conversion of phosphoenolpyruvic acid to pyruvic acid is one such irreversible step. This stumbling block must be bypassed by a reaction unique to gluconeogenesis: the conversion of oxaloacetic acid to phosphoenolpyruvic acid, catalyzed by the enzyme phosphoenolpyruvate carboxykinase (Sec. 8.3). This is also one of the reactions that is used to control the rate of gluconeogenesis.

In mammals, the conversion of lactic acid to glucose is an important process. After vigorous muscular activity, the amount of lactic acid in the bloodstream may attain high levels (Sec. 7.4) and can even cause substantial muscle pain. The cyclical process by which lactic acid is converted to glucose in liver cells and eventually reappears as muscle glycogen is known as the *Cori cycle* (Fig. 9.3). In this way, the partial degradation product, lactic acid, can be used to restore glucose levels and the full complement of glucose energy obtained.

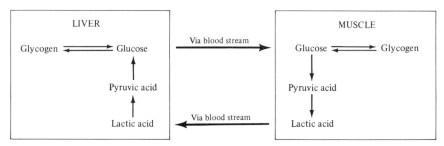

Figure 9.3 The Cori cycle is the body's way of recycling lactic acid. In the liver, lactic acid is converted to pyruvic acid. Pyruvic acid is carboxylated to form oxaloacetic acid, which is then used to make glucose.

9.4 Insulin and the control of sugar transport into cells

Insulin is a peptide hormone formed in the pancreas; Sanger's determination of the amino acid sequence of this molecule in the early 1950s was a landmark in protein biochemistry (Sec. 3.5).

In 1889, von Mering and Minowsky discovered that when the pancreas was removed surgically from dogs, the animals developed a condition similar to diabetes in man. It was not until 1921, however, that Banting and Best obtained insulin from extracts of pancreas and demonstrated that the hormone would alleviate the symptoms of diabetes in man. Surprisingly, although insulin has been the subject of much biochemical study, its primary site of action on metabolism is not yet known.

Stimulation of glucose uptake from blood into cells is *one* of the primary results of the administration of insulin. This probably occurs by stimulation of the activity of the hexokinase that converts glucose into glucose-6-phosphate (Sec. 7.1), thus "trapping" more sugar in the cell. Although this explanation of the way in which insulin controls cellular supplies of carbohydrates is simple and logical, it should be pointed out that insulin has other effects on metabolism which are not easily explained by a stimulation of hexokinase.

In human beings, one of the results of a deficiency of insulin is a reduction in the transport of glucose from the bloodstream into cells and tissues. A diabetic state is therefore characterized by high blood sugar (hyperglycemia) and a high rate of sugar excretion in the urine; this can be counteracted by the injection of insulin, which results in an increase in the transport of sugar into cells and a fall in blood sugar levels. An overdose of insulin leads to a drastic depletion of the blood sugar (hypoglycemia), which, in turn, may cause insulin shock, which results from reduction of the supply of sugar to the brain. (Brain cells do not store much carbohydrate and rely almost entirely on a continuous supply of glucose from the blood.)

In recent years many diabetics have profited from the introduction of oral "antidiabetic" agents, such as tolbutamide, chlorpropamide, and tolazamide. These drugs are not substitutes that act in the same way as insulin; their effectiveness is believed to lie in their ability to stimulate the production and/or release of insulin. There are clearly two possible reasons for a diabetic's failure to make sufficient insulin:

1. A defect in the control of insulin synthesis; the pancreas does not respond to changes in blood glucose levels.

2. A defect in or loss of the ability to synthesize insulin, as would occur if the pancreas were removed.

It should be clear that only class 1 diabetics can respond to oral "antidiabetic" agents; class 2 diabetics cannot make insulin under any circumstances, and their condition can only be alleviated by providing an external supply of the hormone.

A number of other hormones serve to counterbalance the effects of insulin, as we will see below. Hormone imbalances that overstimulate glycogen breakdown cause hyperglycemia, while those which understimulate lead to hypoglycemia. Therefore, defects in insulin-controlled glucose transport or in the metabolism of glycogen can lead to the same result (Table 9.2).

9.5 Glucagon and epinephrine control glycogen breakdown

The control of glycogen synthesis and breakdown is obviously a major factor in the control of blood glucose levels; both processes are subject to the action of regulatory hormones released by the endocrine glands. Two hormones are of particular importance in this respect: *glucagon*, produced in the pancreas, and *epinephrine*, made in the adrenal gland. Glucagon is a peptide hormone which acts on liver cells. Epinephrine

TABLE 9.2 Hormonal Effects on Carbohydrate Metabolism

Blood glucose level	Caused by
High sugar (hyperglycemia)	Lack of insulin; adequate sugar does not enter cells (diabetes)
	or
	Too much glucagon or epinephrine; overstimulation of glycogen breakdown
Low sugar (hypoglycemia)	Too much insulin; excessive sugar transport into cells
	or
	Too little glucagon or epinephrine; insufficient glycogen breakdown

is a small molecule chemically related to the amino acid tyrosine; it acts on liver and muscle cells. Another name for epinephrine is adrenalin, the "fight, flight, or fright" hormone that always flows at just the right times through the bloodstreams of heroes and heroines of novels and movies. Although these two hormones act similarly with respect to stimulating glycogen breakdown, they have other, independent, physiological effects.

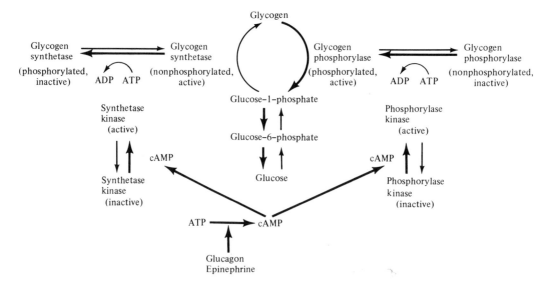

Figure 9.4 The relationships between glycogen and glucose and the controlling function of cyclic AMP. The figure indicates the scheme of glycogen–glucose interconversion and its regulation by phosphorylation of the synthetase and phosphorylase kinases. These phosphorylations are, in turn, controlled by cAMP, which is made from ATP when triggered by the hormones glucagon and epinephrine. The thick arrows indicate conversions stimulated by the two hormones.

Glucagon and epinephrine control glycogen metabolism by altering the activity of two enzymes: glycogen phosphorylase and glycogen synthetase (Sec. 9.1 and Fig. 9.4). These enzymes exist in two forms, one enzymatically active and one inactive; the interconversions between the two forms require modification by phosphorylation of certain specific amino acid residues in the proteins (the hydroxyl groups in the side chains of serine and threonine). The phosphorylation reactions are controlled by the action of the hormones, but, somewhat surprisingly, the hormones do not act directly on the phosphorylating enzymes. This control is mediated by stimulation of the production of small molecule "messengers," which, in turn, regulate the enzymes that catalyze the phosphorylation reactions. The "messenger" molecule in the case of glycogen metabolism is cyclic adenosine monophosphate (cyclic AMP or cAMP).

A schematic presentation of the way in which glycogen synthesis and breakdown are controlled is shown in Fig. 9.4. This may look somewhat complex, but the principles behind it are simple; when glucagon or epinephrine stimulate glycogen breakdown to feed glucose into glycolysis and the Krebs cycle, the synthesis of glycogen is simultaneously inhibited. Conversely, when glycogen is being made, the flow of glucose from glycogen through glycolysis and into the Krebs cycle is reduced. Thus, cells contain regulatory systems that, no matter what the circumstances, function to shunt intermediates into or away from glucose in order that immediate metabolic needs can be met.

9.6 Photosynthesis: the synthesis of carbohydrates using solar energy

In discussing how cells use carbohydrates as energy sources, we have ignored the question of the primary source of carbohydrates. The question in reality should be rephrased as "What is the primary source of energy?" and the answer is, naturally, the sun. How, then, is the sun's light energy converted into chemical bond energy, to be used to provide energy for cell work? Light energy is used to make carbohydrates from CO_2 and water by photosynthesis; this process is outlined in Fig. 9.5.

The first part of photosynthesis is collectively called the *light reaction*; it can occur only when light (normally from the sun) is available. In this process the light energy is captured by receptor molecules (chlorophylls) in the chloroplasts in plant and algal cells and converted into chemical energy in the form of ATP. When particles of light (photons) hit chlorophyll molecules, electrons are activated (i.e., the electrons gain energy). These activated electrons are then passed through a series of electron acceptors, with small amounts of energy being given off with each passage. The energy given off during this electron transfer is captured for the coupled synthesis of ATP. A group of complex metalloproteins such as cytochromes and ferredoxin are the electron acceptors in reactions basically similar to those that take place in the electron transport/oxidative phosphorylation system of mitochondria (Sec. 8.4 and Fig. 8.6).

Water is an essential reactant in photosynthesis in green plants and algae. In overall terms, water is split into three components during photosynthesis: replacement

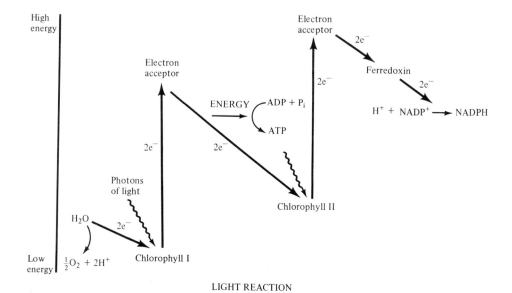

LIGHT REACTION

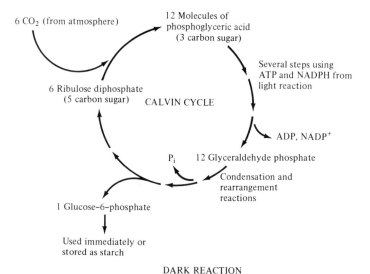

DARK REACTION

Figure 9.5 Photosynthesis. These figures illustrate the two series of reactions of photosynthesis that occur in green plants and algae. The reactions are indicated only in outline; a number of intermediates have been omitted. Similar electron transfer reactions occur in nitrogen fixation (Sec. 11.1) and ferredoxin also plays a role in this process.

Carbohydrate Formation and Storage: How Cells Control Energy Supplies

electrons to allow the continuation of the light reaction, hydrogen for the NADPH needed in the dark reaction, and the vital by-product oxygen. All the oxygen in the atmosphere comes from photosynthesis. Hence, harnessing of the sun's energy not only provides carbohydrates as energy sources, but also the atmospheric oxygen, which provides other life forms with the means to oxidize energy sources. The value of green plants and algae in man's welfare is inestimable. Evidence suggests that molecular oxygen is a relatively recent addition to the earth's atmosphere; therefore, we can conclude that aerobic organisms and, indeed, the mitochondrial electron transport system, are not as old in the evolutionary time scale as photosynthetic organisms, anaerobic organisms, and the glycolytic system. The more complex photosynthetic organisms must have evolved from simple organisms and released molecular oxygen in the atmosphere before aerobic organisms could have developed.

The second part of photosynthesis is called the *dark reaction*—a series of biosynthetic reactions that can take place in the presence or absence of light, provided that energy in the form of ATP is present and that the cell has an adequate supply of NADPH (usually obtained from the light reaction). The ATP and NADPH are required to drive a cycle of reactions that begins with the addition of atmospheric carbon dioxide to a five-carbon monosaccharide acceptor. We refer to this as *carbon dioxide fixation*, and it provides photosynthetic organisms with the wherewithal to assimilate CO_2 for making six-carbon sugars. The two-step fixation reaction is catalyzed by ribulose diphosphate carboxylase, an enzyme that is present in large amounts in chloroplasts.

Ribulose-1,5-diphosphate → 3-Phosphoglyceric acid

A series of reactions then converts 3-phosphoglyceric acid to glucose (Fig. 9.5), which plants store in the polysaccharide, starch. This is named the *Calvin cycle*, after the Nobel prize winner in whose laboratory the nature of many of the biochemical reactions of photosynthesis was established. The use of radioactive compounds in "tagging" various intermediates played an important role in these studies. For example, the use of radioactive carbon dioxide enabled the detection and identification of the early intermediates in the Calvin cycle.

The elucidation of the light and dark reactions of photosynthesis, as outlined in Fig. 9.5, represents a very substantial biochemical achievement involving the work of many scientists. However, our knowledge of photosynthesis remains incomplete; some of the steps are still unknown and the exact details of others are yet poorly understood.

9.7 From plants to animals

The link in the chain between the energy of sunlight and the maintenance of cellular activity in animals occurs when they eat plants. For example, the starch made during photosynthesis is the major nutrient of most human beings.

Other plant polysaccharides are also important in the food chain. For example, ruminants such as cattle, sheep, and giraffes can live on grass because bacteria present in their rumens degrade cellulose to small molecules, such as lactic and acetic acids, by anaerobic processes. These acids then pass into the animal's bloodstream and are eventually combined with nitrogen compounds (also from plants) to make protein, which is later eaten by man. The mediation of bacteria is thus crucial in providing man and other carnivores with essential nutrients such as carbohydrate and protein. Man cannot live on grass because he possesses neither enzymes capable of digesting this plant, nor the bacteria capable of doing it for him. Good, fresh vegetables taste better, anyway!

Another interesting example of the intervention of microbes in carbohydrate utilization is seen in the renowned ability of termites to destroy wooden buildings. These insects owe their ability to digest wood to protozoans that inhabit their alimentary tract. The protozoans produce cellulases, allowing termites to wreak damage and obtain nourishment at the same time.

SUMMARY

The simple carbohydrates (e.g., glucose) needed for cell energy come from several sources: they can be provided in the diet, by the degradation of storage polysaccharides inside the cell, or by the *de novo* synthesis of glucose from other metabolic intermediates.

Carbohydrates come in several metabolically related forms: simple monosaccharides, disaccharides, and polysaccharides. At the center of carbohydrate metabolism is the six-carbon monosaccharide glucose. Glucose can be degraded by glycolysis and the Krebs cycle to provide energy for cell work. Glucose is the monomer of storage polysaccharides, such as glycogen and starch, and of the structural polysaccharides of plants. Glucose is also the precursor of the five-carbon riboses needed for nucleic acid synthesis. When glucose (and glycogen) are in short supply, cells can make glucose from amino acids by gluconeogenesis.

Carbohydrate Formation and Storage: How Cells Control Energy Supplies 161

Glucose is the primary source of energy for most cells. Because it is so essential, animals have developed complex hormonal systems for balancing the rates of glucose catabolism, glycogen synthesis and breakdown, and gluconeogenesis. Epinephrine, glucagon, and insulin are some of the hormones that act to keep blood glucose levels within a narrow range, suitable for the maintenance of life.

The ultimate source of carbohydrates is photosynthesis, a process in which green plants use light energy to synthesize glucose from atmospheric CO_2 and water. The glucose is stored in plant polysaccharides, which are subsequently consumed by animals and microorganisms.

PRACTICE PROBLEMS

1. Green plants make glucose in the dark reaction of photosynthesis. Give two reasons why plants need glucose when they can convert light energy into the chemical energy of ATP.

2. How many ATPs can a cell gain by the aerobic catabolism of 1 molecule of maltose? (Maltase cleaves maltose to two molecules of glucose.) How many from a linear glycogen molecule containing 100 glucose residues?

3. Believe it or not, toothpaste containing chlorophyll was a rage in the 1950s. It was advertised as the ultimate in stopping "bad breath." What color was this toothpaste? Would people who used this toothpaste be capable of photosynthesis?

4. Indicate which of the following statements apply to (a) insulin or (b) glucagon or both by placing the appropriate letter or letters before each statement.
   ____ (1) Decreases blood glucose by enhancing glucose transport into cells.
   ____ (2) Stimulates glycogen phosphorylase activity.
   ____ (3) Addition can cause hypoglycemia.
   ____ (4) Addition can cause hyperglycemia.
   ____ (5) Lacking or nonfunctional in a diabetic person.
   ____ (6) Secreted by the pancreas.
   ____ (7) Elevates cAMP levels in the cell.
   ____ (8) Is a polypeptide hormone.

5. During acute starvation, animals eventually degrade muscle proteins as a source of energy. Outline the pathways by which the carbon atoms of two different amino acids can be used to synthesize the glucose required by nerves and other cells. In what organ of the body does this gluconeogenesis occur? What vitamin(s) are needed for the pathways you have outlined?

6. Man, other animals, and plants depend on each other for many nutrients. Complete the cycle shown here by writing the following substances over the appropriate arrows: starch, cellulose, oxygen, protein-rich products, carbon dioxide.

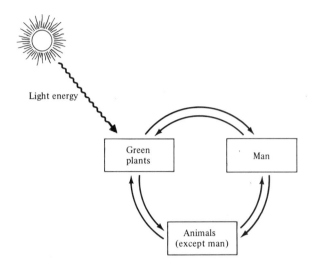

SUGGESTED READING

ALBERSHEIM, PETER, "The Walls of Growing Plant Cells," *Sci. American*, 232, No. 4, p. 80 (1975). Cellulose fibers, bound together by more complex polysaccharides, form the rigid, yet extendable, walls of plant cells. Describes the chemical analyses of polysaccharides.

BASSHAM, J. A., "The Path of Carbon in Photosynthesis," *Sci. American*, 206, No. 6, p. 88 (1962). Describes the intricacies of the Calvin cycle.

DIMLER, ROBERT J., "Dextran Helps Save Lives," in *The Yearbook of Agriculture*, U.S. Department of Agriculture, House Document No. 239, 90th Congress 2d session, p. 314. U.S. Government Printing Office, Washington, D.C., 1968. Outlines the efforts of researchers dedicated to making this polysaccharide available to shock victims.

GOVINDJEE and RAJNI GOVINDJEE, "The Absorption of Light in Photosynthesis," *Sci. American*, 231, No. 6, p. 68 (1974). A somewhat complicated discussion of the chemistry and physics of the light reaction.

JANICK, JULES, CARL H. NOLLER, and CHARLES L. RHYKERD, "The Cycles of Plant and Animal Nutrition," *Sci. American*, 235, No. 3, p. 75 (1976). The carbon cycle is one of several nutritional cycles which link man to other animals, plants, and microorganisms.

KRETCHMER, NORMAN, "Lactose and Lactase," *Sci. American*, 227, No. 4, p. 71 (1972). Many adults lack the enzyme needed to digest the disaccharide lactose, the major carbohydrate in milk.

LEHNINGER, ALBERT L., *Bioenergetics*, W. A. Benjamin, Inc., New York, 1965. Chapter 5 describes photosynthesis.

PASTAN, IRA, "Cyclic AMP," *Sci. American*, 227, No. 2, p. 97 (1972). Many animal hormones, including those which control blood glucose levels, act by means of cAMP.

SHARON, NATHAN, "Glycoproteins," *Sci. American*, 230, No. 5, p. 78 (1974). Some proteins have carbohydrates linked to them.

10 Lipids

In our daily lives we encounter many lipid-rich substances. We cook with butter, olive oil, vegetable shortening, and lard. We expect the butcher to trim the fat off the meat. We soften our skin with lanolin and glycerin. Less obvious are the steroid hormones, including many compounds used in drug therapy, and the fat-soluble vitamins.

Our awareness of the importance of lipids is increased by a growing concern about the relationships between these compounds and our health. We are constantly bombarded with warnings about being "too fat," the effects of eating too many saturated fatty acids, and the consequences of elevated levels of cholesterol and triglycerides in our blood.

Why are so many seemingly different compounds classified together? *Lipids are defined as organic molecules which are not soluble in water, but are soluble in organic solvents such as ether or chloroform.* Unlike other macromolecules, which are defined in terms of their chemical composition and structure, lipids are defined by this single physical property, and this definition is not very revealing in terms of the structure or function of the molecules. Once biochemists began to analyze and study water-insoluble cellular material, it became clear that lipids were a very heterogeneous group of compounds and can be as different from one another as, for example, proteins are from nucleic acids.

Table 10.1 Major Types of Lipids

All lipids contain water-insoluble chains or rings of —CH_2— groups.

	General Structure	Example
FATTY ACIDS		
Saturated	$CH_3(CH_2)_n COOH$	$CH_3(CH_2)_{14} COOH$ Palmitic acid
Unsaturated	$CH_3(CH_2)_n \overset{H}{C}=\overset{H}{C}(CH_2)_n COOH$	$CH_3(CH_2)_7 CH=CH(CH_2)_7 COOH$ Oleic acid
COMPLEX LIPIDS		
Triglycerides	$H_2C-O\overset{O}{\overset{\|}{C}}(CH_2)_n CH_3$ $HC-O\overset{O}{\overset{\|}{C}}(CH_2)_n CH_3$ $H_2C-O\overset{O}{\overset{\|}{C}}(CH_2)_n CH_3$	$H_2C-O\overset{O}{\overset{\|}{C}}(CH_2)_{14} CH_3$ $HC-O\overset{O}{\overset{\|}{C}}(CH_2)_{14} CH_3$ $H_2C-O\overset{O}{\overset{\|}{C}}(CH_2)_{14} CH_3$ Tripalmitate
Phosphoglycerides	$H_2C-O\overset{O}{\overset{\|}{C}}(CH_2)_n CH_3$ $HC-O\overset{O}{\overset{\|}{C}}(CH_2)_n CH_3$ $H_2C-O-\overset{O}{\overset{\|}{P}}-O-X$ $\qquad\;\; O^-$	$H_2C-O\overset{O}{\overset{\|}{C}}(CH_2)_{14} CH_3$ $HC-O\overset{O}{\overset{\|}{C}}(CH_2)_7 CH=CH(CH_2)_7 CH_3$ $H_3C-O-\overset{O}{\overset{\|}{P}}-O-CH_2CH_2\overset{+}{N}(CH_3)_3$ $\qquad\;\; O^-$ Phosphatidyl choline
STEROLS[a]	(steroid ring system with R group)	Cholesterol

[a] In chemical shorthand, a six-membered ring of —CH_2— groups is written as ⬡ ; each point on the hexagon represents one —CH_2— group.

Lipids

The major physical property common to all lipids is their insolubility in water, which results from a common chemical structure. All lipids contain long chains or rings of —CH_2— (methylene) groups:

$$\begin{array}{c} \text{H H H H H H H} \\ | \; | \; | \; | \; | \; | \; | \\ -C-C-C-C-C-C-C- \\ | \; | \; | \; | \; | \; | \; | \\ \text{H H H H H H H} \end{array}$$

and relatively few charged groups (i.e., few —NH_3^+, —COO^-, —PO_4^{2-}). These long uncharged chains are very insoluble in water.

10.1 The major lipids: fatty acids, glycerides, and sterols

(a) FATTY ACIDS ARE STRUCTURALLY THE SIMPLEST LIPIDS

Fatty acids have the general formula

$$H_3C-CH_2-CH_2-CH_2-CH_2-CH_2-CH_2-COOH$$

Since writing all the CH_2's is cumbersome, long chains are abbreviated $CH_3(CH_2)_n \ldots$, where n is the number of repeated methylene units:

$$H_3C-(CH_2)_n-COOH$$

Fatty acids can be either saturated or unsaturated (Table 10.1); all saturated lipids contain the maximum possible number of hydrogen atoms, while unsaturated compounds contain fewer hydrogens and at least one unsaturated (double) bond between carbons. Unsaturated fatty acids can contain one (monounsaturated) or more than one double bond (polyunsaturated). Plants produce relatively high percentages of both types of unsaturated fatty acids. Animals synthesize lesser amounts of polyunsaturated fatty acids; human beings are unable to make certain polyunsaturated fatty acids, and at least three of these compounds are essential components of their diet (Table 10.2).

An interesting physical property of lipid mixtures is that their melting point is roughly proportional to their content of saturated fatty acids. The greater the saturated fatty acid content, the higher the melting point of the mixture. This explains why mixtures of animal lipids (e.g., butter, lard) are solids, while mixtures of plant lipids

TABLE 10.2 Essential Unsaturated Fatty Acids

Human beings cannot synthesize fatty acids that contain more than one unsaturated bond, and therefore these polyunsaturated fatty acids must be provided in the diet.

Linoleic acid	$CH_3(CH_2)_4CH=CHCH_2CH=CH(CH_2)_7COOH$
Linolenic acid	$CH_3CH_2CH=CHCH_2CH=CHCH_2CH=CH(CH_2)_7COOH$
Arachidonic acid	$CH_3(CH_2)_4(CH=CHCH_2)_4(CH_2)_2COOH$

(e.g., olive oil, peanut oil) are liquids at room temperature. You may be familiar with "oleomargarine" and "vegetable shortening," both of which are solids at room temperature and may seem to contradict this rule. However, if you read the label on these products, you will find that they contain "partially hydrogenated vegetable oil"; this means that some of the double bonds in the unsaturated fatty acids have had hydrogen added, making the molecules more saturated and giving the final mixture a melting point above room temperature.

(b) FATTY ACIDS ARE USUALLY PART OF COMPLEX LIPIDS

We have described simple fatty acids, usually called "free fatty acids," in which the carboxyl group of the molecule is not attached to any other molecule. In cells, however, most fatty acids are not "free," but are part of complex lipids, often triglycerides or phosphoglycerides (Table 10.1). These combination molecules consist of a backbone, which is an alcohol, joined to one or more fatty acid molecules.

The alcohol backbone of both triglycerides and phosphoglycerides is glycerol. Glycerol, which can be formed from glucose (Chap. 7), is a three-carbon alcohol with three hydroxyl groups. In the glycerides, *fatty acids are linked to glycerol by an ester bond* between the hydroxyl groups of glycerol and the carboxyl groups of the fatty acids.

$$\begin{array}{c}
H O \text{—Ester bond} \\
| \| \\
H-C-O-C-(CH_2)_nCH_3 \\
| O \\
\| \\
H-C-O-C-(CH_2)_nCH_3 \\
| O \\
\| \\
H-C-O-C-(CH_2)_nCH_3 \\
| \\
H
\end{array}$$

In triglycerides, all three of the hydroxyl groups of glycerol are linked to a fatty acid. Cells also contain smaller amounts of diglycerides (glycerol linked to two fatty acids) and monoglycerides (glycerol linked to one fatty acid). The three fatty acids in glycerides may be identical or different.

In mammals, triglycerides are the major lipid in adipose tissue (Fig. 10.1). When the diet contains more calories than are needed to meet immediate energy needs, carbohydrates (and lipids) are converted to fatty acids [Sec. 10.3(a)] and stored as triglycerides in adipose tissue; that is, we get fatter. When the caloric intake is insufficient to meet energy needs, the fatty acids are released from triglycerides and degraded to provide energy in the form of ATP [Sec. 10.3(b)]; that is, we get thinner.

Although triglyceride degradation, "lipolysis," can occur as a direct response to energy demands, this degradation is controlled by several hormones. As an example, insulin decreases the rate of lipolysis, in part because it enhances the cellular uptake of glucose, making fatty acids unneeded as an energy source. Glucagon and epinephrine have the opposite effect; these hormones stimulate lipolysis. It is important to compare the effects of these hormones on the dispersion of lipid stores with their effects on carbohydrate stores, that is, glycogen (Secs. 9.4 and 9.5).

Phosphoglycerides, the second major group of complex lipids, have a more complicated structure than triglycerides. As in triglycerides, the alcohol backbone is glycerol; two fatty acids are joined to this molecule by ester bonds. The third hydroxyl of the glycerol backbone is linked to a phosphate group, which in turn is joined to another molecule, called X in Table 10.1. X can be the amino acid serine, the sugar inositol, or the amine choline. When X is choline, the resulting phosphoglyceride is called a lecithin.

The primary cellular role of phosphoglycerides is in the construction of cell membranes and lipoproteins (Sec. 10.5). Phosphoglycerides have a water-insoluble "end," composed of fatty acid residues, and a water-soluble, charged "end," the phosphate group and its attached X. Because of this "two-headedness," these molecules act as links between proteins and more water-insoluble lipids, such as triglycerides and sterols, which explains their importance in cell membrane structure. Their ability to solubilize other lipids makes phosphoglycerides good emulsifying agents; lecithins are used for this purpose in many commercially prepared foods.

(c) STEROLS

The third major group of lipids is the sterols (Table 10.1), water-insoluble compounds in which the ends of the —CH$_2$— chain are joined to form ring structures; most of the rings contain five or six methylene groups, and may be saturated or unsaturated. Some sterols have attached R groups, such as the 8-carbon side chain of cholesterol (Table 10.1), the major animal sterol.

Cholesterol is a component of animal cell membranes and is the precursor of many other steroids, including the adrenal and sex hormones [Sec. 10.6(b)].

10.2 Lipid metabolism in the whole animal

Most body tissues can degrade fatty acids to acetyl CoA [Sec. 10.3(a)] and hence use these lipids as energy sources. But, in animals, certain other pathways of lipid metabolism are localized in specific body organs. Some "background" information about the role of the organs prominent in lipid metabolism is necessary, not only to understand normal metabolism, but also to understand the complications that arise when a particular organ or pathway malfunctions.

The three organs most important in lipid metabolism are the intestine, the liver, and adipose tissue (Fig. 10.1). The intestine is the site of lipid absorption and excretion. Adipose tissue is the storehouse for triglycerides, whose fatty acids are released when energy is needed. The liver is the major center for phosphoglyceride and triglyceride synthesis and is the site of lipoprotein synthesis and secretion. The liver also produces cholesterol and bile acids (Sec. 10.4).

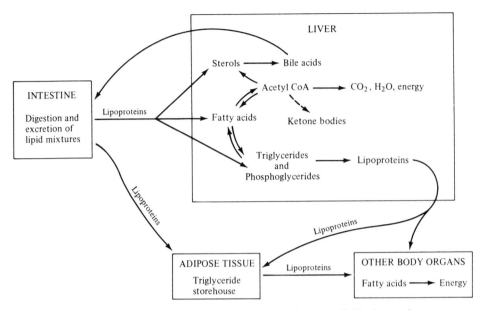

Figure 10.1 Lipid metabolism in animals. Lipoproteins carry lipid mixtures between organs.

The localization of lipid metabolism is of significance because lipids, being insoluble in water, are also insoluble in blood, urine, and other body fluids. Specialized transport proteins (lipoproteins, Sec. 10.5) are needed to carry lipids from one organ to another. As you study the metabolic pathways in succeeding sections, keep in mind that the end products of these pathways cannot always move freely through the body, and that many aspects of lipid metabolism have evolved to absorb, move, or excrete these water-insoluble compounds.

10.3 Fatty acids start and end as acetyl CoA

(a) FATTY ACID SYNTHESIS

Fatty acids are built up two carbon units at a time; a six-step reaction cycle is repeated each time two more carbon atoms are added to the growing chain. The two carbon fragments originate as excess acetyl CoA, which comes from the catabolism of dietary carbohydrates.

Fatty acid synthesis begins when two two-carbon units condense to form a four-carbon fatty acid (Fig. 10.2); the second turn of the cycle adds two more carbons, converting the four-carbon fatty acid to a six-carbon linear molecule. The six-carbon chain grows to eight carbons, and this stepwise addition continues until the fatty acid

Lipids

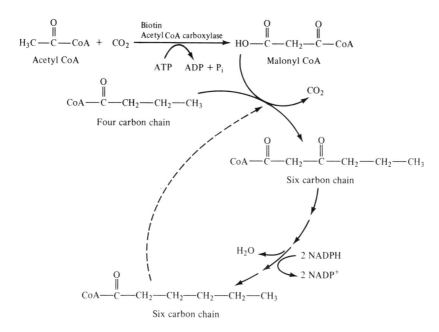

Figure 10.2 Fatty acid synthesis. Fatty acids are built up from two-carbon units that originate in acetyl CoA. Each time this cycle turns, the —CH_2— chain of the fatty acid grows by two carbon atoms. This diagram shows the —CH_2— chain growing from four to six carbons; in the next turn of the cycle (dashed line), the six-carbon chain will react with malonyl CoA to form an eight-carbon chain.

typically contains 16 or 18 carbon atoms. Special reactions are needed to form the rare fatty acids that contain odd numbers of carbon atoms.

In biological systems, acetyl CoA molecules do not simply condense to form a linear —CH_2— chain, since this is not feasible in energetic terms. To overcome this energy barrier, acetyl CoA first reacts with carbon dioxide to form a three-carbon compound, malonyl CoA. It is malonyl CoA, an "activated" molecule, which donates two carbons to the growing fatty acid chain each time the anabolic cycle of fatty acid formation turns.

The formation of malonyl CoA from acetyl CoA and CO_2 is interesting for two reasons. First, this reaction requires ATP. Thus, fatty acid synthesis, like other anabolic pathways, uses energy. Second, formation of malonyl CoA requires the vitamin biotin, a cofactor in a number of enzymatic reactions in which carbon dioxide is "fixed", that is, reactions in which atmospheric CO_2 is converted to the carboxyl group of an organic molecule.

When malonyl CoA condenses with the growing fatty acid chain, only two of its three carbon atoms are incorporated into the linear chain; the third carbon is released as carbon dioxide. Studies using ^{14}C-labeled malonyl CoA have shown that the released CO_2 contains the carbon atom from the CO_2 which was added onto acetyl

CoA to form malonyl CoA in the first step. So, although CO_2 is required for the synthesis of fatty acids, it does not become a permanent part of these lipids. In essence, CO_2 is a catalyst for fatty acid synthesis.

The first condensation product of malonyl CoA and the growing fatty acid chain is not a new, longer fatty acid, but a molecule with a —C=O group (a carbonyl group) in the carbon chain (Fig. 10.2). Three enzymatic reactions are needed to reduce this carbonyl group to the methylene group of the finished fatty acid. The hydrogens needed for this reduction are donated by NADPH (Chap. 8). This series of reactions is repeated until the desired fatty acid chain is made. Synthesis of fatty acids takes place in the cytoplasm as distinct from fatty acid degradation that takes place in the mitochondria.

Nearly all aerobic organisms, including plants and animals, produce some unsaturated fatty acids by the removal of hydrogens from the corresponding saturated fatty acids. These enzymatic reactions require molecular oxygen. In this way, for example, palmitic acid,

$$CH_3(CH_2)_{14}COOH$$

is converted to palmitoleic acid,

$$CH_3(CH_2)_5CH=CH(CH_2)_7COOH$$

Free fatty acids are attached to glycerol to form triglycerides (and phosphoglycerides). The fatty acid must be activated to its CoA derivative (these reactions require ATP) prior to the formation of the ester bond; glycerol must be in the form of glycerol phosphate. In summary:

$$3 \text{ fatty acyl CoA} + \text{glycerolphosphate} \longrightarrow \text{triglyceride} + H_2O + P_i$$

(b) FATTY ACID CATABOLISM

When the body's caloric intake is less than its caloric demand, the fatty acids released from triglycerides (by the action of lipases) are broken down in order to satisfy the ATP deficit. The breakdown of the long —CH_2— chains of the fatty acids provides a hefty supply of acetyl CoA (and hence ATP). The catabolism of fatty acids is in many respects a reversal of their synthesis, although different enzymes are used; the end product of fatty acid degradation is acetyl CoA, the same molecule that served as the starting material for their synthesis. However, the pathway is not an exact reversal, since malonyl CoA is the additive component used in synthesis. Also, it is important that synthetic and degradative pathways of molecules such as fatty acids differ, so that the two pathways can be regulated independently.

The breakdown of fatty acids is shown in Fig. 10.3. The first step (which occurs only once for each fatty acid molecule) is the "activation" of the fatty acid by conversion to its coenzyme A derivative. This reaction requires ATP and is analogous to the ATP-using steps in the first stage of glycolysis (Sec. 7.1); the cell must often use a little of its ATP to produce a lot more (like priming a pump). Because the ATP is

Lipids

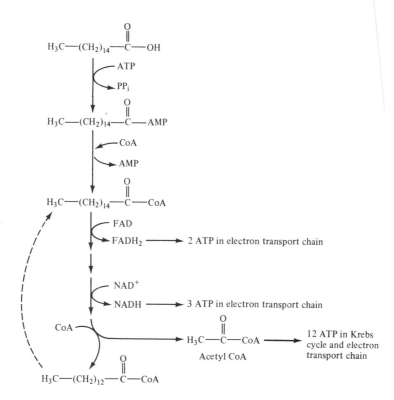

Figure 10.3 Fatty acid degradation. Before degradation, fatty acids must be activated by conversion to the CoA derivative. This two-step process releases AMP and PP_i and uses the energetic equivalent of 2 ATPs.

Fatty acyl CoAs are degraded by the sequential removal of two-carbon units as acetyl CoA. For the complete degradation of palmitic acid (16 carbons), the four-step cycle must be repeated seven times.

converted to AMP, this reaction uses up the (energetic) equivalent of two ATP molecules.

The fatty acyl CoA molecule then enters a cycle of four reactions, which results in the release of one molecule of acetyl CoA and a fatty acyl CoA molecule that is shortened by two methylene groups. In two of the four reactions, hydrogen atoms are released and are accepted by FAD in the first reaction and by NAD^+ in the third reaction. The reduced hydrogen acceptors, $FADH_2$ and NADH, in turn donate the hydrogens to the electron transport chain (Chap. 8), where their energy is used to produce ATP. The coenzyme A derivative of the shortened fatty acid is now recycled and loses two more methylene groups as acetyl CoA. This four-step pathway is repeated again and again until the fatty acid is entirely degraded to acetyl CoA.

A special shuttle system is needed to transport the fatty acid from the cytoplasm into the mitochondrion, where fatty acid degradation takes place. This compartmentalization of reactions means that the acetyl CoA, $FADH_2$, and NADH are formed at the place where they can be utilized for ATP production by the Krebs cycle and the electron transport system.

Carbohydrate and lipid metabolism are closely interrelated. This is well illustrated by a phenomenon known as *ketosis*, a metabolic imbalance that results when high rates of fatty acid degradation are accompanied by a low intake, or an underutilization of carbohydrates and depleted glycogen stores. This metabolic disorder occurs during diabetes and starvation. When organisms need to obtain energy from stored fats, fatty acids are enzymatically cleaved from triglycerides in adipose tissue; the released fatty acids are transported to the liver (Sec 10.2), where they are catabolized to acetyl CoA. During diabetes or starvation, this lipid catabolism may be so extensive that the liver's capacity to degrade acetyl CoA is pushed to its limits. The Krebs cycle machinery of the liver cannot keep expanding to degrade this acetyl CoA, particularly since only very limited amounts of carbohydrates are available for oxaloacetic acid production. Since, in addition, carbohydrates are unavailable for glycerol phosphate synthesis, the acetyl CoA cannot be reincorporated into fatty acids and complex lipids. In short, liver cells are glutted with acetyl CoA. This excess acetyl CoA is converted to four-carbon keto acids: acetoacetic acid and β-hydroxybutyric acid (ketone bodies). Acetoacetic acid is degraded to acetone, giving a distinctive, unpleasant odor to the breath of ketotic individuals.

As long as the production of ketone bodies remains low, peripheral tissues, especially muscle, can use them as an energy source, but as their level in the bloodstream increases, the excess ketone bodies are excreted in the urine. This excretion requires large amounts of fluid, and untreated ketosis can lead to severe dehydration. Because ketone bodies are negatively charged, their excretion is accompanied by the excretion of positively charged ions, usually sodium, which may deplete the body's supply of cations to a dangerous, and even fatal, extent.

10.4 Cholesterol: from acetyl CoA to gallstones

Cholesterol (Fig. 10.4) is one of the lipids in which the now-familiar sequence of —CH_2— groups is not linear but links up to form four rings. With the exception of the D vitamins, this basic ring structure is found in all other lipids derived from cholesterol (Fig. 10.4), such as other sterols, the bile acids, the sex hormones, and the adrenocortical hormones.

All the carbons of cholesterol originate in acetyl CoA. In a complicated series of reactions, 18 molecules of acetyl CoA are combined to form the 30-carbon unsaturated molecule squalene. This anabolic pathway releases carbon dioxide and requires hydrogen atoms from NADPH and energy from ATP. Squalene (Fig. 10.4) is a linear chain of carbon atoms; it has no ring structure. One of the most extraordinary reactions of cellular metabolism converts squalene into a sterol. In a single step, the squalene molecule folds upon itself and the correctly paired carbon atoms are enzymatically joined to form the four rings characteristic of sterols. This reaction requires

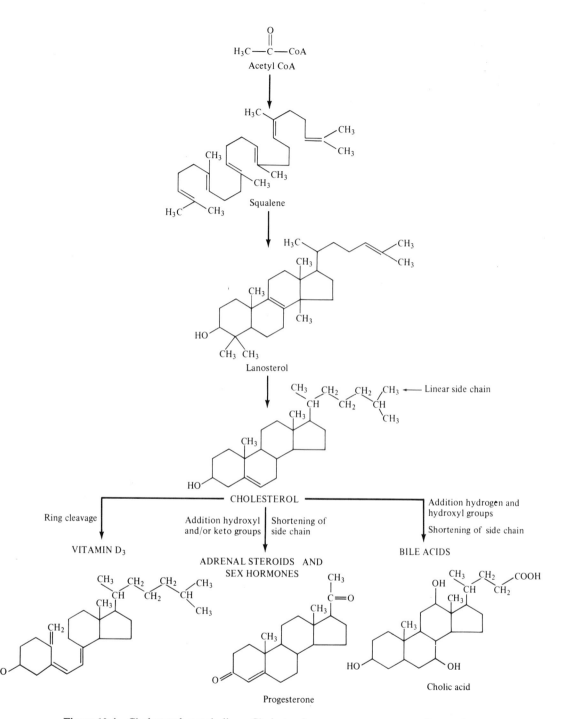

Figure 10.4 Cholesterol metabolism. Cholesterol serves as a precursor of several important classes of compounds. Each arrow represents a multistep pathway. The physiological roles of these compounds are described in Sec. 10.6. Cells cannot degrade cholesterol to acetyl CoA; therefore, it cannot serve the cell as an energy source.

molecular oxygen.* The product of the condensation of squalene is lanosterol, which is converted to cholesterol by a few, relatively minor, enzymatic modifications. Although most animal cells contain only traces of lanosterol, it is a major component of lanolin, the fatty coating of sheep's wool, which is an ingredient of many soaps and lotions.

In animals, cholesterol is the precursor of all other sterols (Fig. 10.4). The linear eight-carbon side chain of cholesterol can be modified and/or shortened; hydroxyl or keto groups can be substituted for hydrogen on the rings. But, in animal cells, neither cholesterol nor its derivatives can be catabolized to a straight-chain molecule or to acetyl CoA. Therefore, in contrast to other lipids, sterols cannot be used as an energy source by cells.

The bile acids (primarily cholic acid) are the most plentiful derivatives of cholesterol. Bile acids, made in the liver, are then joined to an amino acid to form bile salts; unlike their precursors, bile salts are soluble in water. These molecules are stored in the gallbladder prior to excretion into the intestine. Bile salts play a significant role during digestion because they dissolve and solubilize (emulsify) lipid material in the intestine. All lipids must be emulsified before they can be absorbed, and this includes the essential fatty acids and the fat-soluble vitamins.

Bile salts not only help in lipid uptake, but are also required for cholesterol excretion. Because the body cannot degrade cholesterol, there are only two ways it can be eliminated: first, as bile salts, in excess of the amount needed for lipid absorption, and second, as free cholesterol. Free cholesterol is insoluble in water (and urine) and must be solubilized in order to be excreted from the liver into the gallbladder and then into the intestine. During this passage, it is the bile salts that keep the cholesterol in solution. If the concentration of cholesterol in the gallbladder is too high for the available bile salts to solubilize, the cholesterol precipitates out of solution, forming gallstones. These spheres of almost pure cholesterol may sometimes be as large as half an inch in diameter.

10.5 Membranes and lipoproteins: structure and function

It has been emphasized repeatedly that lipids are insoluble in water. This physical property is a "double-edged sword." On the one hand, it means that lipids are ideal for the construction of cell membranes. On the other hand, it means that organisms need highly sophisticated mechanisms for the absorption and transport of these essential molecules.

Triglycerides are almost totally insoluble in water; phosphoglycerides have mixed solubility characteristics. The two long chains of methylene groups contributed to the

---

* Although we usually think of oxygen as being needed only for mitochondrial energy production, a few other enzymatic reactions catalyzed by oxidases also require this molecule.

phosphoglyceride by the fatty acids are completely insoluble in water. Because of this aversion to an aqueous environment, we call this part of the phosphoglyceride molecule "hydrophobic." Hydrophobic structures such as these tend to aggregate with each other to avoid water; such "huddling together" is called "hydrophobic bonding." You may recall that certain amino acid R groups associate in this way. The remainder of the phosphoglyceride, the charged phosphate group and its attached amine or amino acid (the X, Table 10.1), does not share this dislike of water. In fact, the charged end of the phosphoglyceride molecule prefers an aqueous environment and, for this reason, is termed "hydrophilic." In terms of solubility, then, phosphoglyceride molecules are "two-headed": one end of the molecule associates with water or charged molecules; the other end, comprising the —CH_2— chains, avoids water by associating with other long chains of methylene groups (i.e., other lipids).

This "two-headedness" of phosphoglycerides is crucial to the construction of cell membranes. The cell membrane acts as a barrier between the cell machinery and the external environment. This membrane must keep out unneeded and poisonous compounds while confining the critical metabolic intermediates inside the cell. At the same time, the cell membrane must allow the selective transport of essential nutrients into, and waste products out of, the cell. Similarly, internal membranes (e.g., nuclear and mitochondrial membranes) are used to define cell compartments.

How is the "two-headedness" of phosphoglycerides used in the construction of cell membranes? When placed in water (the normal cellular environment), phosphoglycerides tend to aggregate so that the hydrophobic ends of the molecules associate away from the aqueous environment while the hydrophilic portions face outward into the water (somewhat like the legendary pioneer wagon trains forming their circle). One arrangement of phosphoglyceride molecules that satisfies this requirement is the *phospholipid bilayer* (Fig. 10.5). In cell membranes, the phosphoglycerides form such a bilayer. Membranes also contain triglycerides and cholesterol buried in the inner hydrophobic region of the bilayer.

However, cell membranes do not consist of lipid only; they are composed of about 40% lipid and 60% protein (by weight). What is the spatial relationship between the proteins and the lipids? The proteins, which include structural proteins, transport proteins, and certain enzymes, can either "sit" on the phosphoglyceride bilayer, extend part way into the bilayer, or completely penetrate the lipid layer (Fig. 10.5). In any case, those parts of the protein molecules that lie within the hydrophobic region of the lipid bilayer must themselves by hydrophobic, that is, must be rich in amino acids with uncharged side chains containing many methylene groups, such as isoleucine, leucine, and the aromatic amino acids (Table 2.2). The major structural proteins of cell membranes act as a glue holding the membrane together, and are almost insoluble in water.

Hydrophobic bonds between phosphoglycerides and between phosphoglycerides and proteins are the only bonds that keep the membrane structure intact. Individually, these are very weak bonds; the strength of the cell membrane lies in their enormous numbers. It is truly amazing that this structure, universal to all living cells, contains no covalent bonds between its member molecules.

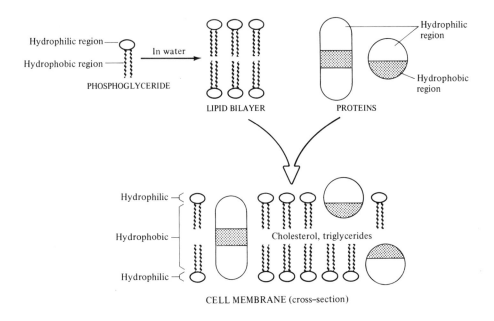

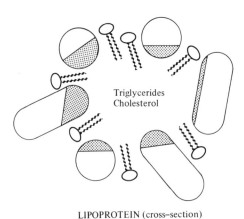

Figure 10.5 Structures of cell membranes and lipoproteins. In cell membranes and lipoproteins, the hydrophobic portions of the lipid molecules aggregate together away from water. This hydrophobic region is surrounded by a shell of hydrophilic molecules.

If the utility of lipids lies in their insolubility in water and their propensity to form hydrophobic bonds, this physical property is also responsible for many of the complications of lipid metabolism. These molecules are, of course, insoluble in blood and other body fluids. Hence, particularly in higher animals, special mechanisms are needed to transport lipid molecules. This is the function of *lipoproteins*. Like cell membranes, lipoproteins consist of an inner hydrophobic core surrounded by a water-soluble hydrophilic "shell" (Fig. 10.5). The innermost core consists of cholesterol and triglycerides in association with the hydrophobic amino acid R groups of the proteins. The outer shell consists primarily of small, charged amino acid R groups of the proteins. The "two-headed" phosphoglycerides serve as a bridge between the core and the outer shell. Lipoproteins vary in both lipid and protein composition, depending on what lipids are being transported and whether transport is from intestinal, liver, or adipose tissue.

In animals, the liver (Fig. 10.1) is especially important in the formation of lipoproteins. In this organ, the protein part of these molecules is synthesized and most lipoprotein molecules are assembled. Any impairment of lipoprotein synthesis or assembly leads to fat deposits in the liver, which impairs liver function and can progress to cirrhosis (in which fibrous scars fill the liver). "Fatty liver" is caused by protein starvation, high-fat diets, carbon tetrachloride poisoning, or prolonged alcohol consumption.

10.6 Many other lipids play specialized physiological roles

We have discussed the roles of triglycerides, phosphoglycerides, and cholesterol, which together account for the bulk of the lipids in most cells. This section contains a brief description of some other lipids which, although present in small amounts, play essential roles in cell and organism metabolism.

(a) FAT-SOLUBLE VITAMINS

The "fat-soluble" vitamins are so named because, like other lipids, they are insoluble in water but will dissolve in lipid solvents and in lipid mixtures. The properties of these four vitamins are summarized in Table 10.3.

Unlike any other "vitamin," human beings and other animals can synthesize vitamin D; it may be argued that vitamin D is actually a hormone. It is made from cholesterol (Fig. 10.4) when the skin is exposed to sunlight. In colder climates (e.g., Madison, Wisconsin) and regions where sunlight is limited, the amount synthesized is not sufficient to meet metabolic requirements, and vitamin D must be present in the diet. Fish, most of which live in darkness, make vitamin D by a pathway that does not require sunlight; their production of vitamin D (and A) explains the nutritional value of cod-liver oil.

Table 10.3 Fat-soluble Vitamins

Four fat-soluble vitamins are required by human beings. Each of these vitamins exists in several forms, which have slightly different chemical structures but identical physiological properties. The alternative forms are designated by numbered subscripts (e.g. K_1, K_2, K_3...). The chemical structure of one form of each vitamin is given below.

		Symptoms of:	
	Role	Deficiency	Overdose
Vitamin A (A_1)	Maintenance of epithelial tissue and mucus secretions; cofactor in physiologic mechanism of vision	Night blindness; epithelial tissue damage in lungs, sinuses; kidney and bladder disorders	Painful joints, thickening of long bones; hair loss
Vitamin K (K_1)	Needed for synthesis of prothrombin, a protein required for blood clotting	Uncontrollable hemorrhaging	Injection in infants leads to jaundice and mild anemia
Vitamin D (D_1)	Increases absorption of calcium and phosphate from intestine; plays role in calcification of bone	Rickets, a skeletal abnormality	Calcification of soft tissue, including lungs and kidney
Vitamin E	Unknown in man; may prevent destruction of other unsaturated lipids	Unknown in man; necessary for reproduction in rats; may lead to anemia in human beings	?

178

Deficiencies of the fat-soluble vitamins can result not only from dietary insufficiency, but also from any generalized failure of lipid absorption, particularly from the disruption of the flow of bile which results from liver disease or bile duct obstruction. In this age of "megavitamin" therapy, when high doses of certain vitamins are claimed to cure or prevent many human disorders, it is important to recognize that large doses of the fat-soluble vitamins may be toxic, especially to children. As with most vitamins, necessary and sufficient quantities of these compounds are present in sensible varied diets.

(b) FAT-SOLUBLE HORMONES: STEROIDS AND PROSTAGLANDINS

Hormones are chemical messengers that coordinate the metabolic activities of different body tissues in higher organisms. The molecular basis of their regulatory activities is discussed more fully in Chapter 14.

In terms of their chemical structure, there are two classes of lipid hormones: the *steroid hormones* (sex hormones and adrenal steroids) and the *prostaglandins*. All the steroid hormones are made from cholesterol; all contain the four rings characteristic of the parent compound. The synthesis of these hormones involves many multistep pathways, usually including the removal of all or most of the linear side chain of cholesterol and the addition of hydroxyl or keto groups (Fig. 10.4). The prostaglandins are derived from 20-carbon unsaturated fatty acids. The structure of prostaglandin E_1 is shown in Table 10.4; the other prostaglandins are structurally related.

There are two functional classes of adrenal steroids: the mineralocorticoids and the glucocorticoids. The role of the mineralocorticoids is to promote sodium ion (Na^+) retention by increasing its reabsorption. Both the intracellular and extracellular sodium ion concentration are critical in maintaining the proper volume of water in body compartments.

The glucocorticoids act to maintain the proper glucose and glycogen levels. Their release triggers a group of reactions which, taken together, increase both blood glucose and liver glycogen levels. Their overall effect is to stimulate gluconeogenesis (Sec. 9.4) in the liver. The glucocorticoids also decrease the uptake of amino acids by most nonhepatic cells and stimulate protein catabolism; the amino acids are transported to the liver, where they are converted to glucose, which is stored as glycogen. Simultaneously, fats are mobilized from adipose tissue and the glycerol released may also be converted to glucose. Perhaps unrelated to their gluconeogenic properties, the glucocorticoids also possess remarkable antiallergic and anti-inflammatory effects and are used medically to treat allergies and arthritic conditions.

In mammals, the gonads (ovaries and testes) produce the steroid sex hormones (Table 10.4). The "male" hormones, the androgens, are responsible for the maturation and maintenance of the male reproductive tract; their positive effect on protein anabolism contributes to muscular growth and development during puberty in males. The "female" hormones, the estrogens and progestogens, are responsible for the development and maintenance of the female reproductive tract. The role of these

TABLE 10.4 Chemical Structures of the Fat-soluble Hormones

All the steroid hormones are made from cholesterol. Prostaglandins are derivatives of unsaturated fatty acids.

Class	Important example	
SEX HORMONES Androgens	Testosterone	(structure)
Estrogens	Estradiol	(structure)
Progestogens	Progesterone	(structure)
ADRENAL STEROIDS Mineralocorticoids	Aldosterone	(structure)
Glucocorticoids	Cortisol	(structure)
PROSTAGLANDINS	PGE_1	(structure)

180

hormones in determining male and female behavior remains controversial. Progesterone inhibits ovulation and its synthetic analogs are used effectively as contraceptives. Diethylstilbesterol (DES), a potent estrogen, was formerly given to beef cattle because it promotes more efficient weight gain. Its use as a feed additive has been curtailed because it has been shown to produce tumors when administered to mice in high concentrations, which raises questions about its possible side effects on humans. More disturbing is the fact that daughters born to women given DES as drug therapy have a high incidence of cervical cancer.

Steroid hormones are not the exclusive property of mammals. The molting hormone of insects, ecdysone, is a steroid, and similar compounds have been identified in crustaceans.

The prostaglandins, which have come into prominence recently, are a large group of fat-soluble hormones with a wide range of physiological activities. Nearly 20 different prostaglandins are now known, each with its own unique effects on metabolism. Some prostaglandins regulate the activity of smooth muscle, others control gastric secretions, some regulate endocrine secretions, and still others affect blood pressure. Several prostaglandins have striking effects on the female reproductive tract; certain of these hormones stimulate uterine contractions and are used to induce labor, and still others are effective agents for abortion. The palliative effects of aspirin are believed to be, in part, due to an effect on prostaglandin levels in the body. Although their precise mechanism of action is not yet known, the prostaglandins seem to act as a link between other hormones and the sympathetic nervous system. The routine use of prostaglandins as drugs must await further study of their total effect.

(c) COMPLEX LIPIDS OF BRAIN AND NERVE TISSUES

Structurally unique complex lipids are required for the construction of the membranes that surround brain and nerve cells. Although these lipids are heterogeneous, all contain a basic unit consisting of the long-chain amino alcohol sphingosine, combined with a fatty acid. Most nerve and brain lipids also contain sugars: the basic unit (shown in the rectangular area below) is linked to a monosaccharide (glucose or galactose) or to a short chain of these two sugars or their derivatives. An exception is sphingomyelin, in which the basic unit is joined to phosphorylcholine.

These lipids have gained recent medical notoriety because, when accumulated in excess in nerve cells, they cause a group of fatal degenerative diseases known as the

Sphingosine ⟶ | $CH_3-(CH_2)_{12}-CH-CH=C-CH-CH_2-O$ | ⟶ Sugars

with OH above the central C, H below, and NH below connecting to:

Fatty acid ⟶ $CH_3(CH_2)_{16}-C(=O)-NH$

"lipid-storage diseases." Most lipid-storage diseases result from a lack of one of the enzymes which specifically degrades a particular complex lipid. Lacking the enzyme, the cells produce more and more lipid but have no way of removing it. As the lipid accumulates, the cells become less and less functional and, depending on which nerve cells are affected, this leads to a loss of muscle control and/or to a decline in mental processes. Most of the lipid-storage diseases are inherited as autosomal recessive traits (Sec. 12.10).

Tay-Sachs disease, which affects the nerve cells of the brain, is the best known of the lipid-storage diseases. The symptoms progress from muscular weakness to mental retardation, with seizures and blindness, with death usually occurring by the age of three. Children with this disease lack the degradative enzyme hexosaminidase A, which cleaves one sugar molecule from the complex lipid called ganglioside GM_2. Because very sensitive assays for hexosaminidase A activity (or the lack of it) are now available, a prenatal diagnosis of Tay-Sachs disease can be made when this enzyme is found to be missing from amniotic fluid; Tay-Sachs carriers (Sec. 12.10) can be identified because they produce only half the normal amount of hexosaminidase A (although they are themselves free of the symptoms of the disease).

(d) WAXES

Waxes are surface lipids secreted by both plants and animals. These compounds provide a protective and water-repellent film on skin, feathers, and leaves; the wax layer on leaves helps to conserve water within the plant. Waxes also serve as a means of storing energy in many aquatic animals.

Each wax molecule contains a long-chain (10 to 30 methylene groups), monohydroxy alcohol joined to a fatty acid by an ester bond. Secreted waxes are usually mixtures of molecules with different-size alcohol and fatty acid components.

10.7 Lipids, diet, and heart disease

To begin on the positive side, we must remember that lipids are a necessary part of our daily diet. Not only are they an excellent energy source, but both polyunsaturated fatty acids and the fat-soluble vitamins must be present in our diet. Triglycerides, phosphoglycerides, and cholesterol, which can be synthesized by our cells, are crucial for energy storage and in the formation of all cell membranes.

However, the ongoing debate over the effects of dietary cholesterol and unsaturated fatty acids cannot be ignored. The "cholesterol debate" begins with the question of whether or not there is a *cause-and-effect* relationship between serum cholesterol levels and heart disease, particularly atherosclerosis. In atherosclerosis, lipid-rich deposits form in the arterial walls and continue to increase in size until the blood vessel is closed off. It is known that individuals with atherosclerosis often have elevated blood cholesterol levels and that this sterol, along with other lipids, is found in atherosclerotic deposits. It is less clear, however, that a high level of circulating cholesterol actually *causes* lipid deposits.

For those who think that cholesterol *is* the villain, the logical step is to recommend that everyone lower their blood cholesterol level in the hope of preventing atherosclerosis. This is done, first of all, by eating less cholesterol. Since cholesterol is made only by animals (plants make sterols, but these are poorly absorbed by human beings), this means reducing meat and egg consumption, switching to low-fat milk and getting more of our protein from plant products. The effectiveness of such a reduction in dietary cholesterol in reducing serum cholesterol is not clear. The average American male eats about 0.3 gram of cholesterol per day while he synthesizes over three times this amount (1 gram/day) from acetyl CoA; the contribution of the latter to circulating cholesterol is not easily controlled by dieting.

This is where polyunsaturated fatty acids (PUFA) enter the picture. It has been suggested that an increase in one's intake of PUFA (and a concomitant decrease in dietary saturated fatty acids) will promote the esterification of cholesterol. (The formation of an ester bond between the hydroxyl group of cholesterol and the carboxyl group of the PUFA.) Such esterified cholesterol tends to be more readily taken up by cells than is unesterified cholesterol, thus removing it from the bloodstream. A seldom-asked question is: "Where does cholesterol go when it leaves the bloodstream?" Some investigators have suggested that it is simply stored intact in body tissues, which may or may not be a desirable situation. Because plant lipids have a higher ratio of PUFA to saturated fatty acids than do animal lipids, it has been recommended that we substitute plant products for animal products in our diet. No one is recommending drastic changes in our diet, and certainly no one recommends a great increase in total fatty acid intake. In their book, *The Cholesterol Controversy*, Pinckney and Pinckney describe people who became ill by substituting several tablespoons of polyunsaturated lipids for the meat and eggs of their normal diet.

Whether or not such dietary manipulations can successfully prevent the onset of atherosclerosis is not known. There are so many factors that influence heart disease (as well as general health) that it is nearly impossible to sort them all out. The following have been implicated in heart disease: obesity, lack of exercise, high blood pressure, softened drinking water, gender, heredity, vitamin B_6, magnesium, calcium, and emotional stress. Cigarette smoking is also a contributing factor and it has been suggested that deaths due to heart disease could be reduced by one-fourth if smoking were banned. Clearly, prevention of heart disease is no simple matter; and consideration of one component of our diet is unlikely to provide a solution to the problem, as popular "health" books would like to suggest.

SUMMARY

Lipids are a diverse group of water-insoluble metabolites with a common chemical feature: chains or rings of $—CH_2—$ groups. Cells produce a seemingly endless array of such compounds: straight chains, branched chains and ring structures, saturated and unsaturated compounds, simple and complex molecules. These variations in structure are associated with a wide variety of physiological roles.

The simplest lipids are the fatty acids. Fatty acids are made from acetyl CoA, particularly when caloric intake exceeds immediate metabolic needs. Free fatty acids are subsequently attached to glycerol by ester bonds to form triglycerides, which are stored in adipose tissue. When the diet no longer contains enough carbohydrates and lipids to satisfy energy requirements, the fatty acids, released from triglycerides, are degraded to acetyl CoA, which yields ATP in the Krebs cycle and electron transport systems.

Fatty acids, esterified to alcohol backbones, also form the basis of other complex lipids: phosphoglycerides, waxes, and the lipids of brain and nerve tissue. Many of these compounds are used for the construction of cell membranes.

Sterols are lipids containing four rings of —CH_2— groups. Like fatty acids, these molecules are synthesized from acetyl CoA. Cholesterol, the major animal sterol, is a component of cell membranes, and is the precursor of the bile acids, vitamin D, and the steroid hormones.

The water insolubility of lipids is both a problem and a blessing to organisms. Because of this physical property, lipids cannot move freely in body fluids. Lipids have to be transported from one organ to another in lipoproteins, mixtures of lipids and proteins with hydrophobic interiors and hydrophilic exteriors. On the plus side, the water insolubility of lipids makes them ideal for the construction of cell membranes. Layers of lipid, surrounded by proteins, form semipermeable barriers that separate the aqueous environments inside and outside the cell.

PRACTICE PROBLEMS

1. Prove to yourself that the complete catabolism of the 16-carbon fatty acid, palmitic acid, $CH_3(CH_2)_{14}COOH$, to CO_2 and water provides the cell with a net yield of 129 molecules of ATP.

2. Hydrogens from NADPH are needed in the reduction reactions of fatty acid synthesis. What reactions in animal cells are the source of these hydrogens? What reactions in plant cells serve this purpose?

3. Name two vitamins needed to make fatty acids from acetic acid and describe the reactions in which they participate.

4. Briefly outline the pathway(s) by which:
    (a) Glucose is converted to fatty acids.
    (b) Alanine is converted to cholesterol.
    (c) Triglycerides are degraded to CO_2, water, and energy.

5. Which of the following kinds of lipids can serve as cellular energy sources?
    (a) Triglycerides.
    (b) Estrogens.
    (c) Phosphoglycerides.
    (d) Free fatty acids.
    (e) Bile acids.

Lipids

6. Compare the effects of insulin and glucagon on liver glycogen stores and on the triglycerides stored in adipose tissue.
7. The liver from a well-fed rat was found to synthesize fatty acids provided that the following compounds were present: carbon dioxide, glucose, NADPH, ATP, and biotin. If each of these compounds was radioactively labeled with ^{14}C, which would lead to the production of radioactive fatty acids?
8. Explain the following:
   (a) The disease abetalipoproteinemia is characterized by the absence of an important class of lipoproteins in the blood. Affected individuals accumulate triglycerides and phosphoglycerides in the liver and intestine. Why?
   (b) Drugs are known which block the biosynthesis of cholesterol at various stages. The use of these drugs has been sharply curtailed because of their harmful side effects. What metabolic problems would you expect these drugs to create?
   (c) Oral preparations of vitamin D are not effective in individuals with bile-duct obstruction, that is, when the bile duct is blocked and cannot empty into the intestine. Why not?
9. Assume that membrane proteins can contain any amino acid. If this were the case, which of the amino acid R groups in the proteins would be sticking out into the water? Which of the amino acid side chains would form hydrophobic bonds with the phosphoglycerides?
10. Excess carbohydrates can be converted to lipids; sugars are degraded to acetyl CoA, which is incorporated into fatty acids. On the other hand, in animals a dietary excess of fatty acids (or glycerides) does not produce a net increase in carbohydrates; the products of fatty acid degradation cannot be converted to simple sugars. Describe the metabolic problem(s) involved in converting fatty acids to glucose and show why an excess of these lipids cannot cause an increase in the body level of glucose or other carbohydrates.

SUGGESTED READING

BENDITT, EARL P., "The Origin of Atherosclerosis," *Sci. American*, 236, No. 2, p. 74 (1977). Mutations in arterial wall muscle cells, rather than blood lipids, may be the cause of atherosclerotic plaques; excellent photographs and diagrams.

BRADY, ROSCOE O., "Hereditary Fat-Metabolism Diseases," *Sci. American*, 229, No. 2, p. 88 (1973). Shows structures of complex lipids of brain and nerve tissue and explains how failure to degrade these compounds leads to lipid-storage diseases.

CAPALDI, RODERICK A., "A Dynamic Model of Cell Membranes," *Sci. American*, 230, No. 3, p. 26 (1974). Good description of the roles of both lipids and proteins in cell membranes; excellent diagrams.

LIEBER, CHARLES S., "The Metabolism of Alcohol," *Sci. American*, 234, No. 3, p. 25 (1976). Prolonged alcohol consumption leads to adaptive changes in the liver, including an increased production of fatty acids and lipoproteins.

LOOMIS, W. F., "Rickets," *Sci. American*, 223, No. 6, p. 77 (1970). A lively chronicle of the research which proved that vitamin D production in animals requires sunlight. Rickets is most prevalent in industrialized northern cities where air pollution is severe.

PINCKNEY, EDWARD R., and CATHEY PINCKNEY, *The Cholesterol Controversy*, Sherbourne Press, Nashville Tenn., (1973). Emphasizes the negative side of the cholesterol debate.

SCRIMSHAW, NEVIN S., and VERNON R. YOUNG, "The Requirements of Human Nutrition," *Sci. American*, 235, No. 3, p. 50 (1976). Three essential fatty acids and the fat-soluble vitamins must be present in the human diet.

WILLIAMS, CARROLL M., "Third-Generation Pesticides," *Sci. American*, 217, No. 1, p. 13 (1967). Insect hormones, which are derivatives of branched-chain, unsaturated fatty acids, hold promise as insecticides.

YOUNG, VERNON R., and NEVIN S. SCRIMSHAW, "The Physiology of Starvation," *Sci. American*, 225, No. 4, p. 14 (1971). During starvation, the degradation of fatty acids yields ketone bodies, which the brain uses as an energy source.

11
The Metabolism of Amino Acids & Other Nitrogen Compounds

Up to this point we have focused our attention on pathways by which cells build and degrade compounds containing primarily carbon, hydrogen, and oxygen. A fourth element, nitrogen, is equally important in living systems; it is found in proteins, nucleic acids, and certain phospholipids and sugars.

Amino acids are at the center of nitrogen metabolism (Fig. 11.1); not only are they the monomers for proteins, but they also serve as the precursors for purine and pyrimidine bases and literally hundreds of other nitrogen-containing metabolites. The structures and chemistry of the amino acids have been described in Chap. 2.

11.1 The nitrogen cycle

Relatively few deposits of nitrogen compounds are found in the earth's crust, because most nitrogen compounds are either soluble in water, are lost from the earth's surface as a gas (i.e., are volatile), or are readily decomposed to soluble or volatile forms. Hence, the nitrogen content of soil tends to decrease and must be constantly replenished to meet the metabolic needs of microbial, plant, and animal life. The process of alternate nitrogen loss and nitrogen capture by living systems is known as the *nitrogen cycle* (Fig. 11.2).

Nitrogen from living or once-living material passes into the atmosphere as ammonia and other volatile nitrogen compounds. Ammonia is produced by the

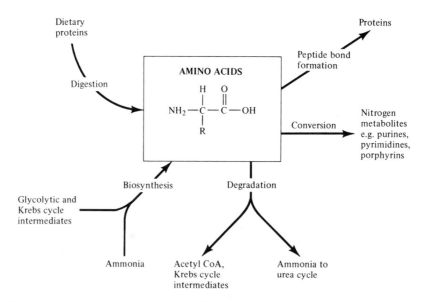

Figure 11.1 Amino acids are at the center of nitrogen metabolism. Basically, amino acid metabolism is the same in all cells. All cells make proteins and nucleic acid bases; all cells can degrade amino acids to energy-yielding compounds and ammonia. However, other aspects of amino acid metabolism are highly organism-specific. Cells differ with respect to which amino acids they can synthesize, which nitrogen compounds they excrete, and in the variety of metabolites they produce.

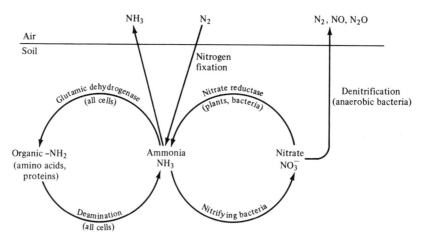

Figure 11.2 Nitrogen cycle. Nitrogen is an essential element for the cellular production of proteins, nucleic acids, and many other biologically active organic molecules. Nitrogen lost by deamination and denitrification is replaced by nitrification, nitrogen fixation, and nitrate reduction. Animals are dependent on bacteria and plants to capture and reduce inorganic nitrogen compounds for them.

deamination of amino acids, by the degradation of urea and by other reactions common to cells; the ammonia passes into the atmosphere or is converted to nitrate (NO_3^-) by bacteria. Under anaerobic conditions, as are often found in decaying organic matter, denitrifying bacteria convert nitrate to nitrogen (N_2), nitrous oxide (N_2O), and other nitrogen gases that most cells cannot use. These processes result in a net loss of nitrogen from the biosphere.

Nitrogen enters the biosphere as ammonia or nitrate in rainwater, as nitrogen gas (N_2), or as nitrate, ammonia, or urea in fertilizers.* In general, these compounds are converted to ammonia, which becomes biologically active when incorporated into amino acids by means of the glutamic dehydrogenase reaction (Sec. 11.3) found in all cells.

Two biological processes, nitrification and nitrogen fixation, convert inorganic nitrogen compounds to ammonia. *Nitrification* is the reduction of nitrate to ammonia by nitrate reductase, a molybdenum-containing enzyme complex found in plants and a few microorganisms. *Nitrogen fixation* involves the reduction of nitrogen gas (N_2) to ammonia (NH_3). Nitrogen is chemically inert and its reduction is an energy-consuming process; 12 ATPs are used each time a molecule of nitrogen is reduced to two molecules of ammonia. Nitrogen fixation is catalyzed by nitrogenase, an iron- and molybdenum-containing enzyme complex and is unique to certain procaryotes: a few free-living bacteria, some blue-green algae, and the familiar symbiotic bacteria that fix nitrogen only when present in the nodules of specific host plants. For those of you who are gardeners, a bagful of nitrogen-fixing bacteria can be purchased at any seed store; the inoculation of host seeds (beans, peas, peanuts, etc.) with bacteria prior to planting will reduce the need for nitrogen fertilizer.

11.2 Cells need a basic set of amino acids

In order to make proteins and to produce many hormones, nucleic acid bases, and other nitrogen-containing molecules, all cells need a basic set of amino acids (Table 2.2). These may be obtained in two ways: (1) by the synthesis of the amino acids from sugars, ammonia, and other simple precursors; or (2) by the acquisition of intact amino acids from the environment. Human beings and many other animals can make only about half of the amino acids they require, the "nonessential" amino acids; the rest, the essential amino acids, must be supplied in the diet (Table 11.1). Higher plants are more versatile; they synthesize all the amino acids they require, as can most microorganisms.

---

* Plants use nitrates readily, but nitrates are water-soluble and wash away in rain water. Ammonia, when injected as a gas, adheres fairly well to the soil. Urea is hydrolyzed to ammonia and CO_2 by enzymes in bacteria and other soil microbes.

TABLE 11.1 Nonessential and Essential Amino Acids in Human Beings

Human beings can synthesize about half of their required amino acids; the remainder must be supplied in the diet. Our amino acid requirements change somewhat with age and general condition; for example, arginine, a product of the urea cycle, is required only by children. For those who find memory aids useful, "PVT. TIM HALL" can help one remember the first letter of each essential amino acid.

Nonessential	Essential
Alanine	Arginine
Aspartic acid	Histidine
Cysteine	Isoleucine
Glutamic acid	Leucine
Glycine	Lysine
Proline	Methionine
Serine	Phenylalanine
Tyrosine	Threonine
	Tryptophan
	Valine

The 18 amino acids listed in Table 2.2 are all used as building blocks for proteins. In addition, proteins contain four "derived" amino acids which are made by enzymatically modifying the R groups of members of the basic set. The four derived amino acids (Table 11.2) are glutamine and asparagine, amide derivatives of glutamic and aspartic acids, and hydroxyproline and hydroxylysine, made by adding a hydroxyl group to the R group of the parent amino acid.

Human beings obtain their essential amino acids from the meat, milk, and vegetable proteins in their diet. Most cells in the body are impermeable to intact proteins; hence, all dietary proteins must be degraded to amino acids before they can be absorbed and used to make proteins and amino acid derivatives in human beings. This digestion is carried out by enzymes, such as pepsin, trypsin, and chymotrypsin, which are secreted into the gastrointestinal tract.

The human amino acid requirements are as important as the need for vitamins, minerals, and unsaturated fatty acids. One of the most serious nutritional diseases found in the world is protein malnutrition (kwashiorkor). Victims are starved, not for calories, but for specific amino acids. This disease is endemic to many parts of the world; soon after weaning, children develop characteristic symptoms of thin limbs, bloated bellies, and lethargy. Overcoming protein malnutrition requires changes and expansion of the diet to include foods higher in essential amino acids. As one approach toward this end, plant breeders have developed "high lysine" corn to replace indigenous varieties low in this essential amino acid.

TABLE 11.2 Derived Amino Acids.

Modification of R group of parent amino acid	Derived amino acid	Structure at pH 7
Hydroxylation	Hydroxyproline	$\text{H}_2\overset{+}{\text{N}}-\overset{\overset{\text{H}}{\vert}}{\text{C}}-\overset{\overset{\text{O}}{\|}}{\text{C}}-\text{O}^-$ with CH₂–CH(OH)–CH₂ ring closing to N
	Hydroxylysine	$\overset{+}{\text{H}_3\text{N}}-\overset{\overset{\text{H}}{\vert}}{\text{C}}-\overset{\overset{\text{O}}{\|}}{\text{C}}-\text{O}^-$, side chain $-\text{CH}_2-\text{CH}_2-\text{CH(OH)}-\text{CH}_2-\overset{+}{\text{NH}_3}$
Amide formation	Glutamine	$\overset{+}{\text{H}_3\text{N}}-\overset{\overset{\text{H}}{\vert}}{\text{C}}-\overset{\overset{\text{O}}{\|}}{\text{C}}-\text{O}^-$, side chain $-\text{CH}_2-\text{CH}_2-\text{C}(=\text{O})-\text{NH}_2$
	Asparagine	$\overset{+}{\text{H}_3\text{N}}-\overset{\overset{\text{H}}{\vert}}{\text{C}}-\overset{\overset{\text{O}}{\|}}{\text{C}}-\text{O}^-$, side chain $-\text{CH}_2-\text{C}(=\text{O})-\text{NH}_2$

11.3 The synthesis of nonessential amino acids from glycolytic and Krebs cycle intermediates and ammonia

Amino acid biosynthesis has two phases: formation of the carbon skeleton, and incorporation of the amino group. The carbon skeleton is often derived from Krebs cycle intermediates; the α-amino group comes, ultimately, from ammonia.

This section outlines the relatively straightforward pathways for the synthesis of nonessential amino acids in human beings. These pathways are similar in all cells of all organisms. The pathways used in plants and microbes for the biosynthesis of essential amino acids will not be described.

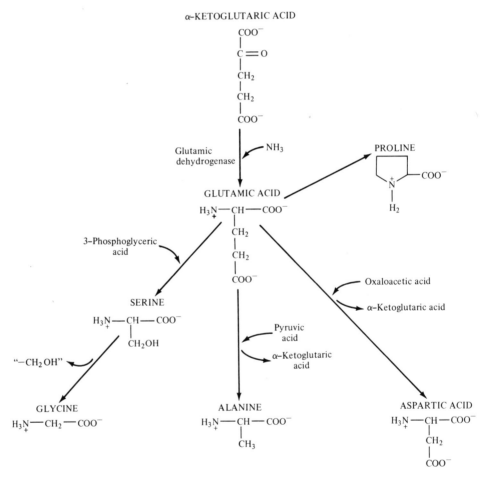

Figure 11.3 Synthesis of nonessential amino acids. Most nonessential amino acids are made from glycolytic or Krebs cycle intermediates and ammonia.

The Metabolism of Amino Acids and Other Nitrogen Compounds

The production of nonessential amino acids (Fig. 11.3) can be seen as starting with the incorporation of ammonia into the Krebs cycle intermediate, α-ketoglutaric acid:

$$\begin{array}{cc}
\text{O=C-C(=O)-OH} & \text{H}_2\text{N-CH-C(=O)-OH} \\
| & | \\
\text{CH}_2 & \text{CH}_2 \\
| & | \\
\text{CH}_2 & \text{CH}_2 \\
| & | \\
\text{C(=O)-OH} & \text{C(=O)-OH}
\end{array}$$

$$\text{NH}_4^+ + \alpha\text{-Ketoglutaric acid} + \text{NADPH} \xrightleftharpoons[]{\text{Glutamic dehydrogenase}} \text{Glutamic acid} + \text{NADP}^+$$

This is an extremely important reaction, in which the α-amino group is made directly from ammonia.* Glutamic acid is the source of the amino group and carbon atoms for the synthesis of other amino acids. Alanine and aspartic acid are made by transamination reactions (Sec. 8.3) in which the amino group is exchanged between the respective α-keto acids (pyruvic and oxaloacetic acid) and glutamic acid. The synthesis of proline, serine, and glycine from glutamic acid involves further modifications of the carbon skeleton.

In addition to the pathways outlined in Fig. 11.3, cells make cysteine from methionine and serine. Tyrosine is produced by the addition of a hydroxyl group to the 6-carbon ring of phenylalanine (Fig. 11.7).

The synthesis of two of the "derived" amino acids, hydroxypyroline and hydroxylysine (Table 11.2), is of special interest because these amino acids are found primarily in the insoluble structural proteins (Chap. 2) of skin, bones, teeth, and connective tissue of higher animals. Free hydroxylated amino acids cannot be incorporated into growing polypeptide chains, as there is no transfer RNA and no codeword (Sec. 13.5) for these amino acids; the hydroxyl groups are added to the R group of the parent amino acid *after* the proline or lysine has been incorporated into the polypeptide chain.

An example of this hydroxylation process is found in the synthesis of collagen, the structural protein that constitutes the organic matrix of human bones, teeth, and cartilage. Collagen synthesis starts with the construction of a long polypeptide; the R groups of specific intrachain proline and lysine residues are then hydroxylated. The enzyme that hydroxylates the intrachain proline residues requires vitamin C (ascorbic acid) as a cofactor, and the collagen formed in the absence of this vitamin is deficient in its biological function. Vitamin C-deficient individuals develop scurvy, a disease characterized by fragile bones, loose teeth, and skin sores.†

---

* Ammonia (NH_3) forms ammonium ions (NH_4^+) in water.

† Scurvy was common among sailors in the sixteenth, seventeenth, and eighteenth centuries. When British physicians recognized that the consumption of lemons or limes could prevent this disease, these fruits were provided on long voyages. Hence, the name "limey" was applied to British sailors.

11.4 The degradation of amino acids feeds the Krebs and urea cycles

A large proportion of a cell's amino acids are used for protein synthesis or serve as precursors for other small nitrogen metabolites (Sec. 11.5). Amino acids can also be used as energy sources, although, in contrast to sugars and fatty acids, this is not their predominant metabolic fate.

In animals, the amino acids that are used as energy sources most often originate from dietary proteins. Digested proteins provide a mixture of amino acids; once the requirements for protein synthesis are met, "leftover" amino acids can be degraded to supply ATP energy. Not surprisingly, under normal conditions, human beings do not degrade their own body proteins to obtain energy (e.g., muscle is not an energy "store"); this occurs only during extreme starvation, after glycogen and lipid stores have been depleted.

The degradation of amino acids, like their synthesis, consists of two parts: the removal of the amino group, and the degradation of the carbon skeleton. The amino group is transferred by transamination to a keto acid (Sec. 8.3), removed by deamination, and eventually excreted. By a number of complex pathways, the carbon skeletons of amino acids can be converted to acetyl CoA, pyruvic acid, acetoacetyl CoA, and/or Krebs cycle intermediates (Figs. 8.4 and 11.4). Hence, amino acid catabolism provides energy (Sec. 8.3) and also precursors for the synthesis of fatty acids, glucose, and the less desirable ketone bodies (Chap. 10).

Through a network of interconversions and transaminations, the α-amino groups of most amino acids find their way to the α-amino group of glutamic acid. Glutamic acid is then directly deaminated to produce α-ketoglutaric acid and ammonium ion:

$$\text{glutamic acid} + \text{NADP}^+ \xrightleftharpoons{\text{Glutamic dehydrogenase}} \text{NH}_4^+ + \alpha\text{-ketoglutaric acid} + \text{NADPH}$$

We have seen how glutamic dehydrogenase catalyzes the reverse reaction during amino acid synthesis (Sec. 11.3). The direction in which the glutamic dehydrogenase reaction runs is very important to cell metabolism, for it determines whether the cell is making amino acids or breaking them down. When a cell is "energy-poor" (i.e., the levels of ATP and GTP are low while ADP and GDP are high), amino acids need to be degraded to replenish the energy supply. The nucleoside diphosphates ADP and GDP activate the enzyme to deaminate glutamic acid and produce α-ketoglutaric acid; α-ketoglutaric acid enters the Krebs cycle and ATP is made. When the cell is again "energy-rich," the higher levels of ATP and GTP inhibit the deamination reaction and reduce amino acid degradation. This is another example of how the cell can emphasize or de-emphasize a particular metabolic reaction in response to changing intracellular and extracellular conditions.

Serine and threonine can also be deaminated, giving ammonia and pyruvic or α-ketobutyric acids, respectively. Such deamination reactions increase the intracellular NH_4^+ concentration and, if not reincorporated, NH_4^+ might reach concentrations in the blood which are toxic to tissues and organs. Hence, cells must have a system to get

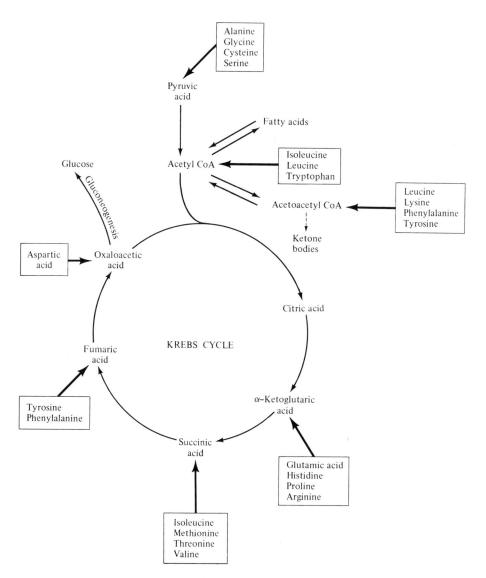

Figure 11.4 Degradation of the carbon skeletons of amino acids. During amino acid catabolism, the carbon skeletons enter the energy-producing pathways of glycolysis and the Krebs cycle. The amino acids that feed into the Krebs cycle can also be "raw material" for gluconeogenesis.

rid of excess ammonium ions, even if considerable energy must be expended in the process.

The mode of amino nitrogen excretion varies and depends on the "life-style" of the organism, particularly whether it is a land or a water creature. Many aquatic animals excrete NH_4^+ itself; terrestrial vertebrates, including man, excrete urea. Birds and terrestrial reptiles, which have a limited water intake, excrete semisolid suspensions of uric acid, a purine derivative (Sec. 11.12). Spiders excrete the purine guanine rather than uric acid. Dalmatian dogs excrete their excess ammonia as uric acid.

The urea cycle is shown in Fig. 11.5. Ammonium ions and carbon dioxide react to form carbamoyl phosphate, which then condenses with ornithine to form citrulline. Although not found in proteins, ornithine, citrulline, and argininosuccinic acid are amino acids; the urea cycle is essentially a series of reactions that converts the R group

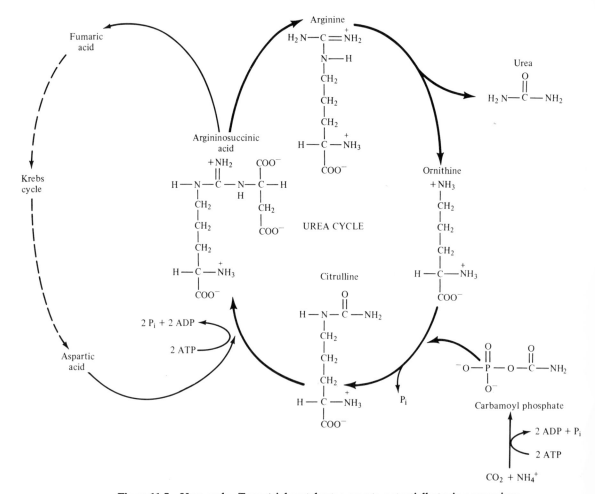

Figure 11.5 Urea cycle. Terrestrial vertebrates excrete potentially toxic ammonium ions (NH_4^+) as urea; four molecules of ATP provide energy for each turn of the cycle.

of ornithine into the R group of arginine. In the final reaction, the R group of arginine is cleaved to produce urea and ornithine. The urea is excreted and the ornithine is used to start another turn of the cycle and produce more urea.

The urea cycle and the Krebs cycle are interconnected. Four ATPs, products of the Krebs cycle and the electron transport system, are consumed in making each molecule of urea. Aspartic acid, formed from the Krebs cycle intermediate, oxaloacetic acid, is the donor of one of the amino groups of urea. Thus, in its use of aspartic acid, urea production "drains" the Krebs cycle. This is not a serious drawback, however, because the fumaric acid released in the urea cycle feeds back into the Krebs cycle, keeping the cycles balanced in a functional sense.

11.5 Nitrogen enters most other biological compounds from amino acids

We have already encountered two biological uses of amino acids: they are the monomers for protein synthesis and they can be degraded to ammonia and energy-yielding compounds. The third possible fate of amino acids is their conversion to or incorporation into various small, but biologically important, nitrogen-containing molecules. Nitrogen metabolites are a chemical and functional "mixed bag." They range from the ubiquitous purines and pyrimidines to the antibiotics produced only by a small group of bacteria and fungi. Nitrogen compounds can be simple molecules such as the hormone epinephrine, or complex molecules such as porphyrins (Fig. 11.6). Some examples of nitrogen-containing metabolites are given in Table 11.3.

We shall describe only two pathways which are vital to many organisms: the synthesis of purines and pyrimidines (Sec. 11.8) and the synthesis of porphyrins; the latter are cofactors that bind metal ions in the important transport molecules: hemoglobin, cytochromes, and chlorophyll.

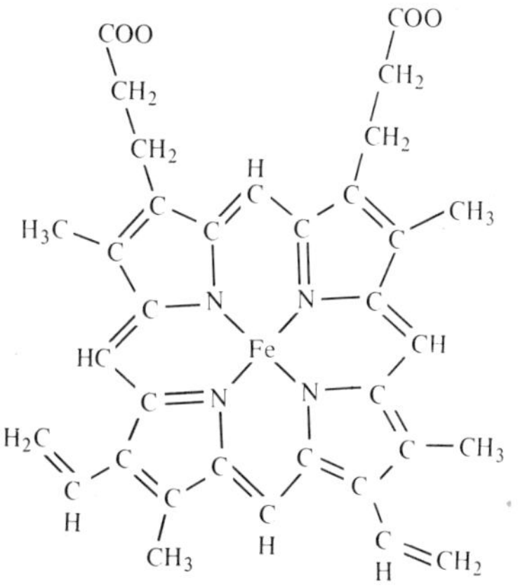

Figure 11.6 Structure of heme.

TABLE 11.3 Nitrogen Compounds Derived from Amino Acids

Amino acids are the precursors of a wide variety of metabolically important nitrogen compounds. Purine and pyrimidine biosynthesis is common to all cells.

	Example	
Function of nitrogen compound	Amino acid precursor	Nitrogen-containing derivative
Information carrying molecules (RNA, DNA)	Glycine, glutamine, aspartic acid	Purines
	Glycine	Pyrimidines
Complex lipids of nerve tissues	Methionine	Choline
	Serine	Sphingosine
Hormones	Tyrosine	Epinephrine, thyroxine
	Tryptophan	Indolacetic acid
Pigments	Tyrosine	Melanin
Cofactors for enzymes and transport proteins	Tryptophan	Niacin (a precursor of NAD^+, $NADP^+$)
	Glycine	Porphyrins
Alkaloids[a]	Proline	Cocaine
	Glutamic acid	Nicotine
	Tyrosine	Morphine, codeine
	Tryptophan	Quinine, strychnine
Vasodilators	Histidine	Histamine
Antibiotics	Valine, cysteine	Penicillin

[a] Nitrogenous plant products having marked physiological action on animals.

Chemically, porphyrins contain four pyrrole rings, arranged in a circle. A pyrrole is a heterocyclic compound with the structure

$$\begin{array}{c} HC\text{---}CH \\ \| \quad \| \\ HC \diagdown_{\displaystyle N}\diagup CH \\ H \end{array}$$

The pyrrole rings are synthesized from the amino acid glycine and the Krebs cycle intermediate, succinyl CoA; these two metabolites condense to form δ-aminolevulinic acid.

$$\underset{\text{Glycine}}{\begin{array}{c} NH_2 \\ | \\ CH_2 \\ | \\ COOH \end{array}} + \underset{\text{Succinyl CoA}}{\begin{array}{c} COOH \\ | \\ CH_2 \\ | \\ CH_2 \\ | \\ C\text{---}CoA \\ \| \\ O \end{array}} \xrightarrow{\text{δ-Aminolevulinate synthetase}} \underset{\text{δ-Aminolevulinic acid}}{\begin{array}{c} COOH \\ | \\ CH_2 \\ | \\ CH_2 \\ | \\ C\text{---}CH_2\text{---}NH_2 \\ \| \\ O \end{array}} + CO_2 + CoA$$

Two molecules of δ-aminolevulinic acid condense to form a pyrrole; four such monopyrroles form the tetrapyrrole ring. Various enzymatic modifications of the R groups around the tetrapyrrole ring create the different porphyrins.

Defects in porphyrin biosynthesis, either inherited or acquired, are the basis of several human diseases, for example the porphyrias. In the porphyrias, the patient *overproduces* one or more of the porphyrin precursors; these accumulate in the blood and are excreted in the urine. An example is acute intermittent porphyria, the inherited disease responsible for the occasional madness of England's King George III, whose unreasonable demands led the American colonies to fight for their independence. Attacks of acute intermittent porphyria, often triggered by mild infections, result from hyperactivity of δ-aminolevulinate synthetase, leading to the overproduction of δ-aminolevulinic acid, other porphyrin precursors, and porphyrin. When their concentrations in the bloodstream reach high levels, these compounds may have toxic effects on the entire nervous system, causing limb weakness, acute abdominal pain, visual difficulties, insomnia, convulsions, and finally paralysis and delirium. Not the sort of disturbances one wants to have in a king!

The four central nitrogen atoms of porphyrins (Fig. 11.6) are present in a specific spatial arrangement and bind positively charged metal ions such as iron, magnesium, cobalt, and copper. This chemical property makes porphyrins ideally suited to serve as a kind of "glue" between metal ions and proteins; protein–porphyrin–metal ion complexes serve many biological functions. Examples include iron-containing hemoglobin and myoglobin used for oxygen transport, iron-containing cytochromes of the electron transport system, and magnesium-containing chlorophylls that carry electrons during photosynthesis. A porphyrinlike structure serves a similar role in vitamin B_{12}. This cobalt-containing molecule is needed for red blood cell formation; decreased ability to absorb vitamin B_{12} in the intestine leads to pernicious anemia.

The degradation of porphyrins is an important physiological process in human beings. The average red blood cell (which is basically a bag of hemoglobin) circulates for about 120 days and then is broken down in the spleen. The globin protein is degraded to amino acids, and the iron is recycled. The porphyrin ring is not reused; rather, it is converted to a linear tetrapyrrole, bilirubin. Bilirubin is transported through the bloodstream to the liver, where it is joined to sugar molecules to produce a more soluble compound or conjugate, a bile pigment, which is excreted through the gallbladder into the intestine.

When, for whatever reason, the breakdown of red blood cells exceeds the capacity of the liver to conjugate or excrete bilirubin, this yellow compound increases in the blood and may be deposited in the skin, mucous membranes, or whites of the eyes; bilirubin deposition leads to a condition named jaundice. Jaundice in adults is often the result of liver disease or a physical obstruction of the excretory ducts of the liver. Mild jaundice is common in newborns, particularly the premature, because their livers are deficient in the enzymes that form conjugates and remove bilirubin. More severe jaundice is found, even in full-term babies, when there is a blood-group incompatibility between mother and child, causing the baby's red blood cells to break down more rapidly than normal. Jaundice is usually not harmful, but *very* high blood levels of bilirubin, most often a problem in newborns, can lead to bilirubin deposits in brain cells and irreparable brain damage. Treatments for severe newborn jaundice include blood-exchange transfusions and phototherapy; high-intensity ultraviolet light breaks down bilirubin to a form that can be excreted.

11.6 The aromatic amino acids are important intermediates in human metabolism

To illustrate the wide-ranging functions of amino acids in human metabolism and physiology, we shall examine more closely the pathways involving the aromatic amino acids, phenylalanine and tyrosine (Fig. 11.7). Cells produce tyrosine from phenylalanine, an essential amino acid. Both these amino acids are, of course, monomers for protein synthesis. In addition, tyrosine is the precursor of several hormones and the skin pigment, melanin. The degradation of tyrosine (and phenylalanine) produces fumaric acid and acetoacetic acid, energy-yielding compounds that can be introduced into the Krebs cycle (Fig. 11.4).

One reason for singling out the pathways involving the aromatic amino acids for special mention is that defects in the enzymes catalyzing these pathways have been recognized as being responsible for several inherited diseases in man (Fig. 11.7). A prime example is phenylketonuria (PKU), a recessive disorder in which an enzymatic defect leads to mental retardation. Individuals with PKU lack the hydroxylase that converts phenylalanine to tyrosine. Hence, dietary phenylalanine, when present in concentrations above that needed for protein synthesis, cannot be degraded to Krebs cycle intermediates. The excess phenylalanine accumulates and "overflows" into the normally insignificant pathway for the production of phenylpyruvic acid and other aromatic keto acids. Tests for PKU, now done routinely on newborns, detect the presence of these unusual metabolites in blood or urine. Low-phenylalanine diets started soon after birth can minimize the brain damage of children with PKU.

The reasons why over- and underproduction of metabolites such as certain amino acids or sugars can lead to mental retardation are not known. Since the brain is totally dependent on the bloodstream for nutrition, it is reasonable to suppose that substantial changes in concentrations of blood components may upset the delicate regulatory balances, particularly in the developing brain, and lead to malfunction.

As with other inherited enzyme defects, there is no real cure for PKU; the missing enzyme cannot be replaced. Although early diagnosis of PKU infants and consequent restriction of phenylalanine levels in their diets will reduce the mental retardation, it is perhaps more sensible in the long run to prevent the conception of such children through the identification and counseling of PKU carriers (Sec. 12.10) who are the parents and potential parents of affected children. Carriers, who have lower-than-normal levels of phenylalanine hydroxylase, can be identified as having increased levels of phenylalanine in their blood and a lowered capacity to remove intravenously injected phenylalanine.

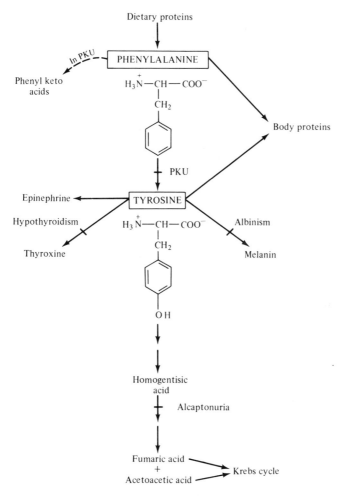

Figure 11.7 Metabolism of phenylalanine and tyrosine in human beings. A slash indicates the enzymatic block in a specific inherited metabolic disease.

Alcaptonuria is another hereditary defect that results from a failure to degrade aromatic amino acids completely; homogentisic acid, a normal intermediate in tyrosine breakdown, cannot be metabolized and is excreted in the urine. On excretion and exposure to air, homogentisic acid is oxidized to a black compound. It is the ink-colored urine, rather than any more serious symptom, that has brought this benign disorder to the attention of physicians and biochemists.

Additional inherited diseases involve other pathways of tyrosine metabolism. Melanin is a pigment in human skin which gives varying degrees of protection against sunlight; a defect in the conversion of tyrosine to melanin leads to albinism. Low levels of the enzymes needed to produce thyroxine from tyrosine are one cause of the hypothyroid diseases such as cretinism (Sec. 14.5).

11.7 Purine and pyrimidine nucleotides

One of the most important groups of small molecules derived from amino acids are the cyclic nitrogen bases, the purines and pyrimidines (Fig. 11.8). All cells synthesize these bases. The nitrogen bases are important components of the nucleotides (ATP, ADP, AMP, GTP, and so on), which act as intracellular energy carriers, as regulatory molecules, and as monomers for the synthesis of the informational macromolecules, RNA and DNA (Table 11.4).

Use of a shorthand notation for the bases will simplify and shorten our discussion. The bases are abbreviated by their first letter: guanine is G, adenine is A, and so on. A *nucleoside* consists of one base attached to a five-carbon sugar, either ribose or deoxyribose. A *nucleotide* is a nucleoside with up to three phosphate groups linked to the sugar residue. Depending on the number of phosphate groups a nucleotide can be a nucleoside monophosphate (NMP), a nucleoside diphosphate (NDP), or a nucleoside triphosphate (NTP). A lower case "d" indicates that the sugar is deoxyribose (e.g., dATP contains deoxyribose, ATP contains ribose).

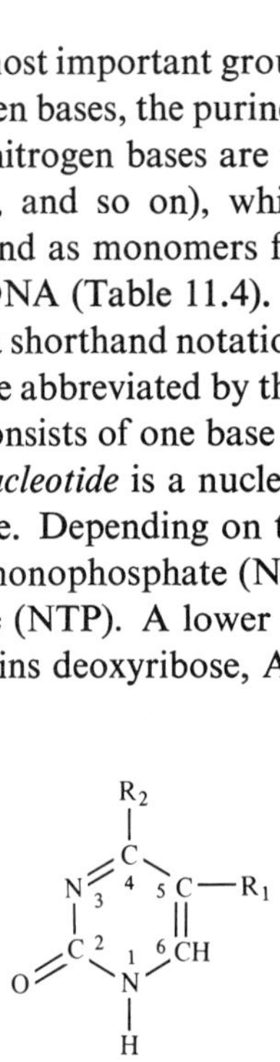

	R_1	R_2
Cytosine	— H	— NH_2
Uracil	— H	= O
Thymine	— CH_3	= O

(a)

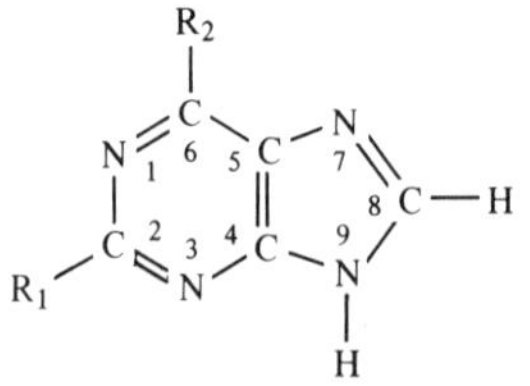

	R_1	R_2
Adenine	— H	— NH_2
Guanine	— NH_2	= O

(b)

Figure 11.8 Structure of the nitrogen bases. (a) Pyrimidines. All pyrimidines have the same six-membered ring structure. The nature of the R groups protruding from the ring distinguishes the different pyrimidines. (b) Purines. All purines consist of the same five-membered ring and a six-membered ring which can have a variety of R groups.

TABLE 11.4 Nucleotides Serve Many Areas of Metabolism

Nucleotide(s)	Role
ATP, GTP, CTP, UTP	Monomers for RNA synthesis (Sec. 13.3)
dATP, dGTP, dCTP, dTTP	Monomers for DNA synthesis (Sec. 12.3)
ATP	Major "middleman" in intracellular energy cycle (Sec. 6.1)
GTP, CTP, UTP, ATP	Activate monomers for macromolecular synthesis (e.g., UTP in glycogen production, Sec. 9.1); GTP supplies energy for protein synthesis
cAMP	"Second message" in hormone action (Sec. 14.5); regulates bacterial RNA polymerase (Sec. 14.1)
ATP	Precursor of enzyme cofactors: NAD^+, $NADP^+$, FAD,[a] coenzyme A

[a] Nicotinamide adenine dinucleotide, nicotinamide adenine dinucleotide phosphate, and flavin adenine dinucleotide.

11.8 The biosynthesis of purine and pyrimidine nucleotides

Each nucleotide is a composite of three distinct parts: phosphate ion(s), a five-carbon sugar, and a purine or pyrimidine base. The order in which these parts are assembled depends on whether the nucleotide contains a purine or a pyrimidine. In both cases, however, the precursor of the pentose–phosphate residue is phosphoribosylpyrophosphate (PRPP) (Fig. 11.9).

Purine nucleotide synthesis (Fig. 11.9) begins with PRPP; the purine ring is built directly onto the ribose residue of PRPP. Nitrogen atoms come from the α-amino groups of glycine and aspartic acid and from the R group of glutamine. The carbon atoms of purines are derived from glycine, CO_2, and methylene groups donated by tetrahydrofolic acid (CH_2—THFA). The first purine nucleotide formed is inosine monophosphate (IMP); IMP is the precursor of the major purine nucleotides, AMP and GMP.

The role of tetrahydrofolic acid as the donor of one-carbon units (e.g., —CH_3, —CH_2—) is not unique to purine synthesis; this carrier molecule shuttles one-carbon units between many diverse anabolic and catabolic pathways. Human cells can modify folic acid but cannot synthesize it from small molecule precursors; we require folic acid which our cells convert to dihydro- and then to tetrahydrofolic acid. Human beings usually obtain sufficient quantities of folic acid from microorganisms inhabiting the intestinal tract, so deficiencies of this vitamin are rare. One step in the *bacterial* pathway for synthesizing folic acid from small molecule precursors is inhibited by the sulfa drugs (Secs. 4.12 and 16.6), and this is the basis for the selective action of these antimicrobial agents.

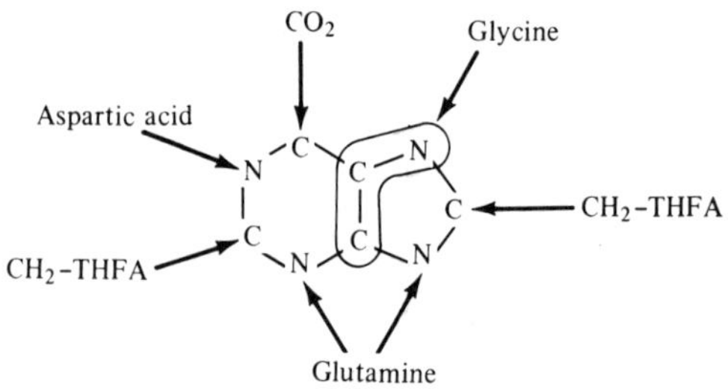

Figure 11.9 Biosynthesis of purine nucleotides. (*a*) Origin of atoms in purine rings. (*b*) Outline of biosynthetic pathway.

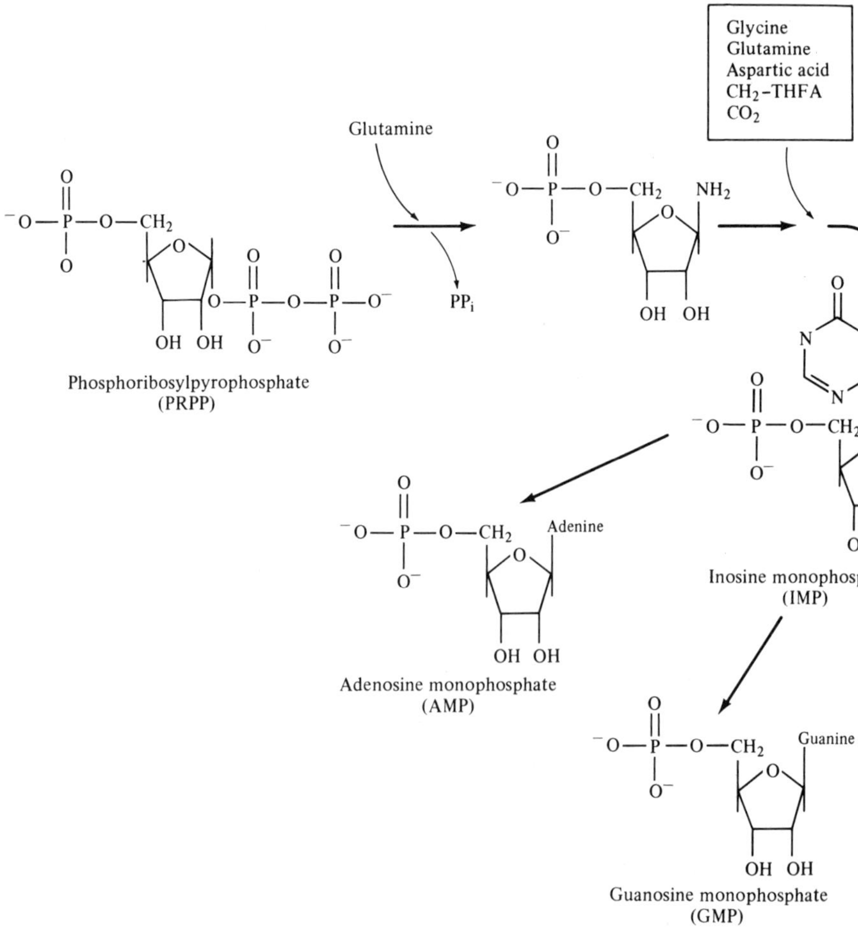

Pyrimidine nucleotide synthesis (Fig. 11.10) begins with the condensation of aspartic acid and carbamoyl phosphate (Fig. 11.5). A free pyrimidine (orotic acid) is formed, which is joined to the ribose residue of PRPP to produce the nucleotide precursor of UMP. UMP is the starting material for the synthesis of the other pyrimidine nucleotides, CTP and dTMP.

Nucleoside monophosphates are converted to the corresponding di- and triphosphates by the stepwise transfer of high-energy phosphate groups from ATP or

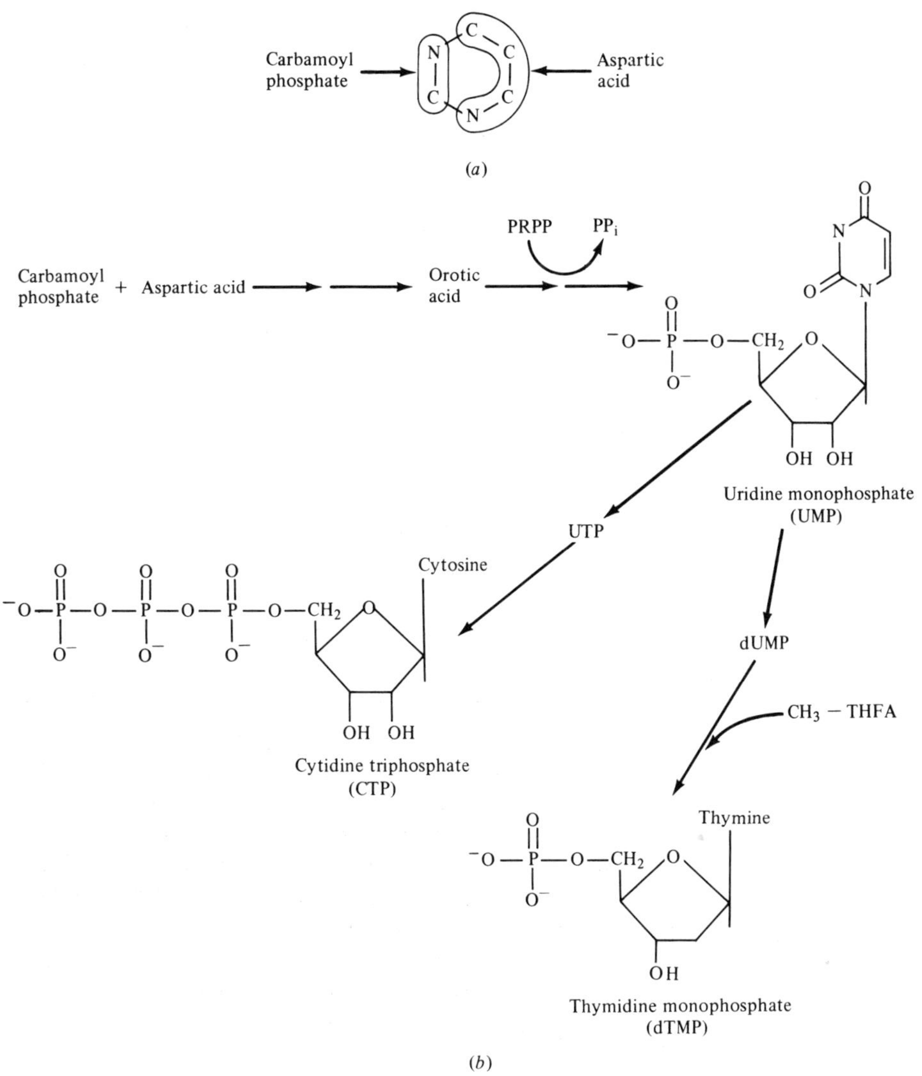

Figure 11.10 Biosynthesis of pyrimidine nucleotides. (*a*) Origin of atoms in pyrimidine ring. (*b*) Outline of biosynthetic pathway.

other nucleoside triphosphates. For example, UMP is converted to UTP by the reactions:

$$\text{UMP} + \text{ATP} \xrightleftharpoons{\text{Kinase}} \text{UDP} + \text{ADP}$$

$$\text{UDP} + \text{ATP} \xrightleftharpoons{\text{Kinase}} \text{UTP} + \text{ADP}$$

These exchange reactions are reversible and can occur between most pairs of deoxyribo- and ribonucleotides with different phosphorylation levels. These exchange reactions do not result in a net increase of high-energy phosphate bonds; the number of such bonds is increased when ATP is formed by substrate-level phosphorylation, as in glycolysis, or by the oxidative phosphorylation of the electron transport chain.

To synthesize DNA, the cell needs deoxyribo- rather than ribonucleotides. In the three cases where the bases are common to both RNA and DNA, the ribonucleoside diphosphates proved to be the precursors of the deoxyribonucleoside diphosphates. dADP, dGDP, and dCTP are formed by the reduction

NDP + NADPH + H$^+$

dNDP + NADP$^+$ + H$_2$O

The fourth DNA precursor, thymidine monophosphate, is made by the methylation of dUMP (Fig. 11.10); the methyl group is donated by CH$_3$–THFA. Cells never make thymidine ribonucleotides, which might be mistakenly incorporated into RNA.

Since all four dNTPs are needed for DNA synthesis (Sec. 12.3), inhibition of the formation of dTMP from dUMP will indirectly, but quickly, stop DNA synthesis. Several drugs used in the treatment of leukemia and other cancers act by inhibiting dTMP formation, the rationale being that the faster growing cancer cells, with their very rapid DNA synthesis, will be inhibited by the shutdown of dTMP production before their slower-growing, noncancerous companions are affected. One set of these drugs are the halogenated pyrimidines, particularly 5-fluorouracil* and its derivative, fluorodeoxyuridine; these base analogs are inhibitors of the enzyme that methylates dUMP. A second group of drugs, including aminopterin and amethopterin, are

---

* 5-Fluorouracil has a fluorine atom, rather than hydrogen, on carbon 5 of the pyrimidine ring (Fig. 11.8); this is the carbon that carries the additional methyl group of thymine.

analogs of dihydrofolic acid and block its conversion to tetrahydrofolic acid; without THFA to act as a methyl donor, dUMP cannot be converted to dTMP.

Cells make purine and pyrimidine nucleotides as monomers for RNA and DNA synthesis and as intracellular energy carriers. The role of nucleotides as precursors of the enzyme cofactors, NAD^+, $NADP^+$, FAD, and coenzyme A, is also critical to cell metabolism. The biosynthesis of the hydrogen carrier, NAD^+, is of particular interest because it depends upon both the production of ATP and the metabolism of the amino acid tryptophan. NAD^+ has the structure

$$\text{From ATP} \begin{cases} \text{Adenine} \\ \text{Ribose} \\ \text{Phosphate} \end{cases}$$

$$\text{From nicotinate ribonucleotide} \begin{cases} \text{Phosphate} \\ \text{Ribose} \end{cases} \text{Nicotinamide}$$

NAD^+

This cofactor is made by linking two nucleotides, ATP and nicotinate ribonucleotide. The nicotinate portion of the latter nucleotide is synthesized from nicotinic acid (niacin), which is derived from tryptophan.

Human beings can synthesize sufficient quantities of niacin from tryptophan if their diet is high enough in this essential amino acid. However, if dietary tryptophan is low, niacin is required. If the diet is low in *both* tryptophan and niacin, the individual may develop the vitamin-deficiency disease, pellagra, which is characterized by dermatitis, diarrhea, and mental disturbances.

11.9 The breakdown and excretion of purines and pyrimidines

Nucleotide degradation begins with the cleavage of the mononucleotide into ribose-1-phosphate and the free base. Ribose-1-phosphate can be converted back to PRPP and used again in the biosynthesis of nucleotides. The free bases have two possible metabolic fates; they are either (1) salvaged for reuse in nucleotide synthesis, or (2) degraded and excreted.

In salvage pathways, the bases are rejoined to PRPP, forming a new nucleotide. For example,

$$\text{guanine} + \text{PRPP} \xrightarrow{\text{Transferase}} \text{GMP} + \text{PP}_i$$

The importance of salvage pathways is illustrated by Lesch-Nyhan disease, an inherited condition characterized by gout, mental deficiency, spasticity, aggressiveness, and self-mutilation (patients literally gnaw away their fingers and lips). Lesch–Nyhan patients lack the transferase which salvages free purines by joining them to PRPP.

Patients have increased serum levels of uric acid, which explains their gout, and of PRPP. There is also an unexplained, but presumably secondary, increase in the levels of several enzymes of purine biosynthesis. It is not known which of the enzymatic lesions are directly responsible for the psychiatric and neurological symptoms of the disease.

The pathway for the degradation and excretion of the free bases depends on whether the base is a purine or a pyrimidine. The six-membered ring of pyrimidines is opened and degraded to NH_4^+, CO_2, and succinyl CoA, which enters the Krebs cycle. In contrast, the ring structure of purines may not be destroyed; purines are converted to uric acid, which, in primates, is excreted in the urine.

$$\underset{\text{Uric acid}}{\begin{array}{c}\end{array}}$$

In certain species, including birds, uric acid formation is the major route for nitrogen excretion; excess amino acid nitrogen is used for purine biosynthesis and then the purines are degraded to uric acid. In lower animals, uric acid may be further degraded to urea or to NH_4^+ and CO_2.

When a human being produces uric acid faster than it can be excreted, the individual may develop gout. When blood levels of uric acid are high, sodium urate crystals are deposited in the joints, triggering painful arthritis; deposition of these crystals in the kidney damages that organ. The causes of gout are numerous. Some patients have an inherited enzymatic defect in purine biosynthesis, leading to an overproduction of purines; low levels of purine salvage enzymes also lead to gout, as in Lesch-Nyhan disease. Other causes of gout may be a decreased capacity to transport uric acid in the serum or to secrete it from the kidneys. Although cartoonists have long associated gout with overindulgence, dietary excesses are not the cause of this disease.

SUMMARY

The major elements in organic compounds are carbon, hydrogen, oxygen, and nitrogen. Nitrogen is unique in that it is not readily available in forms that can be used by animals and plants. Plants depend on bacteria or fertilizer factories to convert atmospheric nitrogen into compounds that they can incorporate into metabolic intermediates. Plants are eventually eaten by animals. The nitrogen cycle is completed when microorganisms convert (decay) the organic nitrogen in plant and animal debris back into volatile inorganic compounds.

Within the cell, amino acids are at the center of nitrogen metabolism. Ammonia (inorganic) is incorporated into glutamic acid (organic), the starting compound for many metabolic pathways, including the production of other amino acids. Organisms differ in their capacity to synthesize individual amino acids; human beings, for example, can make only half of the 19 amino acids required for protein synthesis.

The Metabolism of Amino Acids and Other Nitrogen Compounds

    Amino acid degradation involves the removal of the amino group from the carbon skeleton. The latter usually enters the Krebs cycle, providing cell energy. The nitrogen of amino acids is used either for further amino acid synthesis or is excreted. The mode of nitrogen excretion varies among organisms; human beings excrete primarily urea.

    Amino acids are the monomers for protein synthesis. In addition, they are the precursors of a host of nitrogen-containing metabolites, including porphyrins, purines and pyrimidines and several hormones. Purine and pyrimidine synthesis is crucial for all cells, since these bases are needed for nucleotide and nucleic acid production.

PRACTICE PROBLEMS

1. Rework the Practice Problems following Chapter 2.

2. A rat was fed a diet containing ^{14}C-alanine for 1 hour. Then the rat was killed and its liver removed and analyzed. Which of the following compounds, all found in the liver, would be radioactively labeled? Albumin, citric acid, a bile acid, tryptophan, a 16-carbon fatty acid, phenylalanine hydroxylase, vitamin B_{12}, cytochrome *c*.

3. Give a biochemical explanation for the following:
   (a) Although not a dietary disorder, gout symptoms can sometimes be alleviated by reducing protein intake.
   (b) Individuals with PKU often have light-colored skin and hair.
   (c) Protease inhibitor eaten with every meal will lead to death.
   (d) A deficiency of vitamin B_6, the dietary precursor of pyridoxal phosphate, would be expected to decrease an individual's production of urea.

4. Which of the following compounds, directly or indirectly, contribute nitrogen atoms to urea in humans? Ammonia, glutamic acid, glycogen, casein, ATP, heme. Where appropriate, outline the pathways by which this occurs.

5. A tribe of cannibals, wishing to dominate the other peoples of the world, released large quantities of nitrogenase inhibitor. After a few generations, they found that there was no one left to dominate and that their own population was decreasing. Why did their scheme fail?

6. Oxygen is required to maintain urea production in isolated human liver slices. Why?

7. In terms of evolution, why is it significant that human cells use only nonessential amino acids to synthesize purines and pyrimidines?

8. Why can't thymine nucleotides be incorporated into RNA?

9. Many compounds are intermediates in more than one metabolic pathway. By comparing Figs. 11.4 and 11.5 and Sec. 9.5, find at least one intermediate common to the pathways of:
   (a) The Krebs cycle and the urea cycle.
   (b) The urea cycle and pyrimidine synthesis.
   (c) Glycogen synthesis and PRPP synthesis.
   (d) The Krebs cycle and porphyrin synthesis.
   (e) Purine synthesis and protein synthesis.

10. A nitrogen-fixing bacterium is growing anaerobically on a medium in which glucose is the major nutrient. How many molecules of glucose must this cell catabolize to produce the energy to convert six molecules of nitrogen gas to ammonia?

11. Explain how an intracellular NADPH deficiency can lead to an inhibition of DNA synthesis.

SUGGESTED READING

AXELROD, JULIUS, "Neurotransmitters," *Sci. American*, 230, No. 6, p. 58 (1974). Nitrogen-containing compounds are the chemical messengers by means of which nerve cells communicate. Many psychoactive drugs interfere with neurotransmitter activity.

BRILL, WINSTON J., "Biological Nitrogen Fixation," *Sci. American*, 236, No. 3, p. 68 (1977). Certain procaryotic cells have a complex enzyme system that reduces atmospheric nitrogen to ammonia. Laboratory manipulations of these organisms promise to increase world crop production.

DELWICHE, C. C., "The Nitrogen Cycle," *Sci. American*, 223, No. 3, p. 136 (1970). Atmospheric nitrogen must be combined with hydrogen or oxygen before it can be assimilated by plants, which are later consumed by animals. Both biological and industrial processes are currently used to fix nitrogen; this technology has been so successful that the disposal of nitrogen compounds is becoming a serious problem.

HARPSTEAD, DALE D., "High-Lysine Corn," *Sci. American*, 225, No. 2, p. 34 (1971). New breeds of corn may reduce the incidence of human protein-deficiency malnutrition, particularly where corn is the principal staple.

KERMODE, G. O., "Food Additives," *Sci. American*, 226, No. 3, p. 15 (1972). Many nitrogen compounds are used to enhance the flavor or color, extend the shelf life, or protect the nutritional value of processed foods. A helpful guide for interpreting the ingredients labels on convenience foods.

LONSDALE, KATHLEEN, "Human Stones," *Sci. American*, 219, No. 6, p. 104 (1968). Uric acid (or its salts) sometimes precipitates from acidic urine and forms stones in the organs of the urinary tract.

MACALPINE, IDA, and RICHARD HUNTER, "Porphyria and King George III," *Sci. American*, 221, No. 1, p. 38 (1969). The fascinating story of how misdiagnosis of the disease of a British monarch affected world history.

ROSS, RUSSELL, "Wound Healing," *Sci. American*, 220, No. 6, p. 40 (1969). Collagen synthesis, an important step in wound healing, requires vitamin C.

SCRIMSHAW, NEVIN S., and VERNON R. YOUNG, "The Requirements of Human Nutrition," *Sci. American*, 235, No. 3, p. 50 (1976). Humans require nine essential amino acids, which must come from dietary proteins. This article describes the proteins in various foods and shows which staples complement each other to provide an adequate protein intake.

YOUNG, VERNON R., and NEVIN S. SCRIMSHAW, "The Physiology of Starvation," *Sci. American*, 225, No. 4, p. 14 (1971). In the initial stage of starvation, amino acids are used as energy sources; as starvation continues, the retention of body proteins becomes a major goal of metabolism.

12 Deoxyribonucleic Acid: the Genetic Material

Deoxyribonucleic acid (DNA) is the primary information-carrying molecule in living cells. A cell's DNA defines which other macromolecules (e.g., RNAs, proteins) the cell can produce, thus determining whether the cell is a bacterium or part of an animal or plant. Precise replication of a cell's DNA allows this same information to be passed from one generation to the next.

In DNA, perhaps more than in any other macromolecule, we will see the relationship between structure and function. The base sequence of this long, double-stranded helix carries the information for the amino acid sequences of thousands of different proteins. In addition, the structure of DNA is such that exact copies can be made by the mechanism of semiconservative replication.

12.1 The structure of DNA: single strands and the double helix

A single chain, or strand, of DNA contains a linear backbone of alternating sugar and phosphate groups [Fig. 12.1(a)]. The five-carbon sugar, deoxyribose, has a ring structure with hydroxyl groups at the 3′ and 5′ carbon atoms*; in the DNA backbone,

---

* Read 3′ and 5′ as "three-prime" and "five-prime."

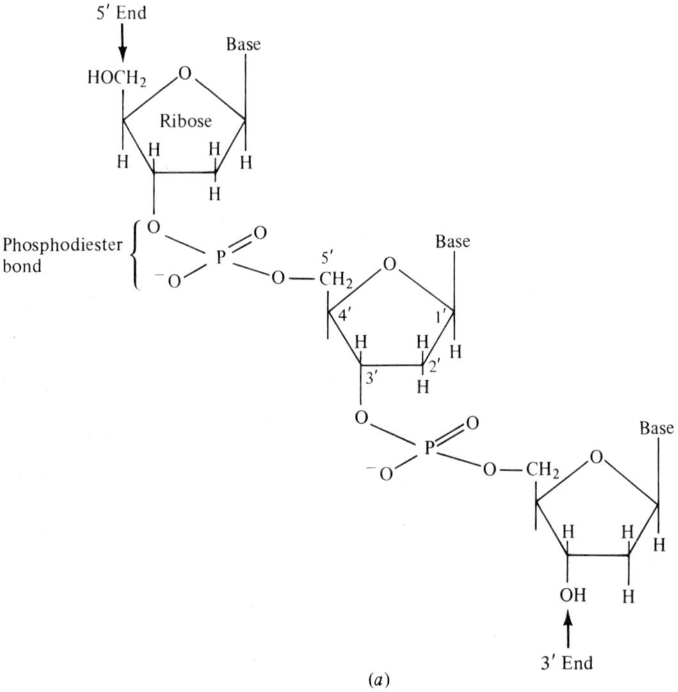

Figure 12.1 Structure of DNA. (*a*) Each strand of DNA consists of a linear sugar–phosphate backbone, with a nitrogenous base attached to each deoxyribose unit. (*b*) The bases in DNA are the purines, adenine and guanine, and the pyrimidines, thymine and cytosine.

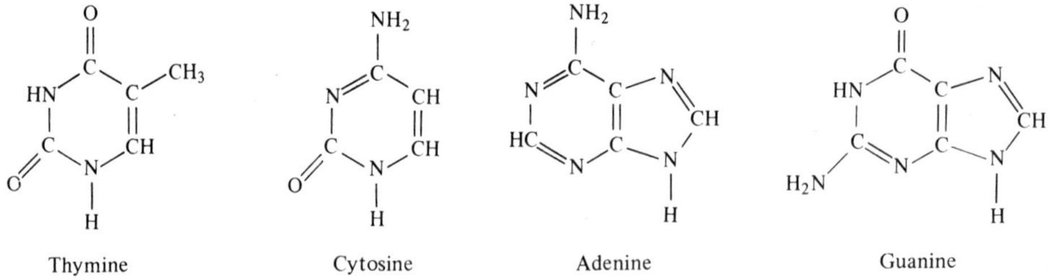

the hydroxyl groups on adjacent molecules are linked together by phosphate bridges (phosphodiester bonds). The ends of this linear chain are chemically different; at one end, the deoxyribose residue has a free 3′ hydroxyl group, while at the opposite end, the sugar has a free 5′ hydroxyl group. These are called the 3′ and 5′ ends of the DNA molecule, respectively.

Heterocyclic nitrogen-containing bases are attached to every sugar residue in the DNA backbone. The base can be a purine, either adenine (A) or guanine (G), or a pyrimidine, either thymine (T) or cytosine (C) [Fig. 12.1(b)]. The metabolism of these bases is discussed in Chapter 11.

Single strands of DNA are often found in biochemist's freezers, but rarely in living cells. *In cells, DNA exists as a two-stranded molecule: the double helix.* To form a double-helical structure, two strands of DNA must line up ("pair") and wind around each other in a very precise and orderly manner. The two strands are aligned and held together by hydrogen bonding (Watson–Crick base pairing) between the bases on opposing strands. As we learned when studying proteins (Sec. 2.4), because of the charged character of amino and carbonyl groups, the hydrogen of an amino group is attracted to the oxygen of a carbonyl group; this attraction is called a *hydrogen bond*. Both purines and pyrimidines have amino and carbonyl groups that can form hydrogen bonds, but because of structural limitations, only two kinds of stable base pairs can be formed; both contain one purine and one pyrimidine [Fig. 12.2(a)]. Adenine always pairs with thymine (AT pair) and guanine always pairs with cytosine (GC pair). This base pairing is not only essential to DNA structure but is the key to the exact replication of DNA (Sec. 12.3), and to the production of messenger RNA (Sec. 13.3). Base pairing of this type is also responsible for the translation of the genetic code in protein synthesis (Sec. 13.4).

To envision the double helix, consider the two DNA strands with their paired bases [Fig. 12.2(b)]. The strands are antiparallel; that is, they run in opposite directions so that the 3' end of one strand is next to the 5' end of its sister strand. The flat base pairs connecting the backbones are stacked in planes more-or-less perpendicular to the long axis of the molecule. A helix is formed when this "ladder" structure is twisted about itself, as would happen if the bottom end of a flexible ladder were held stationary while the top end was rotated in a right-handed direction. The result is a structure reminiscent of a many-storied circular staircase [Fig. 12.2(b)]. Double-helical staircases can be found in several palaces in Europe; however, they are usually left-handed.

The enormous size and the long, threadlike shape of DNA molecules must be emphasized. Most DNA molecules are several hundred times bigger than proteins, ribonucleic acids, or any other macromolecules. For example, the DNA of a mammalian cell contains 5.5×10^9 base pairs. If all the DNA molecules in such a cell were lined up end to end, the resulting "thread" would be 2 meters long! Since it is only about 10^{-9} meters between sister strands in the double helix, the thread would be 2,000,000,000 times longer than it was wide. Even taking into account that this DNA exists in 30–40 pieces (one for each chromosome), the length/width ratio is still impressive. The asymmetry of DNA molecules makes them difficult for biochemists to study, for they are easily broken during laboratory manipulations. Even when they can be isolated intact, determining the linear sequence of so many bases is a formidable task.

Studies of the base composition of DNA have revealed much useful information about this material:

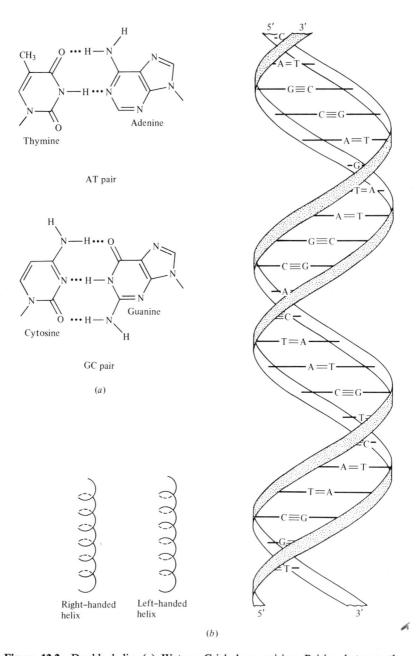

Figure 12.2 Double helix. (*a*) Watson–Crick base pairing. Pairing between the bases in opposing strands of the DNA molecule aligns the two strands and stabilizes their partnership. Only two types of base pairs are permitted in DNA: adenine–thymine pairs, which contain two hydrogen bonds, and guanine–cytosine pairs with three hydrogen bonds. (*b*) Two complementary strands of DNA wind about each other to form the right-handed double helix. The strands are antiparallel. [(*a*) After Stuart J. Edelstein, *Introductory Biochemistry*, Holden-Day, Inc., San Francisco, 1973, p. 26. (*b*) After Stuart J. Edelstein, *Introductory Biochemistry*, p. 24.]

1. The DNA content of every cell of a given animal or plant is identical. Every cell of a given organism contains the same amount of DNA,* and its base composition does *not* vary from cell to cell. DNA from our skin cells is just like our muscle cell DNA, indicating that differentiation does not involve gross qualitative or quantitative alterations in the genetic information.

2. For all DNA molecules, regardless of their source, the amount T always equals the amount A, while the amount G always equals that of C. Once the double helical structure was identified, these equivalences were readily explained by the fact that each base in a DNA strand is paired with ("complemented by") a base in its sister strand; for every A in one strand, there is a T opposite it in the sister strand; for every G, there is a C; and so on.

3. While the percentage of A always equals T and the percentage of G always equals C, the absolute amount of each base pair varies significantly among different species. As examples, total human DNA contains 58% AT pairs and 42% GC pairs. DNA from the bacterium *Pseudomonas aeruginosa* contains 33% AT pairs, and that from wheat 55% AT pairs. The reasons for these variations in base-pair composition are not yet clear; the genetic code (Sec. 13.9) is the same for all organisms. Part of the explanation for these differences comes from the fact that a number of DNA molecules contain long tracts of repeated sequences. The variations must have evolutionary significance; the base-pair ratios of taxonomically related organisms are often similar, and base compositions of DNA can be used as a tool for classification.

12.2 The DNA of higher organisms is organized into chromosomes

In procaryotic cells and viruses, the genetic material usually exists as a single nucleic acid molecule, which is not associated with specific proteins or more complex structures. The DNA of procaryotic cells usually exists as one double helix of DNA; often the ends are joined together to form a circular molecule. Some viruses (Chap. 15), which are intracellular parasites, have exceptional kinds of genetic material; although many viruses contain double-stranded DNA, in others single-stranded DNA, or even RNA, serves as the genetic material.

The situation is more complicated in eucaryotic cells, in which double-stranded DNA molecules surrounded by specific proteins, form the chromosomes (Fig. 12.3). Although these nuclear structures have been well studied by cytologists, we know surprisingly little about their macromolecular organization or how protein–DNA interactions relate to chromosome behavior during cell growth and division. The

---

* Gametes (eggs and sperm) are the exception to this rule; they contain half the amount of DNA found in body (somatic) cells.

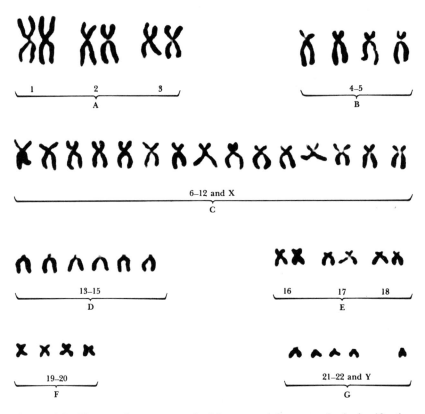

Figure 12.3 Human chromosomes (male) arranged in a standard classification known as a karyotype. Such pictures are made by arresting cell division at metaphase, staining the cells, and then photographing appropriate cells under the microscope. Individual chromosomes are then cut from the photographic print, matched into pairs, and arranged by size. (After V. A. McKusick. *Human Genetics*, 2nd ed., Prentice-Hall, Inc., Englewood Cliffs, N.J. 1969.)

number of DNA molecules inside each eucaryotic chromosome is not known for sure, but evidence is mounting that one uninterrupted DNA molecule runs the entire length of a chromosome. From a comparison of the length of DNA molecules to the size of chromosomes, it is obvious that the DNA must be folded and compacted inside the chromosome, but the nature of this packing is unclear.

Information on chromosomal proteins is also sparse. The best-studied nuclear proteins are the histones; histones are characterized by their positive charge,* which

---

* The amino acid content of these proteins is unusual; lysine and arginine account for about 25% of the amino acids in histones.

causes them to bind to the negatively charged phosphate groups of the DNA. Nuclei contain only a few chemically different histones, but since there are many copies of each histone per nucleus, histones make up a large proportion of the total nuclear protein. Histones probably play a role in the folding of DNA and its packaging into chromosomes; recent work suggests that chromosomes consist of groups of histones ("beads") occurring at regular intervals along the DNA "string." A heterogeneous group of nonhistone nuclear proteins is also present, but studies of these molecules are in their infancy.

Why do we know so little about the chemistry of chromosomes? It should be possible to isolate individual chromosomes, identify their DNA and protein components, and then study how these nucleoproteins are assembled. After all, this has been done for ribosomes, which are also nucleoproteins. The biochemical problems

Figure 12.4 Possible modes of replication of DNA molecules. A basic requirement of DNA replication is that the parent molecule should not be destroyed during the copying process. Both conservative and semiconservative replication meet this requirement. Experiments have shown that all cells replicate their DNA by semiconservative replication. (Reprinted with permission of Macmillan Publishing Co., Inc. from *Genetics* by Monroe W. Strickberger. Copyright © 1968, Monroe W. Strickberger.)

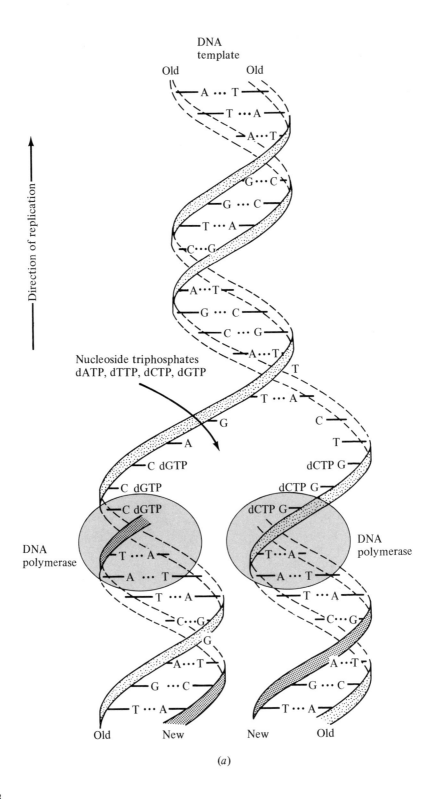

(a)

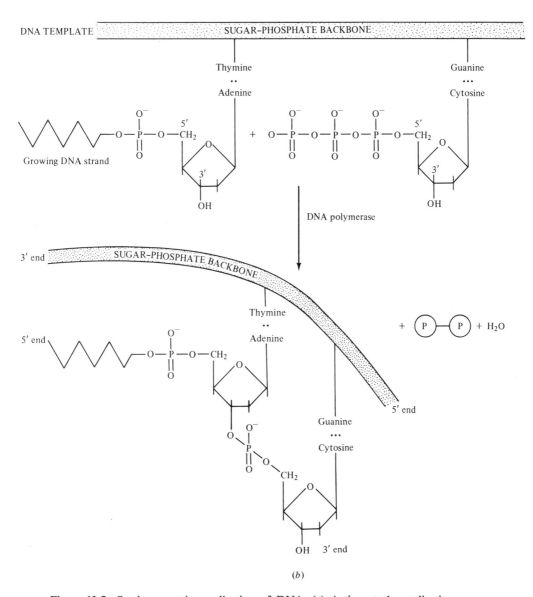

Figure 12.5 Semiconservative replication of DNA. (*a*) Action at the replicating fork. Both strands of the parental double helix are copied simultaneously. Once the (parental) double helix is unwound, Watson–Crick base pairing positions the incoming complementary nucleoside triphosphates along the (parental) DNA template. The dNTPs are added to the growing strand one at a time, in a reaction catalyzed by DNA polymerase. In eucaryotic cells, there are usually many sites of replication along the DNA in each chromosome. In this situation, the replicating fork is at one end of a "bubble" in the DNA molecule (Fig. 12.7).

(*b*) Polymerization reaction. The incoming nucleoside triphosphate is added to the free 3′ hydroxyl of the growing chain. The nucleotide at the 5′ end of the new strand is laid down first and polymerization proceeds toward the 3′ end; this is called synthesis in the 5′ → 3′ direction. As each nucleoside triphosphate is added, one molecule of pyrophosphate is released.

stem from the fact that there is only one copy of each chromosome per cell and it is a very tiny particle compared to the whole cell. It is difficult to isolate intact nuclei and to obtain a large amount of an individual chromosome.

12.3 Semiconservative replication of DNA depends on Watson–Crick base pairing

For genetic information to pass from a cell to its daughter cells, the DNA of the parent cell must be copied (replicated) prior to cell division. The mechanism of DNA replication must be foolproof (Fig. 12.4); just a few copying errors could be lethal to the cell line. The necessary accuracy is achieved by the process of *semiconservative replication*; this mode of replication was a prediction of the original Watson–Crick proposal for the structure of DNA.

DNA replication proceeds by the synthesis of complementary copies of preexisting DNA strands; that is, each DNA molecule acts as a pattern or template for its own replication. Single strands of DNA are not degraded during replication; sugar–phosphate bonds are rarely broken, and the bases are not removed from the backbone.

Replication (Fig. 12.5) begins with the separation of the two strands of the parental DNA molecule; as the double helix unwinds, the base pairs separate and the once-internal bases are exposed. Copying proceeds by Watson–Crick base pairing between the bases of an individual parental strand and incoming nucleoside triphosphates* (dATP, dGTP, dTTP, or dCTP). All four nucleoside triphosphates must be present for replication to occur. As the nucleoside triphosphates pair with their appropriate partners along the template strand, sugar–phosphate bonds are formed enzymatically between the adjacent nucleotides in the new (daughter) strands. In eucaryotes, replication takes place simultaneously at many well-separated places along the DNA; for *Drosophila*, it is estimated that there are 6000 sites of replication (replicating forks) per DNA molecule (i.e., per chromosome).

The process of base pairing and polymerization continues until a complementary copy of the entire parental strand has been synthesized. Both strands of the parental helix are replicated simultaneously; that is, replication gives rise to two identical DNA molecules. Each parental strand and its complement form a new double helix, which, in eucaryotes, interacts with proteins to form a chromosome; some of these associations may occur during the process of replication, before the new molecule is complete.

All the chromosomes in the eucaryotic nucleus are replicated at the same time, so the final result is a doubling of the cell's chromosome number. Chromosome doubling sets the stage for mitosis, when, prior to cell division, the "matched"

---

* A deoxynucleoside triphosphate (dNTP) has the general structure P—P—P—deoxyribose-base. The bases of the dNTPs used for DNA synthesis are thymidine, adenine, guanine, and cytidine.

chromosomes pair and then separate by movement along the microtubules of the spindle (Table 6.1) until two complete and well-separated sets of chromosomes are present in the parental nucleus. When the cell divides, each daughter cell receives one set of chromosomes. Cell divisions in which the chromosome number is unchanged are responsible for the growth of higher organisms and for the proliferation of cultures of microorganisms; changes in chromosome number occur only during the sexual cycle (Sec. 12.10).

12.4 The many enzymes of DNA replication, repair, and recombination

You may have found it strange that our discussion of DNA replication contained no detailed description of the enzymes involved, although they must be needed to form the sugar–phosphate bonds, at the very least. Our understanding of the enzymology of replication is complicated by two factors.

1. The enzymes of DNA replication have been well studied only in bacteria and viruses; even these simple organisms use several different enzymes and non-catalytic proteins to copy their DNA (Sec. 12.5). One can only guess at the additional biochemical complexity of achieving synchronous replication of the 10–50 chromosomes in a eucaryotic cell.

2. Cells contain many enzymes which can catalyze the cleavage or reformation of the sugar–phosphate bonds along the DNA backbone; not all of these enzymes are involved in replication. Different enzymes with similar catalytic activities (Table 12.1) may function in DNA replication, in the repair of damaged DNA, and/or in recombination between homologous DNA molecules. Both repair and recombination are essential processes requiring breakage and subsequent fusion of DNA molecules. Both must occur rapidly and accurately so that the cell's information content is not lost or turned into "gibberish."

Radiation (e.g., x-rays, ultraviolet light) and chemicals (e.g., nitrous acid, nitrosoguanidine, and many carcinogens) damage DNA by breaking the backbone and/or altering the bases chemically. Repair enzymes mend the backbone and "correct" the base sequence by excising the incorrect nucleotide or sequence and replacing it with the correct one.

In recombination, homologous segments of DNA are exchanged between molecules; this occurs when two copies of a particular chromosome are present in the same nucleus. The enzymes of recombination break the two DNA molecules and then join the exchanged pieces together.

Replication, repair, and recombination are clearly different functions in terms of what the cell is trying to achieve, but they are very similar in enzymatic terms. Hence,

TABLE 12.1 Enzymes of DNA Replication, Repair, and Recombination

Three major classes of enzymes work to build or degrade DNA. These enzymes all operate on sugar–phosphate bonds; they *cannot* work on the bonds joining the bases to deoxyribose. All cells contain several enzymes in each class.

Enzymes	Substrate(s)	Reaction
DNA polymerases	dATP, dGTP, dTTP, dCTP, DNA template	Form sugar–phosphate bonds in growing DNA strand (Fig. 12.5)
Deoxyribonucleases Exonucleases	Intact DNA	Break sugar–phosphate bonds Cleave nucleoside monophosphate from end of DNA strand
Endonucleases		Break P—O bonds in middle of DNA strand
DNA ligases	Double-stranded DNA with a break in the backbone of one strand	Fuse broken molecule by forming sugar–phosphate bond

if one has isolated an enzyme that catalyzes sugar–phosphate bond formation, it is difficult to assign the enzyme to one of the three functions.

The importance of DNA repair enzymes can be illustrated by a discussion of the rare inherited skin disease, xeroderma pigmentosum. The skin of affected individuals is abnormally sensitive to the ultraviolet component of sunlight; on exposure, the outer layer of skin atrophies and eventually multiple skin cancers develop. Ultraviolet light causes chemical reactions between adjacent thymine bases along the DNA backbone; the DNA is nonfunctional unless these modified bases are removed. Individuals with xeroderma pigmentosum lack an enzyme needed to excise the altered nucleotides and replace them with unaltered thymidines. It is of interest to note that in any of us, when lying in the sun, DNA is being repaired at a higher rate than usual; that "healthy, glowing" tan may not contribute to your longevity!

The controlled use of certain DNA processing enzymes has recently stirred controversy, for it has become possible to join together genes of unrelated organisms in the test tube. For example, toad genes have been inserted into bacterial DNA, which allows them to replicate inside bacteria. These experiments (Fig. 12.6) rely upon the cleavage of isolated DNA from both species by the same endonuclease. Because the endonuclease cleaves DNA backbones only at specific base sequences, all the DNA fragments produced by this enzyme have similar ends and can be joined together by ligases to form a DNA molecule containing both toad and bacterial genes. The hybrid molecule is then introduced into a bacterium, where it is replicated as if it were a normal bacterial gene. There are many potential benefits from this recombinant DNA research, ranging from the treatment of genetic diseases to the production of rare hormones and the introduction of bacterial nitrogen-fixation genes into plants to decrease the need for fertilizer. Opponents of this type of research argue that there is a

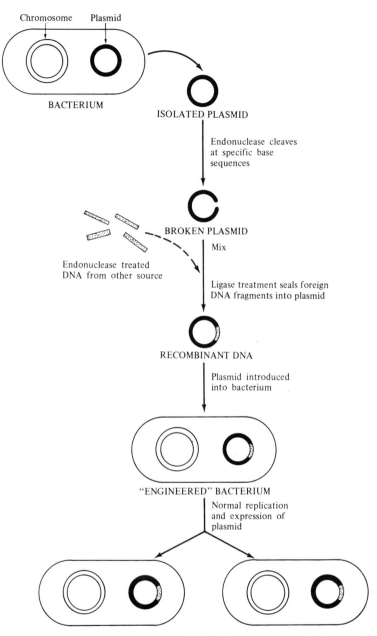

Figure 12.6 Production of recombinant DNA. The genetic material of a bacterium is mostly found in a very large, circular chromosome that is difficult to recover intact. Some bacteria also carry plasmids, small circles of DNA that are copied and passed on to the daughter cells in the same way that chromosomal DNA is replicated.

Plasmid DNA molecules can be purified intact, enzymatically broken and resealed (with or without additional pieces of DNA), and then reintroduced into a bacterial host. This advanced technology allows biochemists to insert nonbacterial genes into plasmids and to create strains of bacteria that can produce copies of foreign genes and foreign gene products.

finite possibility that new pathogenic species could be produced if, for example, genes for cholera toxin, antibiotic resistance, or tumor production were introduced into a currently nonpathogenic, antibiotic-sensitive bacterium.

12.5 DNA replication in bacteria

In Section 12.3, DNA replication was described as occurring by the unwinding of the parental template DNA, the lining up of the complementary nucleoside triphosphates, and the formation of sugar–phosphate bonds by DNA polymerase. While this scheme is correct in outline, studies of the enzymology of DNA synthesis in bacterial systems have shown that semiconservative replication is actually a much more complex process than it might appear. Some of the nuances of DNA synthesis in bacteria are discussed below. The components of bacterial DNA replication are listed in Table 12.2.

(a) DNA SYNTHESIS REQUIRES A PRIMER

The known DNA polymerases cannot start to make a DNA molecule simply by joining the first two dNTPs; rather, they require an initiator molecule, a "primer," to which the new nucleotides are added. In some DNA viruses and possibly in bacteria, the primer appears to be a short RNA molecule (about 100 nucleotides or less). Once the primer is in place at the end of the DNA template, the complementary dNTPs line up and sugar–phosphate bonds are formed by DNA polymerases. Eventually, the RNA primer is removed, leaving only deoxynucleotide residues in the chain.

(b) DNA REPLICATION MUST PROCEED IN TWO DIRECTIONS
    SIMULTANEOUSLY

The two strands of the double helix are antiparallel, yet corresponding segments of both strands are copied simultaneously. Hence, replication must proceed in the

TABLE 12.2 Components of DNA Replication in Bacteria

Component	Role in replication
dATP, dGTP, dTTP, dCTP	Monomers for new DNA
DNA template	Model for new DNA
RNA primer	Initiating material for new DNA strand; removed after DNA is made
DNA polymerase III	Link dNTPs in growing DNA strand
DNA polymerase I	Correct rare mistakes in base sequence
DNA polymerase II	?
Unwinding protein	Separates strands of DNA helix at replicating fork
Cell membrane	Probable site of replication

$3' \to 5'$ direction along one parental strand and in the $5' \to 3'$ direction along the other strand. This became a paradox when enzymologists found that the known DNA polymerases all catalyze synthesis in one direction, from the $5'$ to the $3'$ end of the strand [Fig. 12.5(b)], and cannot work in the opposite direction. This enigma seems to have been resolved with the finding that, on one of the strands, DNA is synthesized in short fragments which are instantly joined by DNA ligase (Table 12.1 and Fig. 12.7). For synthesis to occur in the $3' \to 5'$ direction, DNA polymerase must, in effect, work in the $5' \to 3'$ direction, get off the template, and start again at the newest opening of the replicating fork.

(c) BACTERIAL CELLS CONTAIN THREE DNA POLYMERASES

Studies of the catalytic activities of isolated polymerizing enzymes have not elucidated their individual roles in bacterial metabolism. Three polymerases (I, II, and III) catalyze the attachment of dNTPs to the growing chain in the $5' \to 3'$ direction; all three require both a template and a primer. Isolation of bacterial mutants (Sec. 12.8) in which the level of one of the polymerases is very low has been more revealing. Cells with low levels of DNA polymerase III cannot make copies of their DNA, so this enzyme must be essential for replication. Cells with very low levels of DNA polymerase I are able to replicate their DNA at a normal rate but cannot cope with damage

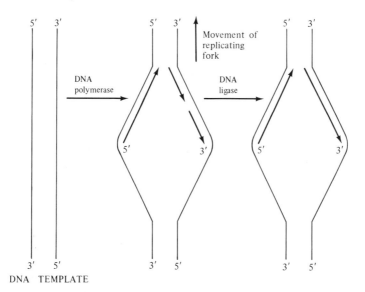

Figure 12.7 Directionality of DNA replication. At the replicating fork, DNA polymerase synthesizes short pieces of DNA in the $5' \to 3'$ direction. Fragments paired to the same template strand are then joined by DNA ligase. The net result is that corresponding segments of both template strands are made simultaneously.

caused by ultraviolet light or chemicals. For these reasons, DNA polymerase I has been assigned a role in DNA repair. The function of DNA polymerase II is still uncertain.

(d) NONCATALYTIC PROTEINS ARE NEEDED FOR REPLICATION

A protein has been found in many organisms that binds preferentially to the unwound, single-stranded DNA at the replicating fork; this "unwinding protein" may act during replication to prevent the template DNA from re-forming the original double helix. Circular DNA molecules, such as are found in bacteria, must be broken in at least one place before they can unwind; the "unwinding protein" may possess this enzymatic activity.

In addition, there is evidence that bacterial chromosomes are associated with the cell membrane and DNA synthesis may take place on the membrane; it is not known which cell membrane components are critical for replication.

If all this seems unnecessarily complicated, remember that DNA is the cell's most precious macromolecule; once DNA is lost or destroyed, its information content can never be retrieved. Over billions of years of evolution, those cells which developed the best mechanisms to ensure the fidelity of replication and to repair damaged DNA have always had a selective advantage.

12.6 Biochemistry meets genetics

We have stated many times that DNA contains the information for making all the macromolecular components of the cell. Most of you are familiar with the concept of genes, the units of inheritance which determine our metabolic potential. Now we will combine our biochemical knowledge with some basic genetic information to see how DNA determines cell chemistry and therefore cell physiology.

First, we need a working definition of a gene. For the present, we can define a gene as a segment of a DNA molecule which carries the information for ("codes for") the amino acid sequence of a particular protein. We have mentioned genes for lipid-metabolizing enzymes (Sec. 10.6), genes for glycogen-metabolizing enzymes (Sec. 9.1) and perhaps you know about genes for hemoglobin. Faulty genes, those which no longer code for their intended proteins, have alterations (mutations) in their DNA, and the resulting physiologically defective cells (or organisms) are called *mutants*.

Justification for our working definition of a gene will be provided by examining the relationship between the structure of DNA and its function as the genetic material:

1. What is the evidence that DNA, rather than protein or lipid or anything else, is the primary information-carrying molecule in cells?

2. What kinds of information does DNA carry? DNA codes for proteins and only indirectly provides information for making polysaccharides, lipids, and other macromolecules.

3. What is the chemical form of the information in DNA? What constitutes the genetic code?

4. What special apparatus does the cell need to utilize DNA's information?

5. What determines whether or not the information in a specific gene will be used by a given cell?

The remainder of this chapter will expand on the first three questions. The mechanism by which information in DNA is used in protein synthesis is described in Chap. 13. Regulation of gene expression is discussed in Chap. 14.

12.7 Evidence that DNA is the genetic material

Two characteristics of the genetic material can be deduced from inheritance patterns. First, the genetic material should be the same, both in amount and in composition, in all somatic cells of an organism, but halved in the gametes. Second, because we know that environmentally induced traits are not inherited, the genetic material cannot undergo changes in response to shifts in the cell's nutritional state or its environment. Analyses of the chemical composition of cells from many sources show that DNA meets both these criteria. This is not, however, sufficient evidence to prove that DNA is the only or even the major genetic material, for other macromolecules, in particular chromosomal proteins, also "fill the bill," in that they appear to be ubiquitous and invariant.

To prove, rigorously, that DNA is *the* genetic material, it had to be shown that *purified* DNA alone could provide information for the synthesis of cell components and that this information is stable and can be passed from generation to generation. Evidence on this point was first obtained in the 1940s by bacterial transformation experiments. Normal virulent pneumococci (S form) are surrounded by a slimy, shiny polysaccharide coat. Mutant, nonvirulent, pneumococci (R form) lack this polysaccharide coat. When an extract from heat-killed S pneumococci was added to a growing culture of R pneumococci, some of the nonvirulent R forms were permanently converted (transformed) to the virulent S form. The nonvirulent R → virulent S transformation was stable for generation after generation. By testing the individual macromolecular components of the S extract, DNA was identified as the active transforming agent; the addition of purified S DNA alone caused inheritable changes in the R pneumococci! Further support for the role of DNA as the genetic material has come from studies of viruses (Chap. 15). Productive infection has been shown to result from the introduction of viral DNA alone.

12.8 Genes code for proteins

It is very difficult to learn what kinds of information DNA carries by studying the biochemical properties of normal cells. It is like trying to tell whether a car runs on windshield washer fluid or on gasoline; one will never know until the car is tested by emptying each tank separately and then trying to run the engine. Both geneticists and biochemists must rely on comparisons between normal cells (wild-type) and cells with defective information or with blocks in their information flow (i.e., mutants).

Although the first suggestion that genes control enzymes came from observations of human beings with inherited disorders in amino acid metabolism, other (simpler) species and their mutants are more amenable to a combined genetic/biochemical analysis. For the biochemist, mutants with defects in simple metabolic functions are easier to study than those affecting gross morphological structure; an arginine-requiring mutant is easier to analyze than a fruit fly with abnormal wings. Microorganisms have several advantages over higher organisms: (1) many generations (10–50) can be observed in a single day, (2) large numbers of mutant strains can be generated with UV light or chemical mutagens, and (3) wild-type strains require only a few, well-defined substances for growth, so that new nutritional requirements can be easily identified. For these reasons, the most revealing subjects for biochemical genetics have been bacteria and fungi and their nutritionally defective mutants.

One biochemical/genetic approach, used originally in the 1940s in studies of the mold *Neurospora crassa*, gave rise to the *one gene/one enzyme hypothesis*. The basic experiment was to treat the microorganism with a mutagenic agent such as UV light and isolate mutant strains that had acquired a single new nutritional requirement such as that for a particular amino acid or vitamin. The mutant strains were then subjected to analysis by sexual crosses to check that the new requirement was indeed inherited and that it resulted from a defect in a single gene. In the vast majority of cases, it was found that the requirement for (i.e., the inability to synthesize) a particular growth substance could be accounted for by a mutation in a single gene.

The study of such "biochemical mutants" together with other knowledge of cell chemistry has led to the following conclusions:

1. All metabolic processes in cells are under genetic control.
2. Metabolic processes consist of a series of discrete steps.
3. Each metabolic reaction is controlled by a single gene (i.e., there is a 1:1 correspondence of gene and biochemical reaction).
4. A single mutation results in the alteration of the cell's ability to carry out a single chemical reaction.

Since enzymes catalyze metabolic reactions, the underlying hypothesis is that each gene controls the specificity and hence the function of one particular enzyme. The one gene/one enzyme hypothesis (Fig. 12.8) states that there is a 1:1 relationship between genes and enzymes.

This hypothesis received support when individual enzymes were found to be missing in particular mutants. Certain arginine-requiring strains of *Neurospora* were found to lack argininosuccinase (Fig. 12.8), the enzyme that catalyzes the final step of arginine biosynthesis. Genetic tests showed that all mutations leading to a loss of this enzyme were in the same gene. Hundreds of analogous nutritional mutants in both fungi and bacteria have been analyzed and found to be unable to perform specific metabolic reactions because they are deficient in the enzymes that catalyze those reactions. All mutations leading to the same enzymatic defect were localized within a single gene. The enzymatic defects in many human disorders have also been pinpointed (Table 12.3). Thus, the 1:1 correspondence between genes and enzymes was established, although many details of the relationship remained to be worked out.

The physiological effects of a mutation (i.e., the mutant phenotype) are not necessarily seen as a nutritional requirement. When a metabolic pathway is blocked, those intermediates formed prior to the step catalyzed by the defective enzyme may be present in very high concentrations. In the pathway A → B → C → D, a defect in the enzyme that converts B to C will lead to an accumulation of intermediate B and perhaps of A. These high concentrations may in themselves be toxic to the cell, or the cell may excrete B and/or A. In either case, the metabolites in question are normal cell constituents, and it is only at very high levels that they disrupt cell function.

Intracellular accumulation of a substance to a toxic level is most often seen when the enzymatic defect is in a degradative pathway. A lack of certain enzymes for degrading macromolecules is the cause of the glycogen-storage and lipid-storage diseases in human beings, in which certain cells accumulate unusable macromolecules (Table 12.3).

Abnormally high excretion of metabolites indicates an enzymatic defect in the biosynthesis or degradation of a small molecule, since these compounds can move

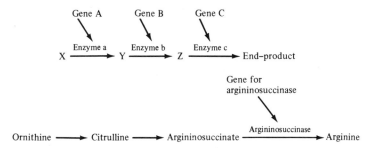

Figure 12.8 One gene/one enzyme hypothesis. Metabolic processes occur as a series of stepwise reactions; each step is catalyzed by a specific enzyme. The structure and hence the reaction specificity of each individual enzyme is controlled by its unique gene.

One example of such a series of reactions is the pathway for arginine biosynthesis in *Neurospora crassa*. The final step is catalyzed by argininosuccinase. Mutants with lesions in the gene for argininosuccinase produce no functional enzyme and hence require arginine.

TABLE 12.3 Human Genetic Diseases

Humans homozygous for a defective gene produce a protein with little or no biological activity. This results in a partial or complete block in the metabolic pathway served by that protein. Since metabolic pathways are interrelated, one single block often disrupts many diverse areas of metabolism. In some cases, the clinical features of the metabolic disease can be traced back to the original block but, particularly in those diseases that affect the nervous system, we cannot as yet relate the (known) protein deficiency with the physiological symptoms.

This table lists only 10 of the hundreds of recognized "inborn errors of metabolism." The 10 were chosen to illustrate the diversity of inherited disorders in terms of the kinds of proteins that can be defective and the areas of metabolism that can be disrupted.

Disease	Malfunctional protein	Metabolic block	Primary consequence of metabolic block	Clinical features
Tay-Sachs	N-acetyl hexosaminidase	Gangliosides cannot be broken down	Complex lipid accumulates in ganglion cells of brain	Mental relapse; loss of motor control
Adrenogenital syndrome	21-Hydroxylase	Cholesterol ⇸ glucocorticoids and mineralocorticoids	Decreased glucocorticoids cause increased output of ACTH, which, in turn, causes increased progestogens and androgens	Masculinization of external genitalia (females); sexual precocity (males)
Galactosemia	Galactose-1-P uridyl transferase	Galactose ⇸ glucose	Toxic derivatives of galactose accumulate	Vomiting, diarrhea follow milk intake
Phenylketonuria (PKU)	Phenylalanine hydroxylase	Phenylalanine ⇸ tyrosine	Toxic derivatives of phenylalanine accumulate	Mental retardation
Lesch-Nyhan	Hypoxanthine-guanine phosphoribosyl transferase	Guanine + PRPP ⇸ GMP + PP$_i$	Overproduction of purines and urate; increased levels of PRPP	Kidney stones, gout; mental deficiency; self-mutilation
McArdle's disease	Glycogen phosphorylase (muscle)	Glycogen ⇸ glucose	Glycogen accumulates in muscle	Strenuous exercise leads to muscle cramps in otherwise normal individual
Sensitivity to primaquine, an antimalarial	G-6-P dehydrogenase in rbc^a (Red blood cell)	G-6-P + NADP$^+$ ⇸ pentose phosphate pathway	With decreased NADPH, —S—S— bridges on rbc surface cannot be broken; rbc's are distorted	Acute anemia when administered drug
Sickle-cell anemia	Hemoglobin	Impaired O$_2$ transport	Deoxygenated HbS precipitates in rbc's	Impaired circulation; anemia
Wilson's disease	Caeruloplasmin, copper transport protein	Defective copper metabolism	Copper accumulates throughout body	Degeneration of brain; cirrhosis of liver
Afibrinogenaemia	Fibrinogen	Defective blood clotting	Failure of blood-clotting mechanism	Severe hemorrhaging after seemingly trivial injury

across the cell membrane. This is of biochemical interest, since it indicates the step where a pathway is blocked, but usually causes no serious problems for the producing cells. In higher organisms, however, small molecules (in the bloodstream) are not always easily removed from the body and may accumulate to toxic levels. This is the case in PKU (Table 12.3) and other human disorders of amino acid metabolism.

The one gene/one enzyme hypothesis can be generalized to the *one gene/one protein hypothesis*. Each and every protein, whether that protein plays an enzymatic, transport, hormonal, or structural role, is controlled by a unique gene. For example, human beings have a gene for the transport protein, hemoglobin; defects in this gene lead to the production of abnormal hemoglobins, many of which cannot carry oxygen as efficiently as the wild-type protein (Sec. 12.10).

Important to our further understanding of the gene–protein relationship was the proof that genes control the amino acid sequences of proteins rather than some other aspect of protein synthesis or activity. Such proof was obtained with the discovery that certain mutant microorganisms, although deficient in enzyme activity, actually produce a nonfunctional enzyme. For example, several tryptophan-requiring mutants in bacteria lack tryptophan synthetase activity; immunological experiments have shown that these strains make a protein similar to the wild-type enzyme but which is unable to function normally in tryptophan biosynthesis. Many such nonfunctional mutant proteins have been purified and their amino acid sequences determined; by comparing the sequences with the analogous wild-type proteins, the exact amino acid alterations caused by the mutations can be identified. These studies led to the conclusion that each gene codes for the amino acid sequence of an active protein; mutations in that gene alter the amino acid sequence and the structure of the protein, and thereby destroy its ability to function in cell metabolism.

As seems to be the case with most generalizations about metabolism, there are always a few exceptions. The one gene/one protein rule does not always apply. For each protein, there is a corresponding gene that carries information for its amino acid sequence, but not all genes govern the production of protein products. The final product of some genes is RNA (Chap. 13) and, in addition, certain segments of DNA known as operators and promoters produce no product at all, but play a regulatory role in gene expression (Chap. 14). Some viruses have genes within genes, such that a given sequence of DNA may code for two proteins, one larger than the other.

12.9 The genetic code: a sequence of three bases signifies one amino acid

Knowing that genes code for the amino acid sequences of proteins, we can ask what chemical features of DNA are responsible for lining up the amino acids in their proper order. Since the sugar–phosphate backbone is identical in all DNA molecules, this structure cannot be responsible for amino acid specificity. It is the sequence of bases in DNA that codes for amino acids.

Logically, there must be at least one unique code word for each amino acid. Since DNA contains only four different bases but proteins are made from 20 amino acids, a sequence of three or more bases is needed to provide each amino acid with its "own" code word. Experiments have shown that the coding ratio is three; *a three-base sequence (triplet) in the DNA codes for one amino acid.*

DNA does not itself direct the physical lining up of amino acids into proteins; that is, amino acids do not assemble into proteins along the chromosomes. As will be described in Chap. 13, DNA information is passed through nucleic acid intermediates for use in protein synthesis. The assignment of base triplets to specific amino acids (i.e., the genetic code) is based on code words in messenger RNA rather than DNA code words. For this reason, the formal presentation of the genetic code will be deferred until the next chapter (Table 13.3).

With three bases per code word, there are 64 possible code words in the genetic "dictionary." All 64 triplets are actually found in DNA. Sixty-one of these triplets have precise amino acid "meanings" and are used, in mRNA form, to direct the insertion into the growing polypeptide chain of one particular amino acid.* All the code words have been matched with their amino acid meanings. There are obviously more code words than amino acids and, in fact, all amino acids except methionine and tryptophan have more than one triplet code word in mRNA.

To direct the synthesis of a complete protein chain, the amino acid code words in each gene must be arranged and "read" in some orderly fashion. How are the triplets read off DNA? The reading is, first of all, nonoverlapping. By this we mean that the base sequence ABCDEFGHI is read ABC DEF GHI, rather than having ABC specify the first amino acid, BCD the second, CDE the third, and so on. Second, the sequence of bases is read sequentially. The triplets for the appropriate amino acids come one right after the other along the DNA; there are no unread bases or spacers within the gene coding for a single protein. This leads to the concept of the *colinearity of gene and protein*; the linear sequence of triplets along the DNA corresponds exactly to the linear sequence of amino acids in the protein product.

An understanding of the genetic code and colinearity of the gene with its protein product provides us with a more detailed understanding of the molecular basis of mutation. A mutation is an alteration in the base sequence of a gene, either a base substitution or a change in the total number of bases. Base substitutions generate "new" triplets. If the new triplet codes for an amino acid (a missense mutation), an incorrect amino acid will be inserted into the protein. If the base substitution creates a "stop" triplet (a nonsense mutation), the protein will be shortened. When the total number of bases is increased or decreased by insertion or deletion of bases, the reading frame of the triplets is shifted, resulting in a "gibberish" protein. In most cases, the protein produced from a mutated gene has no biological activity, although this will depend on the precise nature of the mutation and where it occurs within the gene. The major kinds of mutations and their effects on the amino acid sequences of proteins are shown in Table 12.4.

---

* The other three code words are used as "stop" signals to mark the end of the protein (Sec. 13.5). Such punctuation marks are necessary because each DNA molecule codes for hundreds of proteins.

Table 12.4 Mutations are Alterations in the Base Sequence of DNA

Mutations result in changes in the amino acid (aa) sequence of the protein product of the gene. The mutant protein usually has little or none of the biological activity of wild-type, although this depends on the kind of mutation and its position along the gene.

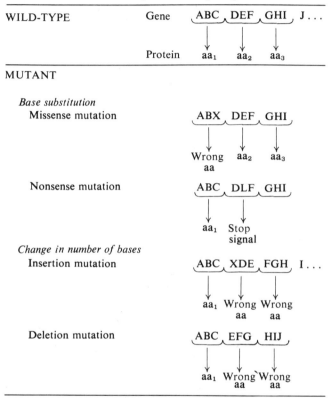

12.10 A look at an inherited metabolic disorder: sickle-cell anemia

To put our combined biochemical and genetic information in perspective, let's look at the inheritance of a single defective gene. Our example will be the metabolic disease known as sickle-cell anemia.

Sickle-cell anemia is caused by a defect in the structure of hemoglobin, the oxygen-transporting protein in red blood cells. The disorder is so named because a (variable) proportion of the red blood cells of affected individuals are elongated, crescent-, or sickle-shaped. These oddly shaped red blood cells contain precipitated sickle-cell hemoglobin (HbS). The abnormal red blood cells become trapped in the small blood vessels and impair circulation; this causes extensive organ damage, particularly in bone and kidney. Sickled red blood cells are unstable; their rapid breakdown leads to severe anemia.

Only *deoxygenated* HbS causes red blood cells to sickle, and this accounts for some of the clinical symptoms of the disorder, including "sickle-cell crisis," in which the proportion of sickled cells is especially high. If the oxygen level decreases and sickling begins, the circulation in small blood vessels decreases to cause an additional local decrease in oxygen levels, more HbS is deoxygenated and precipitates, the blood vessels become more obstructed, and so on.

The molecular difference between HbS and normal hemoglobin (HbA) has been identified. Electrophoretic studies showed that HbS is a more positively charged protein than HbA. This charge difference was found to result from the replacement of a negatively charged amino acid with a neutral one in the β chain (Fig. 2.7); a glutamic acid residue at position 6 in HbA is replaced by a valine residue in HbS.

Although only one amino acid is changed, this is sufficient to cause the abnormal solubility properties of deoxygenated HbS. The valine residue is on the exterior of the HbS molecule. When HbS is deoxygenated, the exterior region is exposed and "sticky," causing HbS molecules to clump together and form the fibrous precipitate that deforms ("sickles") the red blood cells. Oxygenation of HbS masks this "sticky" region and no precipitation occurs. The "sticky" region does not exist in the more negatively charged HbA.

What change in the hemoglobin gene leads to the production of HbS? The mutation is a change in the DNA such that the triplet code word for glutamic acid is replaced by the triplet for valine. These triplets differ only in one base, as can be seen in Table 13.3; sickle-cell anemia results from a single base change in the hemoglobin gene. It is presumed that this change in the hemoglobin gene occurred by chance many generations ago and has been passed from generation to generation through the sexual cycle of meiosis, gamete production, and fertilization (Fig. 12.9).

Sickle-cell anemia is a *recessive* disorder, which means that, to be affected with the disease, an individual must carry two copies of the HbS gene, that is, he must be *homozygous* (HbS/HbS). Individuals with one HbS and one HbA gene (*heterozygotes*) do not have sickle-cell anemia; heterozygotes produce both kinds of hemoglobin and the concentration of HbS in their red blood cells is too low to cause precipitation and sickling.*

Like sickle-cell anemia, most other inherited metabolic disorders in humans are recessive (i.e., the heterozygotes are phenotypically normal). This is because, in most cases, the mutant gene produces a nonfunctional protein but does not interfere with the production or metabolic function of the wild-type protein coded by the nonmutant gene. In other words, heterozygotes produce both active and inactive proteins. Metabolism can go on as usual with the reduced levels of active protein, and heterozygous individuals normally have no physiological difficulties. It is only the homozygously defective individual, who carries no wild-type gene and is thus totally lacking functional protein, who has the symptoms of the inherited disease.

---

* Heterozygotes occasionally show abnormal symptoms under conditions of unusually low oxygen pressure, but this is extremely rare.

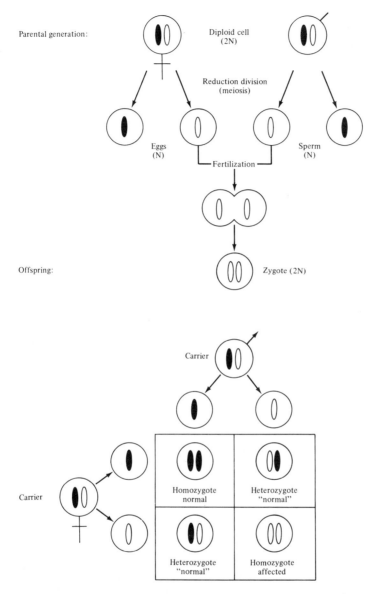

Figure 12.9 Inheritance of metabolic disorders. Human beings are diploid. Their cells carry two copies of each chromosome and hence two copies of each gene. Chromosomes are reassorted and passed from generation to generation, as shown.

An individual will be homozygous for and have the symptoms of an inherited metabolic disorder only if he receives one mutant gene from each parent. Affected homozygotes are usually the offspring of (unaffected) heterozygous parents, called "carriers" of the metabolic disease.

The closed bars indicate the chromosome carrying the wild-type gene; open bars represent the chromosome carrying the mutated gene.

235

To be affected by a recessive metabolic disease, an individual must have received one mutant gene from each parent. Homozygotes are usually the offspring of two (unaffected) heterozygous parents, called *carriers* of the disease (Fig. 12.9). Of course, one or both parents of affected individuals could themselves be homozygous and affected but, since most metabolic disorders are rare and reduce reproductive capability, this is very unlikely.*

Most inherited diseases are quite rare because the (mutated) genes that cause them are deleterious to normal existence and are kept at a low frequency in the human population. The HbS gene is an exception; it is found at a surprisingly high frequency in certain black populations. This fact indicates that some selective force is acting to keep the gene prevalent in the population, counteracting the strong selection against HbS/HbS individuals, who almost never reproduce. It has been found that the parasite which causes malaria cannot propagate as efficiently in the red blood cells of HbS/HbA heterozygotes as it can in the red blood cells of HbA/HbA homozygotes; hence, heterozygotes are less sensitive to malaria than normal (HbA/HbA) homozygotes. Thus, in the tropical regions of Africa, the heterozygote has a selective advantage (i.e., is more likely to reproduce) over both types of homozygote, and the HbS gene is retained in the population at a higher-than-expected frequency.

SUMMARY

Deoxyribonucleic acid (DNA) is the cell's information-storage macromolecule. DNA directs and determines the synthesis of all other kinds of molecules in the cell. The information in DNA passes from one generation to the next to maintain constancy of species.

DNA is a two-stranded macromolecule. Each strand is a linear polymer of alternating deoxyribose and phosphate groups; purine or pyrimidine bases are linked to each deoxyribose residue. The two strands are aligned and held together by hydrogen bonds between the bases: adenine pairs specifically with thymine, and guanine pairs with cytosine. The two paired strands are wound around each other in a double helix. In eucaryotes, the DNA combines with proteins (histones) to form the chromosomes.

During cell division, DNA is copied by semiconservative replication, a mechanism which ensures that DNA information is not lost or damaged. The two strands of DNA separate and free dNTPs base-pair to the individual unwound strands; the dNTPs are then joined together by the action of DNA polymerases to form a new DNA strand that is complementary to the template.

---

* The inheritance of diseases caused by defective genes on the X chromosome is a special case. Males (XY) receive their single X chromosome from their mother, so only the mother need be a carrier of the metabolic disease. Females (XX) with X-linked metabolic disorders usually have affected fathers and carrier mothers.

The information in DNA is contained in its sequence of bases. Three adjacent bases constitute the genetic code word for one amino acid. A gene is the linear sequence of base triplets that codes for the amino acid sequence of a given protein; the amino acid sequence of a protein determines its three-dimensional shape and hence its metabolic function. Each gene is the cell's blueprint from which many copies of one protein can be made so that a specific metabolic job—catalysis, transport, regulation and so on—can be carried out.

A change in the base sequence within a gene will change an amino acid code word and lead to the production of an altered protein. This is often deleterious to the cell, since changing the amino acid sequence of a protein often reduces its ability to function in metabolism. Changes in the base sequence of DNA (mutations) are stable and are passed on from one generation to the next.

PRACTICE PROBLEMS

1. You are given a solution of single-stranded DNA molecules. After this mixture has been allowed to stand for some time, which of the single-stranded molecules are likely to pair with each other and form double-stranded helices? Write down the structures of these pairs.
   (a) AAAAAAAAAAAAAAAA.
   (b) ACACACACACACACAC.
   (c) GGGGGGGGGGGGGGGG.
   (d) TGACTGACTGACTGAC.
   (e) TGTGTGTGTGTGTGTG.
   (f) TTTTTTTTTTTTTTTT.
   (g) CAGTCAGTCAGTCAGT.
   (h) ACTGACTGACTGACTG.
   (i) GTACGTACGTACGTAC.

2. A wild-type bacterium was grown in a medium containing ^{15}N (a "heavy" isotope of nitrogen) so that every nitrogen atom in its DNA is the heavy isotope. The bacterium is then allowed to grow and divide in a medium in which all nitrogen is ^{14}N, the normal "light" isotope. Draw diagrams to support your answers to the questions below.
   (a) After one generation in the ^{14}N medium, how many of the (two) DNA molecules contain ^{15}N?
   (b) After three generations in the ^{14}N medium, how many of the (eight) DNA molecules contain ^{15}N?

3. Glucose is not only a cellular energy source, it is also a precursor for many other molecules. Briefly outline the pathways by which the carbon atoms from glucose get into DNA.

4. A dilute culture containing only S pneumococci was exposed to ultraviolet light. The cells were allowed to divide, and after many generations, the culture contained 90% S pneumococci and 10% R pneumococci. What might have caused this shift in the phenotype of the bacterial population?

5. In cells of eucaryotes, most DNA is found in the nucleus, but mitochondria also contain specific DNA molecules. What proteins might you expect to find coded for by mitochondrial DNA?

6. A family exists in which both parents are phenotypically normal, but four of their six children have an inherited disease, vitamin D-resistant rickets. This form of rickets is not cured by feeding the patients vitamin D.
   (a) Suggest a protein that could be inactive in the individuals with vitamin D-resistant rickets.
   (b) Is this disease recessive or dominant?
   (c) Explain your answer to part (b) by comparing the proteins made by the heterozygote and by both types of homozygotes.
   (d) Which of the parents is a carrier of the disease?

7. Five different mutants of yeast require the amino acid histidine. Each mutant accumulates certain of the intermediates on the pathway of histidine biosynthesis, as listed below.

Mutant	Compounds accumulated
1	A
2	A, C, D, and E
3	C and A
4	A, B, C, D, and E
5	A, E, and C

   (a) By which of the following pathways is histidine synthesized by the yeast cells?
   A → B → C → D → E → histidine
   C → B → D → A → E → histidine
   D → A → E → C → B → histidine
   A → C → E → D → B → histidine
   (b) Now that you have decided on the pathway, indicate which enzymatic step is blocked in each mutant.

8. The enzyme missing in PKU individuals is phenylalanine hydroxylase. This metabolic disease could be cured if the missing enzyme could somehow be supplied to the cells. Briefly discuss the pros and cons of the following suggestions for curing PKU (assume that all are technically possible); direct your answers to both their short- and their long-term usefulness.
   (a) Injecting phenylalanine hydroxylase directly into the cells of the infant.
   (b) Feeding the infant phenylalanine hydroxylase.
   (c) Incorporating the gene (DNA) for phenylalanine hydroxylase into the chromosomes of the infant.

9. You are given a solution containing proteins extracted from liver cells. What two chemical principles could you use to separate the chromosomal proteins from the other proteins in the mixture?

10. Following extensive UV or X-radiation, bacteria excrete deoxynucleotides. Why?

11. How could you prove that the active component in pneumococcal transformation (i.e., the transforming agent) is DNA as opposed to protein? DNA rather than RNA? DNA rather than triglycerides?

12. What is the percent GC in the DNA of *Pseudomonas aeruginosa*? In wheat DNA? (See Sec. 12.1.)

13. Which of the following RNA molecules could form double-stranded structures with the DNA molecules given in problem 1?
    UGUGUGUGUGUGUGUG
    UUUUUUUUUUUUUUUU
    GUCAGUCAGUCAGUCA

SUGGESTED READING

BEADLE, GEORGE W., "The Genes of Men and Molds," *Sci. American*, 179, No. 3, p. 30 (1948). A classic paper tying together the original evidence for the one gene/one enzyme hypothesis.

BRITTEN, ROY J., and DAVID E. KOHNE, "Repeated Segments of DNA," *Sci. American*, 222, No. 4, p. 24 (1970). In eucaryotic DNA, some base sequences are repeated as many as 1 million times.

CRICK, F. H. C., "The Structure of the Hereditary Material," *Sci. American*, 191, No. 4, p. 54 (1954). The Nobel prize winner gives a clear description of the three-dimensional structure of DNA.

CRICK, F. H. C., "The Genetic Code," *Sci. American*, 207, No. 4, p. 66 (1962). Describes experiments which showed that the genetic code consists of nonoverlapping, contiguous base triplets.

FRIEDMANN, THEODORE, "Prenatal Diagnosis of Genetic Disease," *Sci. American*, 225, No. 5, p. 34 (1971). Culturing of embryonic cells obtained by amniocentesis allows for the prenatal diagnosis of many inherited metabolic disorders. These powerful techniques present moral and legal problems for our society.

HANAWALT, PHILIP C., and ROBERT H. HAYNES, "The Repair of DNA," *Sci. American*, 216, No. 2, p. 36 (1967). Cells repair errors in the base sequence of DNA.

MAZIA, DANIEL, "The Cell Cycle," *Sci. American*, 230, No. 1, p. 54 (1974). Chromosome replication sets the stage for division in eucaryotic cells.

McKUSICK, VICTOR A., "The Mapping of Human Chromosomes," *Sci. American*, 224, No. 4, p. 104 (1971). Recombination, seen through family pedigrees, helps investigators assign genes to specific human chromosomes.

SIGURBJÖRNSSON, BJÖRN, "Induced Mutations in Plants," *Sci. American*, 224, No. 1, p. 86 (1971). Irradiation of whole plants or seeds, while generally deleterious, induces a small number of beneficial mutations. Newly created strains of wheat, rice, peppermint, etc., incorporated into sophisticated breeding programs, have been responsible for the "Green Revolution."

YANOFSKY, CHARLES, "Gene Structure and Protein Structure," *Sci. American*, 216, No. 5, p. 80 (1967). Elegant experiments showed the colinearity between the base sequence of DNA and the amino acid sequence of an enzyme.

ZUKERKANDL, EMILE, "The Evolution of Hemoglobin," *Sci. American*, 212, No. 5, p. 110 (1965). Evolution depends ultimately on changes in DNA.

The following articles present both sides of the controversy over recombinant DNA research:

BENNETT, WILLIAM, and JOEL GURIN, "Science That Frightens Scientists, the Great Debate over DNA," *The Atlantic Monthly*, 239, No. 2, p. 43 (1977).

GROBSTEIN, CLIFFORD, "The Recombinant-DNA Debate," *Sci. American*, 237, No. 1, p. 22 (1977). Outlines the guidelines governing recombinant DNA research.

"Recombinant DNA Research, A Debate on the Benefits and Risks," *Chemical and Engineering News*, 55, No. 22, p. 26 (1977). Includes articles by biochemists Bernard D. Davis and Erwin Chargaff and social scientist Sheldon Krimsky.

13 Protein Synthesis: the Translation of Genetic Information into Protein Requires RNA

The information necessary to make, maintain, and propagate each living cell is encoded in the base sequence of its DNA (Chap. 12). The unique sequence of bases in DNA not only gives each cell the potential to become part of a specific organism, but also contains all the information necessary for keeping that cell alive and reproducing. We have established that genes (DNA) code for proteins, and now we will consider how the cell uses specific base sequences in DNA as a blueprint for the synthesis of specific proteins.

A. RIBONUCLEIC ACIDS, MOLECULES OF MANY USES

Analyses of the gross composition of cells show that the majority of the nucleic acid is ribonucleic acid (RNA). It has long been known that addition of either radioactive nucleic acid bases or radioactive phosphate to growing cells leads to rapid labeling of their RNA; RNA is almost always being synthesized. Such experimental results suggest that RNA plays a central role in cell metabolism.

The metabolic role of RNA has now been elucidated: it is the bridge between the genetic information in DNA and the amino acid sequences of proteins. Conveniently for us, the different types of RNA molecules have been given names that tell us what

TABLE 13.1 RNA Molecules

Type of RNA	Approximate size (number of bases)	Function	% total RNA per cell
Transfer (tRNA)	80	Amino acid transfer and recognition in translation	10%
Ribosomal (rRNA)	100, 1500, 3000	Ribosome structure	70%
Messenger (mRNA)	100–5000	Carries information for amino acid sequence of proteins	20%
Viral	4000–30,000	Carries information for synthesis of virus	

they do in protein synthesis (Table 13.1). But before we can consider these roles in detail, we must discuss the structure of this important macromolecule.

13.1 The chemical composition and structure of RNA

The structure of RNA is very similar to that of a single strand of DNA. The major difference is that RNA contains ribose as its five-carbon sugar component, whereas DNA contains deoxyribose. (Ribose has one more hydroxyl group than deoxyribose.) Thus, like DNA, RNA is made up of a sugar–phosphate backbone (Fig. 13.1) with phosphate bridges (phosphodiester bonds) linking the 3′ hydroxyl group of one ribose to the 5′ hydroxyl group of the adjacent ribose.

To each sugar of the ribose–phosphate backbone is attached one of four heterocyclic nitrogen-containing bases; the bases in RNA are adenine, cytosine, guanine, and uracil [Fig. 13.1(b)].* The metabolism of these bases is discussed in Chap. 11. Remember that DNA contains adenine, cytosine, guanine, and thymine.

Since RNA is a linear macromolecule, it is convenient to have a way of naming the two ends. In looking at the structure of RNA in Fig. 13.1, one can see that one end (left) has a free 5′ hydroxyl and the other end (right) has a free 3′ hydroxyl. These are called the 5′ and the 3′ ends of the molecule, respectively.

Most RNA molecules are very long (80–5000 nucleotide residues), with many bases in a seemingly random sequence. In general, DNA molecules are even longer than this; the single molecule of DNA which carries all the information to make a bacterium (about 2 million nucleotide pairs) is many times larger than many RNA molecules.

---

* In certain cases, tRNA for example, some of these bases may be modified by the addition of methyl groups and other side chains to give the "rare" bases (bases of rare occurrence in RNA) such as thymine, inosine, or pseudouracil.

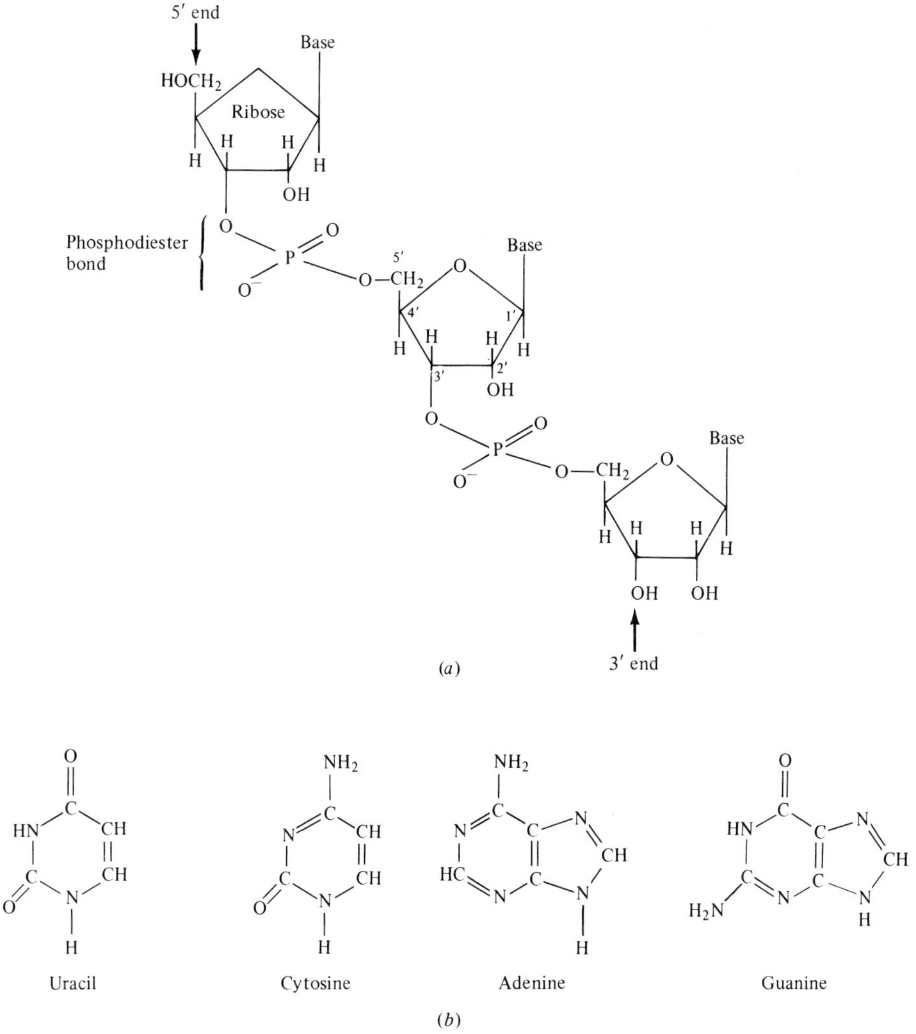

Figure 13.1 Structure of RNA. (*a*) Like DNA, RNA possesses a sugar–phosphate backbone. A purine or pyrimidine base is attached to each ribose unit. Note that hydroxyl groups are present on both the 2′ and 3′ positions of the sugar ring (DNA lacks the 2′ hydroxyl). (*b*) The bases in RNA are uracil and cytosine (pyrimidines), and adenine and guanine (purines).

Because of the free 2' hydroxyl group on each sugar residue in the backbone, RNA is less stable than DNA. The free hydroxyl group makes RNA sensitive to alkali; whereas DNA is stable at pH above 8.0, the RNA backbone is degraded. Obviously, the intracellular pH never gets this high, but this is a useful reaction for distinguishing RNA from DNA in the laboratory.

RNA is degraded by specific nucleases (hydrolyzing enzymes) that cleave the phosphodiester bonds of RNA but do not affect the analogous bonds in DNA. These nucleases distinguish RNA from DNA by recognizing the free 2' hydroxyl groups on

TABLE 13.2 Ribonucleases Are Used to Sequence RNA

(a) Ribonucleases break the sugar–phosphate backbone of RNA adjacent to only certain bases.

Ribonuclease	Base specificity
Pancreatic	U, C
T_1	G
U_2	A, G

Using a combination of these nucleases, RNA sequences can be determined. The DNA is transcribed; then the RNA transcript is isolated, treated with combinations of ribonucleases, and the base sequences of the resulting fragments determined. By lining up overlapping fragments, the base sequence of the intact RNA molecule is determined. The DNA sequence of the gene is simply the complement of the RNA transcript. (For good examples of this technique, see references by Dickson et al. and Bertrand et al. at the end of Chap. 14.)

(b) The lactose operon (Fig. 14.3) codes for the enzymes that degrade the disaccharide lactose; the control region of this operon was sequenced as described above. The promoter region includes the base sequence where the cyclic AMP binding protein attaches (CAP site) and the site for RNA polymerase attachment. On the 5' side of the promoter is the i gene, which codes for the repressor protein that regulates the expression of the operon. On the 3' side of the promoter is the z gene, which codes for the enzyme β-galactosidase. The terminal amino acid sequences of these two proteins can be recognized in the DNA sequence.

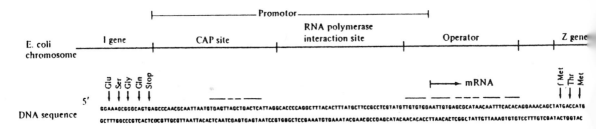

(After Dickson et al., *Science*, 187: 27–35 (1975). Copyright 1975 by the American Association for the Advancement of Science.)

the ribose portions of RNA. Ribonuclease A is a very active enzyme of wide distribution that hydrolyzes RNA; it was the first enzyme for which the complete amino acid sequence was determined, and studies of its action have been important in providing information on how active enzyme molecules are folded and how they work. Some ribonucleases have high specificity for the phosphodiester linkage adjacent to one particular base (e.g., guanine) (Table 13.2); these specific ribonucleases can be used to determine the base sequences of RNA molecules in the same way that chymotrypsin and trypsin are used to determine the amino acid sequences of proteins.

13.2   The three-dimensional structure of RNA

We know that strands of DNA can pair with each other by forming specific hydrogen bonds between complementary bases (Watson–Crick pairing, Sec. 12.1); the bases in RNA share this property. Base pairing plays an important role in maintaining the three-dimensional structure and hence the function of RNA molecules.

In contrast to DNA, most RNAs exist as single-stranded molecules. Pairing between complementary bases located in different parts of the same RNA chain can often lead to localized double-stranded regions. The rules for the pairing of bases in RNA are the same as for pairing in DNA, except that uracil replaces thymine; the base pairs are AU and GC. Thus, a linear structure such as

$$---\text{AGUCCGGAAAGGAGCCGGACC}---$$

might fold up into a hairpin form like this:

```
          AGG
        A     A
        A     G
         G··C
         G··C
         C··G
         C··G
         U··A
      ---AG··CC---
```

Such intrachain structures appear to be essential features of many RNA molecules and seem to play key roles in the function of RNA, especially when it must recognize and interact with other nucleic acids, and with proteins. For example, during protein synthesis, tRNA (Fig. 13.2) interacts with mRNA and with certain protein molecules in a specific fashion. These interactions depend on the tRNA having the proper three-dimensional structure.

Not all RNA molecules in nature are single-stranded; double-stranded RNA molecules are present in certain viruses. In these cases, the double-stranded RNA molecules have a helical structure like that of double-stranded DNA.

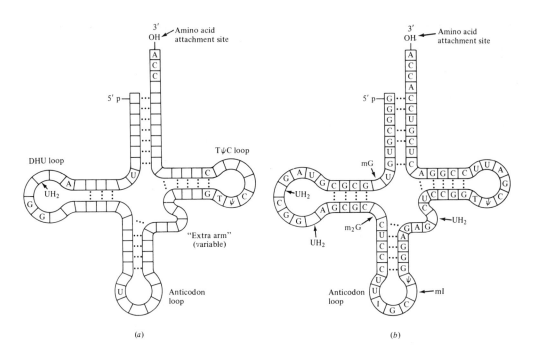
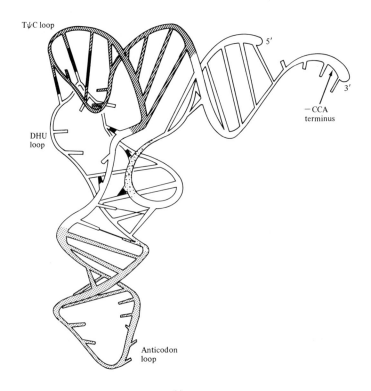

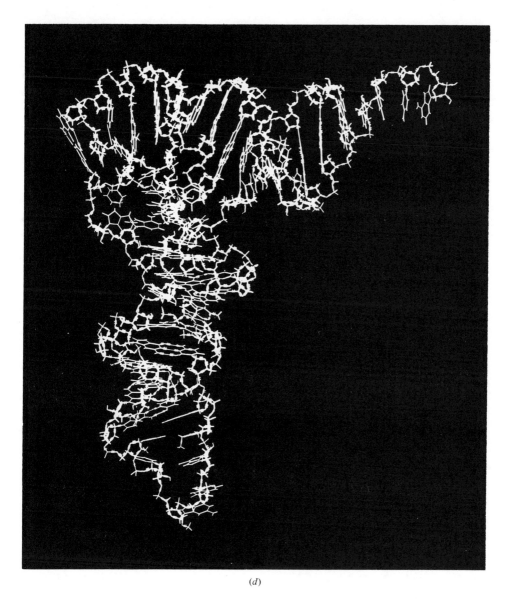

(d)

Figure 13.2 Transfer RNA. Transfer RNA illustrates the complex three-dimensional structure of RNA molecules. (a) Common features of tRNA molecules. The unlabeled squares can be the nucleosides uridine, cytidine, adenosine, or guanosine, depending on the particular tRNA. Abbreviations for nucleosides containing the rare bases are: inosine (I), methylinosine (mI), dihydrouridine (UH$_2$), ribothymidine (T), pseudouridine (ψ), methylguanosine (mG), dimethylguanosine (m$_2$G). (b) Base sequence of yeast alanine tRNA. (c) Diagram of three-dimensional model of yeast phenylalanine tRNA based on x-ray crystallography data. (d) Photograph of a three-dimensional model of yeast phenylalanine tRNA based on x-ray crystallography data. [(a) (b) After Lubert Stryer, *Biochemistry*, W. H. Freeman and Company, San Francisco. Copyright 1975, p. 651. (c) After S. H. Kim et al., *Science*, 185: 435–440 (1974). Copyright 1974 by the American Association for the Advancement of Science. (d) Courtesy Dr. Sung-Hou Kim.]

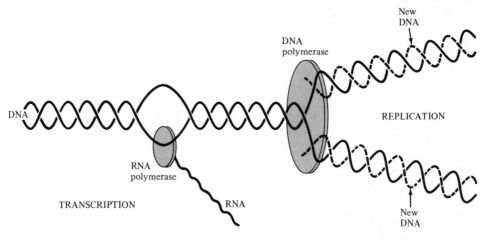

Figure 13.3 Two functions of DNA. DNA serves as the template for the production of RNA molecules needed for the growth and development of the cell and it is also the template for the DNA passed on to the next cell generation.

13.3 Transcription: the synthesis of RNA on a DNA template

The first step in the expression of a gene is the transcription of that gene into RNA.

All nucleic acids (DNA and RNA) in the cell are made by copying a template molecule. In most cases, DNA is the template for the synthesis of both DNA and RNA. This dual use of DNA allows the information in the chromosomes to be

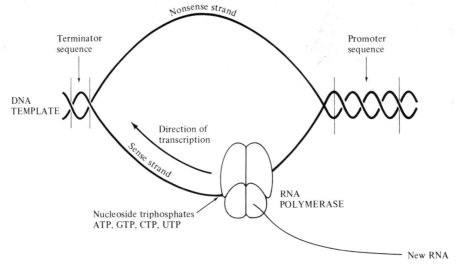

RNA IS MADE ON A DNA TEMPLATE

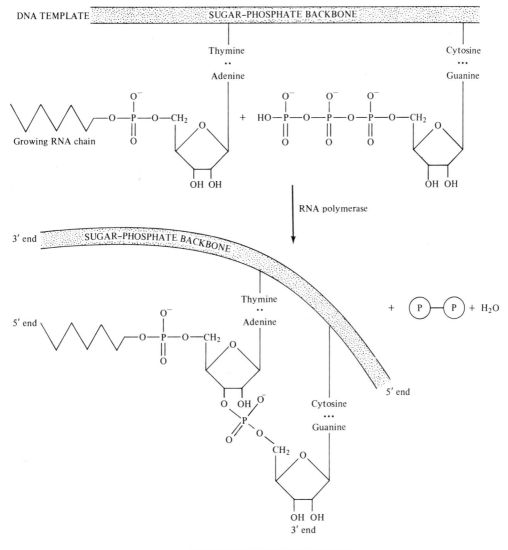

THE POLYMERIZATION REACTION

Figure 13.4 Transcription: the synthesis of RNA. RNA is synthesized on a DNA template. The ribonucleoside triphosphates line up along the DNA by complementary base pairing (CG and UA or AT). The subsequent formation of the chain of phosphodiester bonds is catalyzed by RNA polymerase, a large enzyme with four subunits. For each internucleotide bond formed, one molecule of pyrophosphate is released.

inherited on the one hand, and expressed during the growth and development of the cell on the other (Fig. 13.3).

DNA (the gene) is double-stranded. Since a gene usually codes for one protein only, only one strand of DNA (the "sense" strand) is transcribed into RNA. The other DNA strand is silent during transcription, but, of course, it plays the important role of acting as a template for the sense strand of DNA during DNA replication. Since this silent strand is complementary to the sense strand, it has the same base sequence as RNA (except that thymidine replaces uridine).

When a nucleic acid is to be made, the appropriate nucleotide precursors line up in sequence along the DNA template, by complementary base pairing through hydrogen bonding. Subsequently, the nucleotides are linked together by formation of phosphodiester bonds between the adjacent sugars. This polymerization is catalyzed by enzymes known as polymerases. In the case of DNA, DNA polymerases carry out this synthetic function (Sec. 12.5). In a similar way, RNA molecules are synthesized on DNA template molecules in reactions catalyzed by RNA polymerases (Fig. 13.4). RNA polymerases move (catalyze chain elongation) from the 5' to the 3' end of the DNA template, as do DNA polymerases.

The RNA polymerases of bacteria and higher cells have been extensively studied. In bacteria, a single RNA polymerase catalyzes the synthesis of all the different types of RNA molecules. Cells of eucaryotic organisms contain more than one RNA polymerase and it appears that the different polymerases function in the production of different kinds of RNA molecules; one polymerase may be responsible for mRNA synthesis, another for rRNA and tRNA synthesis, and possibly another is localized in the mitochondria, where it functions in the transcription of genes on mitochondrial DNA.

If we envision RNA polymerase moving along the DNA synthesizing RNA, an obvious problem comes to mind. Cells contain only a few very long DNA molecules, each consisting of thousands of different genes. Since DNA is a continuous molecule, how does the RNA polymerase know where to start and where to stop its progress along the DNA? We cannot have the polymerase starting off in the middle of one gene and ending up in the middle of another! The problem is solved by the presence of special "start" and "stop" sequences in the DNA flanking each gene.

At the 5' end of the gene (where transcription begins), there is a specific DNA sequence to which the RNA polymerase attaches. This sequence of about 10–15 deoxynucleotides is part of the "promoter" sequence that signals the starting point of transcription. The RNA polymerase must attach within this promoter region before the adjacent gene can be transcribed. As a further control of the specificity of the attachment, effectors exist that assist or inhibit the binding of RNA polymerase to the special "start" sequences. One such effector is cyclic AMP, which can influence gene transcription by interacting with a specific protein that binds to the promoter region (Table 13.2).

There are many questions that we can ask about the way in which promoters work: What is the difference between one promoter and another? When RNA polymerase binds to a promoter, does it immediately begin its job of transcribing the

DNA, or is another signal required? The answers to these questions are still being sought.

Having commenced transcription, the RNA polymerase moves along the gene to produce an RNA transcript of the sense strand of the DNA. When it has traversed to the 3' terminus of the gene, the polymerase encounters another specific DNA sequence, called (not surprisingly) the "stop" sequence. Little is known about the mechanism of termination of the synthesis of RNA chains, but it is assumed that when the polymerase reaches the "stop" base sequence at the end of the gene, it falls off the DNA, releasing the complete RNA transcript of the gene.

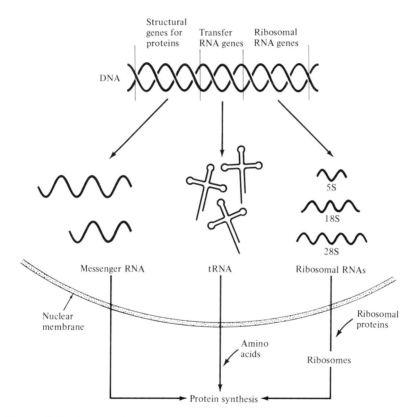

Figure 13.5 Genes code for different kinds of RNA. Messenger, transfer, and ribosomal RNA are all made from a DNA template. Transfer and ribosomal RNA are not translated, so they are the end products of the expression of their genes. Proteins (such as enzymes, transport proteins, hormones, and ribosomal proteins) are the end products of the expression of their structural genes; messenger RNA is intermediate in their production.

In eucaryotic cells, the different kinds of RNA must be transported across the nuclear membrane before they function; in procaryotic cells, the DNA is not enclosed in a membrane.

In eucaryotic cells, RNA is made on the DNA of the chromosomes inside the nucleus and is then transported across the nuclear membrane to the cytoplasm, where it performs its function as messenger, ribosomal, or transfer RNA (Fig. 13.5).

B. HOW PROTEINS ARE MADE—THE TRANSLATION SYSTEM IN CELLS

13.4 An overview

Figure 13.6 shows an outline of the mechanism by which proteins are made—the translation system. With only minor variations, the mechanism of protein synthesis is the same in all organisms. The information for the amino acid sequence of a protein is copied, by transcription, from the gene in the form of mRNA. The *translation system then decodes the information in mRNA into the amino acid sequence of a protein.* The means of decoding is complementary base pairing; each amino acid is first attached to a specific tRNA that will act as a liaison between the amino acid and its code word in mRNA. Translation takes place on the ribosome; successive peptide bonds are formed by enzymes associated with this organelle.

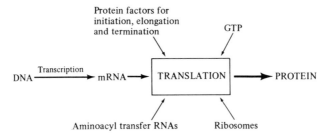

Figure 13.6 Outline of protein synthesis (translation). Many components are required for the conversion of the information in messenger RNA into proteins.

13.5 Messenger RNA and the genetic code

The information for the amino acid sequences of proteins is "written" in the DNA (gene) in the form of a series of triplets of nucleic acid bases; each triplet corresponds to a given amino acid. Messenger RNA, synthesized off the DNA template by RNA polymerase, contains the linear sequence of coding triplets that directs the actual synthesis of a protein of a given amino acid sequence. These mRNA triplets are given the name *codons*; they are the ribonucleic acid code words for amino acids (Table

TABLE 13.3 The Genetic Code

During translation, these mRNA triplets recognize and bind the tRNAs carrying the amino acids indicated. Thus, mRNA directs the insertion of amino acids into protein in the proper order.

The triplet code is the same for all organisms with the exception that the three nonsense (termination) codons, identified in bacteria, may not be used in all organisms.

Second letter

First letter		U	C	A	G	Third letter
U		UUU } Phe	UCU }	UAU } Tyr	UGU } Cys	U
		UUC	UCC	UAC	UGC	C
			} Ser			
		UUA } Leu	UCA	UAA STOP	UGA STOP	A
		UUG	UCG	UAG STOP	UGG Trp	G
C		CUU }	CCU }	CAU } His	CGU }	U
		CUC	CCC	CAC	CGC	C
		} Leu	} Pro		} Arg	
		CUA	CCA	CAA } Gln	CGA	A
		CUG	CCG	CAG	CGG	G
A		AUU }	ACU }	AAU } Asn	AGU } Ser	U
		AUC } Ile	ACC	AAC	AGC	C
		AUA	} Thr			
			ACA	AAA } Lys	AGA } Arg	A
		AUG Met, START	ACG	AAG	AGG	G
G		GUU }	GCU }	GAU } Asp	GGU }	U
		GUC	GCC	GAC	GGC	C
		} Val	} Ala		} Gly	
		GUA	GCA	GAA } Glu	GGA	A
		GUG	GCG	GAG	GGG	G

13.3). A mRNA for a protein of 300 amino acids will be a ribonucleotide chain containing about 1000 bases: 900 bases of amino acid sequence information in the form of codons, plus the "start" and "stop" signals required for translation of this mRNA. As we shall see in Sec. 13.6, tRNA is used to "decode" the nucleic acid triplet messages into their specific amino acids.

13.6 Transfer RNA—the adapter molecule

One of the many problems facing biochemists in their studies of protein synthesis was to explain how a nucleotide sequence in mRNA could be recognized by amino acids during the synthesis of proteins. A specific and error-free mechanism must exist for unique triplet base sequences to direct the insertion of specific amino acids into their correct positions in proteins. Since a highly specific, direct chemical interaction between nucleic acid bases and amino acids was not possible, nature had to evolve an

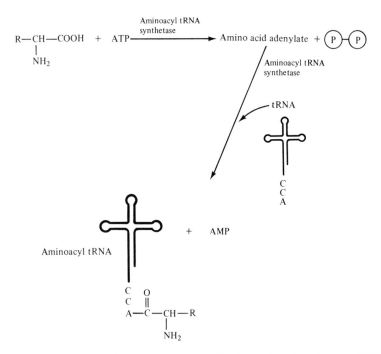

Figure 13.7 Formation of aminoacyl tRNAs. The formation of aminoacyl tRNAs proceeds in two steps. First, the amino acid reacts with ATP to form aminoacyl adenylate. Second, the amino acid is joined to the 3' end (—CCA end, Fig. 13.2) of its specific (cognate) tRNA. These reactions activate the amino acid in a chemical sense so that it can take part in peptide bond formation. Equally important, they give the amino acid a handle by which it can recognize its specific triplet during the decoding of mRNA.

Each amino acid is linked to its cognate tRNA by a specific synthetase. For example, leucine is linked to leucyl tRNA (which carries the anticodon for leucine) by leucyl tRNA synthetase. Because of the specificity of the synthetase, no other amino acid can be attached to $tRNA_{leu}$ nor can leucine be joined to the tRNA for another amino acid.

It is probable that both reactions occur while the molecules of tRNA, amino acid, and ATP are associated with one molecule of synthetase.

alternative way of making nucleotide sequences easily recognizable by amino acids. This recognition is obtained by first joining the amino acids to specific tRNA molecules that contain triplet base sequences (anticodons) that can pair with codons in mRNA. Complementary base pairing between mRNA and the anticodon of the amino acid-carrying tRNA (aminoacyl tRNA) lines up the amino acids in the order called for by the mRNA. The formation of aminoacyl tRNA molecules (Fig. 13.7) has a second purpose; in this reaction the amino acids are activated chemically in preparation for the formation of peptide bonds in proteins.

The roles of tRNA in amino acid activation and triplet recognition are reflected in the structure of a typical tRNA (Fig. 13.2). Three sequences in this molecule are of particular significance. At one end, one always finds a —CCA sequence to which the amino acid is esterified. Near the middle of the molecule, one finds the anticodon loop, containing the sequence of three bases that is complementary to the mRNA coding triplet for the amino acid carried by the particular tRNA. Since there are 20 amino acids, there must be at least 20 different tRNA molecules, each with a different anticodon triplet. A third sequence in each tRNA is responsible for binding to the ribosomes.

One class of tRNA molecules is different from the others; these are initiator tRNAs that are used to start the protein chain by recognition of a specific initiator triplet at the beginning of the mRNA being read into protein. The initiator tRNAs differ somewhat in polynucleotide sequence and in three-dimensional structure from the tRNAs that insert amino acids into internal positions of protein chains.

13.7 The ribosomes—where the action is

The business of protein synthesis does not occur with the mRNA, aminoacyl tRNA, and various required protein factors floating around, free in solution in the cell cytoplasm. Proteins are made on a solid support, the ribosome, that holds the components of the protein synthetic machinery in correct juxtaposition to allow rapid and accurate synthesis. Ribosomes are organelles of very complex structure containing three different RNA molecules and more than 50 proteins (Fig. 13.8). The ribosomes of eucaryotic cells are larger and contain more proteins than those of procaryotic cells.

A popular and convenient description of the role of the ribosome in protein synthesis states that this organelle possesses a number of distinct activity sites on its surface. These include specific binding sites for mRNA, tRNA, and other components needed for protein synthesis; these sites are presumably formed by interactions between rRNA and the various ribosomal proteins in much the same way that an active site is formed on an enzyme. In essence, we can think of the ribosome as a large and complex enzyme with a number of sites capable of holding together the various substrates necessary for the synthesis of proteins.

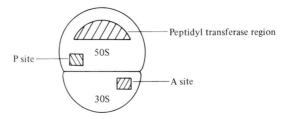

A BACTERIAL RIBOSOME

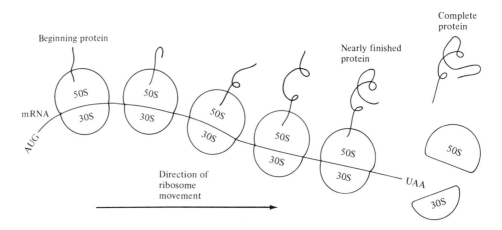

A BACTERIAL POLYSOME

Figure 13.8 Ribosomes are large, complex organelles. Bacterial ribosomes consist of two subunits: a large 50S subunit containing 23S and 5S RNA and about 35 proteins and a smaller 30S subunit containing 16S RNA and about 20 proteins. The intact 50S subunit has the peptidyl transferase region, at which peptide bonds are formed. The 50S subunit also contains the P site, where the tRNA carrying the growing peptide chain is believed to be bound. Messenger RNA binds to the 30S subunit, which also carries the A site, where the mRNA codon recognizes and binds the new aminoacyl tRNA prior to the formation of each peptide bond.

When many ribosomes are simultaneously translating the same mRNA molecule (which is usually the case), the resulting complex is called a polyribosome or polysome.

The ribosomes from all species consist of two parts, a small and a large subunit (Fig. 13.8). The small subunit contains 16S or 18S* ribosomal RNA with its complement of ribosomal proteins; this subunit functions in the initiation of protein synthesis

---

* Large molecules such as RNA are characterized by their behavior in a centrifugal field. The unit of velocity in centrifugation is named after Svedberg, who invented high-speed centrifuges. In centrifugation, S values are proportional to molecular weight. Thus, a small RNA, such as tRNA, has an ultracentrifugation value of 4S and the three rRNAs of procaryotic ribosomes are, in order of increasing size, 5S, 16S, and 23S. In eucaryotic ribosomes the rRNAs are larger: 5S, 18S, and 28S; in addition they contain 5.8S RNA that is exclusive to eucaryotes.

and is the subunit to which mRNA binds. The large subunit contains the 23S or 28S and the 5S rRNAs and possesses the major enzymatic activity of the ribosomes, that of peptidyl transferase. This enzyme is responsible for joining the amino group of one amino acid to the carboxyl group of the next amino acid in the peptide chain. That is, the peptide bond is formed on the large subunit of the ribosome. The two ribosome subunits play separate and complementary roles, and the ability of the ribosome to separate into two subunits is an important feature of protein synthesis.

13.8 Factors and their roles in protein synthesis

One of the most complex, and least understood, aspects of the process of protein synthesis is the involvement of a number of proteins called "factors." They are not enzymes, but are required to interact with ribosomes or tRNA at each stage of translation. The protein-synthesis factors are not classified as ribosomal proteins, since they do not remain attached permanently to the ribosome. They do their job and then leave! The way in which these factors act is not known, but it is thought that they interact with and activate other components of protein synthesis, in much the same way that effectors interact with enzymes to modify their catalytic activity. There are initiation factors, elongation factors, and termination factors, each a different protein with a specific role in translation.

An interesting question to ponder is the evolution of the protein synthetic machinery; more than 70 different proteins (factors, enzymes, and ribosomal proteins) are required for the synthesis of each protein in the cell. Since proteins cannot be made without the mediation of other proteins, the question "Which came first, the chicken or the egg?" becomes even more difficult to answer!

13.9 Getting it all together—the stages of protein synthesis

The steps of protein synthesis are shown in detail in Fig. 13.9. This looks, and is, a very complicated business, but we can simplify things by considering protein synthesis in three stages: initiation, elongation, and termination.

In the first stage, initiation, the mRNA and ribosomes come together in the presence of initiation factors and an initiator tRNA to form an initiation complex. In this complex, the mRNA is "lined up" on the ribosome so that it can be translated from the first codon of a particular protein. The initiation factors and initiator tRNA interact with "start" sequences and ensure that protein synthesis begins at the correct place on the mRNA; it is now "phased" to read the triplets in their correct order. After all, the cell cannot tolerate a situation in which translation starts at any position along an mRNA and produces a random collection of polypeptide chains.

The "start" or "initiation" codon is always AUG, which is a codon for methionine. AUG reads "start" through the mediation of a special methionyl tRNA* that only interacts with the AUG codon in the presence of initiation factors. This would lead one to conclude that all proteins in all cells would have methionine as the N-terminal amino acid; however the methionine (or its derivative) at the N-terminus of the protein is usually removed by the action of proteases.

The amino acid next to the N-terminal initiating amino acid, attached to its specific tRNA which recognizes the codon adjacent to the initiation codon, is now added to the complex. When this "prepeptide" complex is formed, every component is in correct juxtaposition for the formation of the first peptide bond (between the N-terminal methionine and the next amino acid). This reaction requires several elongation factors and also the peptidyl transferase activity that is associated with the large subunit of the ribosome. The energy for peptide bond formation is provided, not by ATP, but by GTP. As shown in Fig. 13.9, with the formation of the peptide bond, the methionine is attached to the amino acid of the second aminoacyl tRNA and the initiating tRNA is ejected by movement of the ribosome along the mRNA. This movement allows the third codon to line up on the ribosome so that it can direct the insertion of the third amino acid. The ribosome then "cranks" along the mRNA, and each amino acid is added, in turn, to the growing peptide chain, its binding to the complex being directed by the codon–tRNA anticodon interaction. As many as 15 ribosomes can move along the same mRNA at one time, each one synthesizing the same protein. The complex of a messenger RNA with an attached group of ribosomes is called a *polyribosome* (Fig. 13.8). The polyribosomes of animal cells often contain more ribosomes than those of bacteria.

In the bacterium *Escherichia coli*, whose lifetime is about 40 minutes, the synthesis of a complete protein takes about 3 minutes (addition rate of one amino acid every 20 seconds). (Compare this with the large effort and inefficiency of the chemical synthesis of much shorter polypeptides in the laboratory.) In eucaryotic cells with longer lifetimes, the rate of addition of amino acids to a growing protein chain is somewhat slower. In the cell many copies of a protein can be made very quickly, because (1) there is more than one copy of the mRNA for this protein, and (2) each mRNA is simultaneously translated by a number of ribosomes.

---

* In eucaryotic cells, methionine is the "starter" or initiating amino acid, but in procaryotic cells a derivative, *N*-formylmethionine, is used.

Figure 13.9 Details of protein synthesis. This scheme illustrates how proteins are made, in stepwise fashion, in all living cells. These steps were worked out by breaking and fractionating cells to produce cell-free protein synthesis in the test tube. By purifying many of the components individually, their precise roles in the sequence have been elucidated. Since certain antibiotics (e.g., tetracycline, chloramphenicol) act by inhibiting specific steps in protein synthesis, these antibiotics can be used to cause intracellular accumulation of intermediates in this complex process, in much the same way that mutants are used to elucidate biosynthetic pathways (Sec. 5.3).

INITIATION

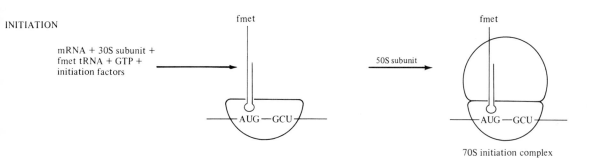

ELONGATION

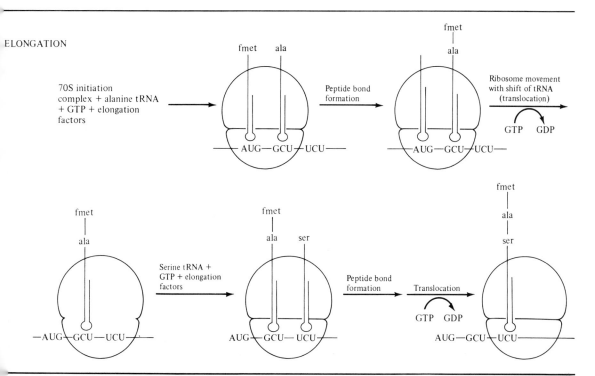

TERMINATION

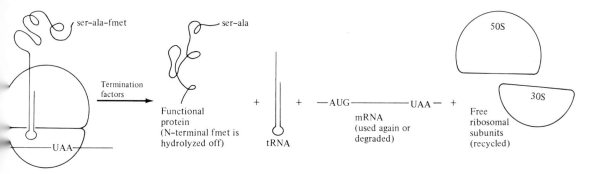

When the 3' end of the messenger RNA is reached, there must be a signal for releasing the completed peptide chain from the ribosome and tRNA to which it is attached. At the end of the sequence of codons in mRNA, there are one or more termination codons (UAA, UAG, or UGA) that, in conjunction with (protein) termination factors, promote the release of the peptidyl-tRNA from the ribosome and the cleavage of the completed peptide chain from the last tRNA. The "empty" ribosomes then fall off the mRNA, separate into large and small subunits, and recommence the protein synthetic process at the beginning (5' end) of another mRNA.

There is an additional aspect of protein synthesis to consider. We have emphasized that proteins have complex three-dimensional structures; how do these shapes form? The information for the folding of proteins is intrinsic to their amino acid sequences. As proteins are being synthesized on ribosomes, they automatically fold into their active conformation, and the completed, folded structure is released from the ribosome when it reaches the "stop" signal.

In the case of proteins that consist of different subunits, each subunit is synthesized on a different mRNA and then released, folded in its active form. The subunits then aggregate to form the active protein molecule.

Now that we understand the way in which mRNA is read into protein, it is not difficult to understand how mutations in DNA can affect protein structure. A mutation is usually a simple change in the base sequence of DNA (Table 12.4). Sometimes a base substitution changes one codon into the codon for another amino acid, with the result that an amino acid change occurs in the protein. For some codons, a base substitution may generate a termination signal (UAA, UAG, or UGA). Base substitutions of the latter type are referred to as *nonsense mutations*, since the net result is premature termination of the synthesis of a protein and the formation of a short, inactive protein. In a form of anemia known as β-thalassemia a defective hemoglobin is formed because of the production of an ineffective β subunit; it is believed that this is the result of a mutation that created a nonsense·codon within the gene for the β subunit of hemoglobin.

SUMMARY

The most abundant nucleic acid in cells is ribonucleic acid (RNA). RNA is a single-stranded molecule with a linear backbone of alternating ribose and phosphate groups; purine or pyrimidine bases are linked to each sugar residue.

The sequence of bases in an RNA molecule is determined by the DNA template (the gene) used in its synthesis. RNA synthesis (transcription) occurs when the appropriate ribonucleotide precursors base-pair with the DNA template; phosphodiester bonds are formed between adjacent nucleotides by RNA polymerase. There are promoter and terminator sequences in the DNA flanking each gene, which "tell" the RNA polymerase exactly where to start and stop transcription.

RNA molecules fall into three classes defined by their role in protein synthesis: ribosomal RNA, messenger RNA, and transfer RNA. Ribosomal RNAs (rRNA) are structural components of the nucleoprotein particles (ribosomes) that are the "workbench" for the process of protein synthesis. Messenger RNAs (mRNA) carry the information (i.e., the sequence of base triplets) for the amino acid sequence of proteins from the genes to the ribosomes. Transfer RNAs (tRNA) are the link (adapters) between the codons in mRNA and amino acids.

The use of the genetic "blueprint" to make an active protein is a multistep process. First, the gene is transcribed into complementary mRNA; the mRNA then binds to a ribosome. The start codon at the beginning of the mRNA is recognized by an initiator methionyl tRNA, thus aligning the first mRNA triplet with the N-terminal amino acid of the protein. Then the second codon in mRNA binds its aminoacyl tRNA. Once the two tRNAs are "side by side" on the ribosome, the formation of the first peptide bond is catalyzed, freeing the initiator tRNA and leaving a dipeptide attached to the tRNA for the second codon. The third tRNA base-pairs with the third mRNA codon and a second peptide bond is made. This translation process continues as the ribosome travels along the mRNA; finally, a stop codon is reached and the completed protein is released.

PRACTICE PROBLEMS

Use Tables 13.2 and 13.3 to answer the following questions. RNA sequences are written with the 5' end at the left and the 3' end at the right.

1. For a protein containing 100 amino acids, calculate the approximate molecular weight of its messenger RNA and its gene (each nucleotide has a molecular weight of about 300). Ignore the sequences required for punctuation (start and stop).

2. In bacteria, messenger RNA is unstable; each molecule is used only a few times and then it is degraded. Can you suggest a reason for this?

3. The following sets of fragments were produced when a short nucleic acid was degraded with three different ribonucleases. Suggest the ribonucleases that were used and determine the sequence of the nucleic acid.
   (a) AG, GU, C, AC.
   (b) A, CCG, UA, G.
   (c) UAG, ACCG.

4. Although most RNA molecules are single-stranded, they are susceptible to cleavage by double-strand specific ribonucleases. Explain.

5. List all the coding triplets that can change to nonsense triplets by a single base change in the DNA.

6. Write the base sequences of mRNA and gene (double-stranded DNA) corresponding to each of the following peptides:
   (a) Leu-ala-val-met-asp-tyr-his.
   (b) Met-ile-cys-trp-glu-lys-cys-ser.
   (c) Phe-(lys)$_3$-(gly)$_2$-thr.
   Is there a unique mRNA sequence for each peptide? If not, why not?

7. What additional triplets would be needed for the mRNAs in problem 6 to function in translation?

8. Indicate whether the following statements are true or false. Explain your answers.
   (a) Only DNA contains thymine.
   (b) RNA polymerase is responsible for the synthesis of all kinds of RNA in bacteria.
   (c) ATP is required for replication, transcription, and translation.
   (d) An inhibitor that blocked RNA synthesis in a bacterium would not inhibit protein synthesis.
   (e) Each amino acid has its own aminoacyl tRNA synthetase.
   (f) The first step in the synthesis of a protein is the binding of alanyl tRNA to the ribosome.

9. The nucleotide sequence around the initiation site on mRNA for a certain protein is

   -GAAGCAUGGCUUCUAACUUUU-

   (a) Which codon specifies the first amino acid in the protein?
   (b) Write the amino acid sequence for the amino terminal section of this protein.
   (c) A codon specifying termination (nonsense), UAA, is indicated. Why doesn't this cause termination?

10. You are given a tube containing a mixture of RNA and DNA. Give one method by which you could destroy only the DNA and two methods to destroy only the RNA.

11. Compare the characteristics of RNA and DNA in eucaryotic cells by filling in the table.

	RNA	DNA
Five-carbon sugar		
Bases		
Template for synthesis		
Location in the cell		
Copies/cell (one vs. thousands)		

12. Canavanine is an amino acid whose chemical structure is similar to arginine. When cells take up canavanine, protein synthesis is inhibited. Suggest a mechanism for this inhibition.

13. DNA molecules from two different species of bacteria contain significantly different percentages of AT and GC pairs; one contains 40% AT pairs and the other 55% AT pairs. Both use the same genetic code. Analyses show that the distribution of amino acids in their total protein is very similar; both contain roughly 5% serine, 4% aspartic acid, and so on. How can the composition of their proteins be so similar if that of their DNA is so different?

SUGGESTED READING

CLARK, BRIAN F. C., and KJELD A. MARCKER, "How Proteins Start," *Sci. American*, 218 No. 1, p. 36 (1968). In bacteria, an initiator tRNA, carrying formylmethionine, recognizes the start codon in mRNA and places its amino acid in the first position in the polypeptide chain.

CRICK, F. H. C., "The Genetic Code: III," *Sci. American*, 215, No. 4, p. 55 (1966). A combination of genetic and biochemical techniques was used to assign each amino acid to its mRNA codon(s).

HOLLEY, ROBERT W., "The Nucleotide Sequence of a Nucleic Acid," *Sci. American*, 214, No. 2, p. 30 (1966). To determine the base sequence of alanyl tRNA, the molecule was digested with ribonucleases and the resulting fragments separated and analyzed.

MILLER, O. L., JR., "The Visualization of Genes in Action," *Sci. American*, 228, No. 3, p. 34 (1973). Excellent electron micrographs of eucaryotic DNA being transcribed into RNA by RNA polymerase. The pictures show transcription and translation as "real" processes rather than as something confined to the test tube or textbook.

NOMURA, MASAYASU, "Ribosomes," *Sci. American*, 221, No. 4, p. 28 (1969). One successful approach to determining the relationship between the structure and function of ribosomes was to dissociate these nucleoprotein particles, examine each component, and then reassemble the parts in various ways.

RICH, ALEXANDER, "Polyribosomes," *Sci. American*, 209, No. 6, p. 44 (1963). Describes biochemical techniques used to show that many ribosomes travel simultaneously along a single mRNA molecule.

RICH, ALEXANDER, and SUNG HOU KIM, "The Three-Dimensional Structure of Transfer RNA," *Sci. American*, 238, No. 1, p. 52 (1978). Determination of the details of tRNA structure has increased our understanding of the role of this kind of RNA in the translation process.

SATIR, BIRGIT, "The Final Steps in Secretion," *Sci. American*, 233, No. 4, p. 28 (1975). Highly specialized organelles transport and secrete those proteins which function outside the cell where they are synthesized.

14 Regulation & Control: How Cells Turn Metabolic Reactions On & Off

All living organisms have the capacity for change. Bacteria adjust their metabolism so that they can survive on an ever-changing supply of nutrients. Animals with varied diets must maintain the blood levels of many substances within a narrow range. Complex higher organisms pass through distinct stages of development: human beings change from child to adult, tadpoles undergo a dramatic metamorphosis to frogs, fruit trees produce their annual crop, and so on. Thus, there are changes that occur in response to immediate needs or environmental stimuli (adaptation) and changes that constitute genetically controlled steps in an organism's life cycle (differentiation and development). These changes may be reversible or irreversible.

Since the genetic information (DNA) of any cell is the same as that of its ancestors and of its progeny, the types of changes mentioned above must be due to differences in the *expression* of genes and not to alterations in the actual gene content (e.g., the loss or acquisition of genes).* Given these facts, the only ways that an organism can change the outward expression of its genes (phenotype) are to regulate the transcription (RNA synthesis) or translation (protein synthesis) of its fixed genetic content (genotype), or to control, directly, the activity of enzymes with activators or inhibitors.

Cells have evolved a multitude of systems for controlling their metabolism; transcriptional, translational, and post-translational control are all used (Table 14.1).

---

* The one exception is discussed in Sec. 14.4.

TABLE 14.1 Mechanisms of Regulation of Gene Expression

There are many ways by which a specific gene or its products can be "turned on" or "turned off" without gross changes in the cell's other metabolic activities.

Level of regulation		Mechanism of regulation
Intracellular	DNA replication	Increasing amount of gene product by multicopying genes (gene amplification); very rare
	Transcription	Inhibition or enhancement of RNA polymerase activity by: (1) changing the strength of the polymerase–promoter binding (2) changing the rate at which the polymerase begins to transcribe the gene (3) chromosome condensation and dispersion
	Translation	Inhibition or enhancement of protein synthesis by: (1) modifications of start or stop signals or ribosomal binding sites on mRNA (2) gene-specific protein synthesis factors or ribosomal components
	Post-translation	Inhibition or enhancement of enzyme activity by: (1) modification of amino acid R groups (2) reversible binding of small effector molecules (3) mass-action effects
Intercellular	Within an individual organism	Hormones synchronize metabolism in diverse organs
	Between organisms	Chemical messages (pheromones) coordinate the activities of physically separated members of the same species

In these ways, metabolic pathways can be "turned on" or "turned off" as the need for a particular end product increases or decreases. In choosing a mechanism for regulating a particular pathway, the cell must strike a balance between economy and speed. Transcriptional control is the most economical; neither material nor energy are wasted making gene products that will not be used. But to achieve a rapid response, it is often necessary to control metabolism at the enzyme level; small molecule activators and inhibitors (effectors) can quickly (and usually reversibly) bind to an enzyme, altering its catalytic activity.

In addition to intracellular regulation, mechanisms are needed to coordinate the highly differentiated organs of complex organisms. Both animals and plants produce a plethora of hormones which balance the specialized metabolic activities of different body parts. In animals, the ultimate integrators of metabolic functions are the brain and nervous system.

14.1 Regulation at the level of transcription

This is probably the most important form of control that exists in nature. By using combinations of regulatory elements, cells can control the expression of genes by turning transcription on or off, at the appropriate times. Transcriptional control occurs during the maturation of larvae, or when a bacterium is presented with an abrupt change in its carbon source; in both cases, the expression of some genes has to be shut off and other genes turned on.

Control is exerted at two points in the transcription process depending upon:

1. Whether or not RNA polymerase can bind to the DNA prior to initiation of RNA synthesis.
2. Whether or not the RNA polymerase, having bound to the DNA, can move along that DNA and transcribe its complementary messenger RNA.

It is easy to visualize mechanisms for regulating these steps of transcription if we consider, not the structure of the gene, but the sequence of bases in the DNA that precedes any particular gene on the chromosome (Fig. 14.1).

All structural genes must have an accompanying promoter site, which includes the binding site for RNA polymerase. Different promoters have different base sequences that have strong or weak affinities for the polymerase. Those structural genes that need to be expressed at a high rate throughout the life of the cell (genes coding for ribosomal components, RNA polymerase, the glycolytic and Krebs cycle enzymes, etc.) would be expected to have "strong" promoters; that is, their promoters have strong affinities for RNA polymerase, so these key structural genes are transcribed frequently and efficiently. Structural genes whose products are needed less often would be expected to be associated with "weak" promoters.

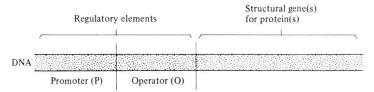

Figure 14.1 Regulatory elements and their structural genes. This figure shows the physical arrangement of a gene and its associated regulatory elements, the promoter and the operator. The regulatory elements produce no gene products; structural genes code for the amino acid sequence of proteins, usually enzymes. RNA polymerase must bind to the promoter and move along the operator before transcription of the structural gene can begin. The base sequence of the promoter associated with the lactose operon (Fig. 14.3) is shown in Fig. 13.2.

In bacteria, the structural genes required for a particular metabolic pathway, such as lactose utilization or tryptophan biosynthesis, may be arranged linearly along the DNA (Fig. 14.2) and coordinately controlled by a single promoter and operator. This genetic unit is known as an operon.

In addition to properties due to the DNA sequence, the affinity of RNA polymerase for promoters may in some cases be increased or decreased by small molecule effectors. In bacteria, one such effector is cyclic adenosine monophosphate (cyclic AMP), which stimulates transcription by increasing the strength of the polymerase–promoter binding. Remember that cAMP is also an important regulatory molecule in glycogen metabolism in mammals (Fig. 9.2 and Sec. 14.5), where it activates protein kinases by interacting with these enzymes. It is interesting that the same molecule can act in such different ways.

While interactions at the promoter affect the binding of RNA polymerase, interactions at an adjacent site, the operator, control whether or not the bound polymerase can begin to move along and transcribe the gene into mRNA. Certain small proteins can bind to operator sequences in DNA and either prevent ("jam up") or enhance transcription (Fig. 14.2); these control proteins, the products of regulatory genes, are known as "repressors" or "activators." Operator-controlled systems of regulation are usually associated with the structural genes for enzymes needed only intermittently during the life of the cell. Many genes, such as those coding for Krebs cycle or glycolytic enzymes, are not under such control because their products are required throughout the cell cycle; these are known as constitutively synthesized enzymes.

(a) NEGATIVE TRANSCRIPTIONAL CONTROL. IS THE LACTOSE
OPERON A MODEL FOR ALL REGULATION?

In 1965, François Jacob and Jacques Monod of the Pasteur Institute received the Nobel prize for their work in unraveling the mechanism by which bacteria adjust their metabolism when they are transferred from a medium containing glucose as carbon source to a medium containing lactose. Mostly from genetic experiments, Jacob and Monod proposed a model of negative regulation that has now been confirmed and

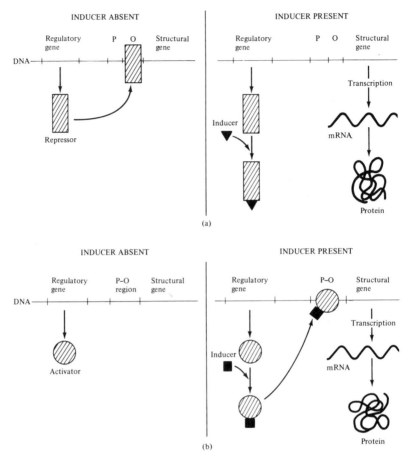

Figure 14.2 Positive and negative control of transcription. Regulatory genes code for control proteins which act at the operator to prevent or enhance the transcription of the structural genes adjacent to the operator—promoter region. The terms "negative" and "positive" refer to the effect of the control protein on transcription (i.e., whether it turns transcription "off" or "on"). The activity of the control protein is often affected by an inducer, a small molecule that is usually a substrate of the pathway whose structural genes are being regulated. In both kinds of control, the addition of inducer stimulates the transcription of the structural genes. (a) Negative control: binding of the control protein (repressor) to the operator blocks transcription. In the absence of inducer, the repressor binds to the operator and prevents transcription of the structural genes. When the inducer is present, it binds to the repressor, changing its configuration so that it can no longer recognize and bind to the operator. With no repressor bound to the operator, transcription of the structural genes can proceed. (b) Positive control: binding of the control protein (activator) to the promoter-operator region enhances transcription. In the absence of inducer, the control protein cannot bind to the DNA and the structural genes are not transcribed. When inducer is present, this small molecule binds to the activator, changing its configuration so that it can recognize the promoter–operator site. With the activator bound, transcription of the structural genes can proceed.

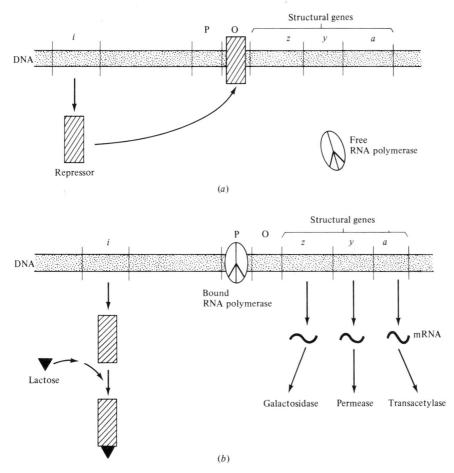

Figure 14.3 Control of the lactose operon in bacteria. (a) When bacteria are growing in glucose, the lactose operon is turned off. The three structural genes, z (galactosidase), y (lactose permease), and a (transacetylase) are not transcribed because (1) RNA polymerase binds poorly to the promoter (P), owing to the lack of cyclic AMP, and (2) bound RNA polymerase cannot move along the DNA because the *lac* repressor, a protein coded by the *i* gene, is bound to the operator (O). (b) When glucose is replaced by lactose, the bacteria must adjust to metabolize this new carbon source or stop growing. This adjustment occurs in two ways. (1) When cells are shifted from glucose to lactose, cyclic AMP is made and this increases the binding of RNA polymerase to the promoter. (2) As lactose enters the cell, it is immediately converted to a metabolite that binds to the repressor protein, preventing the latter from binding to the operator. With no repressor bound, the RNA polymerase is free to move along the DNA and transcribe the structural genes. The permease transports lactose into the cell, where it is cleaved into glucose and galactose by galactosidase and these sugars catabolized by the glycolytic pathway. The function of the transacetylase is unknown.

When glucose is added back to the cells, the synthesis of the enzymes required for the metabolism of lactose is turned off; the repressor would no longer be inactivated and would again bind to the operator and prevent transcription. In addition, cAMP production would be arrested, and this also influences the expression of the *lac* operon.

extended by biochemical studies to include other biosynthetic and degradative pathways. The model is presented in Fig. 14.3. It is called *negative* control because, when the bacteria are growing in glucose, transcription of the lactose genes is turned off by the action of a repressor protein.

Other examples of negative control are known. Most bacteria do not require amino acids for growth; they make their own from simple precursors supplied in the growth medium. If a bacterium is presented with a plentiful supply of amino acids, it would be uneconomical to continue synthesizing these compounds. To avoid this unnecessary waste, the presence of an excess of the required end product (e.g., amino acid) immediately shuts off the transcription of all the genes involved in the biosynthesis of this product. Most bacteria have the capacity to make tryptophan. When the need to make tryptophan is eliminated by provision of this amino acid, tryptophan or a derivative interacts with a control protein in the cell to make an active repressor complex that inhibits transcription of the genes for tryptophan biosynthesis.

(b) POSITIVE CONTROL MAY BE MORE GENERAL THAN NEGATIVE CONTROL

Although the mechanism of negative control of transcription has been well worked out, not many instances of this type of regulation have been found. In fact, positive control mechanisms, in which transcription is turned on by the action of an activator protein (Fig. 14.2), may be more common in many organisms. In this case, the genes are dormant under noninduced conditions; transcription begins only when an inducer is present and the activator protein–inducer complex is bound to the operator. The pathways for the catabolism of maltose and arabinose in *Escherichia coli* are under such positive control; when one of these sugars enters the cell, it binds to an activator protein, forming a complex that binds to the DNA and stimulates the transcription, and hence the synthesis, of the enzymes for the degradation of the sugar.

At present, no one knows how such activator proteins work, but it is possible that they enhance the binding of RNA polymerase to the promoter, or that they bind to DNA and induce shape changes which allow RNA polymerase to transcribe the adjacent structural genes.

(c) OTHER REGULATION AT THE CHROMOSOME LEVEL

Although it seems likely that essentially the same mechanisms of positive and negative control of transcription occur in higher organisms (what is true for bacteria is true for elephants), the greater complexity of the eucaryotic chromosome allows for additional systems of regulation. Eucaryotic chromosomes consist of a complex of DNA and proteins, primarily histones (Sec. 12.2). Chromosomes can exist in a dispersed state, in which form their DNA appears to be readily transcribed; or in a compacted state, in which case their DNA seems to be transcribed less efficiently. In the compacted form, chromosomal proteins may play a regulatory role by "covering up" the DNA and preventing the binding and functioning of RNA polymerase. This implies that, in higher organisms, mechanisms must exist for specific and perhaps sequential attachment and removal of chromosomal proteins, in order that the proper genes be expressed at the correct stages of development of the plant or animal.

Large blocks of genes can be rendered inactive by chromosome condensation during certain stages of development. For example, in human females, one X chromosome is not expressed because it exists in an inactive, condensed state. When the female embryo reaches a few thousand cells, one X chromosome in each cell* condenses to form the genetically inert (and cytologically distinct) Barr body. This inactivation explains why males can survive with one X chromosome, whereas females apparently need two; we all, in fact, have but one active X chromosome.

14.2 Regulation at the level of translation: turning protein synthesis on and off

Control at the level of transcription is economical since energy is not expended in making messenger RNA that is not translated. However, transcriptional control depends on the existence of short-lived mRNAs; that is, turning off the transcription of a gene would be ineffective unless the previously formed mRNAs were rapidly destroyed. In procaryotes, most mRNA molecules are unstable and hence transcriptional control of gene expression is feasible.

However, in certain specialized and/or nondividing cells of eucaryotes, mRNA is long-lived. For example, immature red blood cells produce virtually no new mRNA while they are synthesizing large quantities of hemoglobin; the same mRNAs are used over and over again. Even more striking is the situation in echinoderm (e.g., sea urchin) eggs. The unfertilized eggs carry all the mRNA needed for development into the gastrulation stage; fertilization triggers the translation of these mRNAs and the start of embryogenesis. In these and other similar cases, if gene expression is to be controlled, that control must be exerted at the translation stage or later.

Although it is not clear how and where translation is controlled, there are several steps in the process which are likely candidates as control points:

1. Many mRNA molecules in eucaryotic cells have chemical modifications of the nitrogenous bases at their termini; for example, a methylated guanosine is often found as the 5′ terminal base. These modifications may be important to promote initiation of translation or ribosome binding, since the unmodified mRNA molecules are less active in translation than those with the modification.

2. Since many protein factors are needed for translation (Sec. 13.8), it has been suggested that different factors specific to the translation of certain mRNAs

---

* The X chromosome that is inactivated is random and differs from cell line to cell line.

may exist. For example, some mRNAs may require specific initiation factors that catalyze the formation of initiation complexes for the synthesis of their translation products.

3. Specific ribosomal proteins or their derivatives, produced at appropriate times in the life cycle of a cell or organism, could influence the translation of the mRNA molecules available at that time, by a variety of different mechanisms, through transient changes in the properties of ribosomes.

4. Different transfer RNAs, charged with the same amino acid but differing in their codon recognition because of redundancy in the genetic code, could be available at different times or be produced in response to certain stimuli such that the synthesis of some proteins could be completed, but others not, depending on the codon present in the mRNA made at that time.

14.3 Post-translational control: the regulation of enzyme activity

We have said that regulation at the level of translation is uneconomical. Regulation after the protein is made may seem to be even more wasteful! However, this proves to be the most rapid and perhaps convenient way for a cell to control enzyme activity.

A given reaction can be turned on or off very quickly by activating or inhibiting the enzyme that catalyzes that reaction. There are several ways of accomplishing this; the first is to modify the enzyme protein chemically. Many of the R groups of proteins can be reversibly altered by enzymatic reactions such as acetylation (lysine) or phosphorylation (serine, threonine), and such modifications may have a considerable effect on the activity of an enzyme, especially if the R group being modified is associated with the active site. We have already described how glycogen breakdown is controlled by specific kinases that phosphorylate and activate glycogen phosphorylase, while other kinases phosphorylate and inactivate glycogen synthetase (Fig. 9.2).

Some chemical modifications of enzyme proteins are irreversible. Several digestive enzymes in human beings are synthesized in an inactive form and are activated by the removal of a small number of amino acids. One example is chymotrypsin. The inactive form of the enzyme, chymotrypsinogen, is synthesized in the pancreas and secreted into the small intestine, where the removal of three dipeptides converts chymotrypsinogen to the active chymotrypsin.

The second way in which the activity of an enzyme may be modified is to use a reversible activator or inhibitor (an "effector") of the enzyme. This provides a most sensitive and rapid type of adjustment. Effectors are small organic molecules that bind reversibly to an enzyme at a position distant from the active site; this binding perturbs the three-dimensional shape of the whole protein, changing the structure and hence the functionality of the active site. Effectors can either increase or decrease enzyme activity, depending on how they affect the active site. Recall how activity in the Krebs cycle, glycolysis, and fatty acid biosynthesis can be coordinately controlled

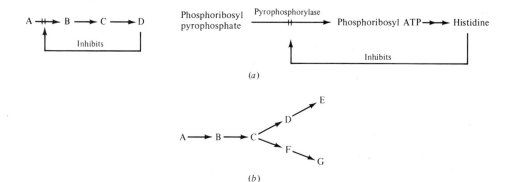

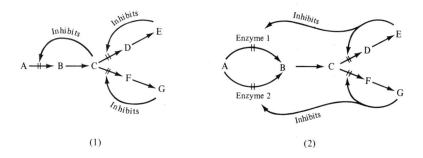

Figure 14.4 Control of biosynthetic pathways by feedback inhibition. Feedback inhibition is one mechanism for controlling the rate of synthesis of an amino acid or other small molecule.

(a) Control of unbranched pathways. When the concentration of end product (D) increases, the excess D inhibits the first enzyme in its biosynthetic pathway, thus stopping its own production. When D levels fall, the pathway starts up again. One example is the bacterial production of histidine.

(b) Control of branched pathways. When two metabolites are made from a common intermediate (C), the feedback loops become more intricate. In such a system, if the end product E builds up, the cell needs to inhibit D production but may or may not want to reduce the production of B and C (precursors of both E and G). An analogous situation occurs if G accumulates. The problem can be solved in several ways, for example:

(1) The branch-point compound feedback inhibits the first common reaction. Both end products inhibit the first reaction unique to their own biosynthesis after the branch point. The branch-point compound (C), which will accumulate only when the cell contains an excess of both E and G, inhibits the A to B conversion.

(2) Multiple enzymes, each sensitive to a different end product, catalyze the first common reaction. As in (1), both end products inhibit the first reaction unique to their own biosynthesis after the branch point. Additionally, each end product can inhibit one of the enzymes responsible for the A to B conversion.

by the concentration of citrate; high levels of citrate inhibit phosphofructokinase (Fig. 8.3) and at the same time stimulate the activity of acetyl coenzyme A carboxylase, an enzyme involved in fatty acid biosynthesis.

Control of enzymatic activity by inhibitors is very important in the regulation of the biosynthesis of amino acids and other small molecules. The final product of a biosynthetic pathway often inhibits the enzyme that catalyzes the first step in the pathway; this is known as *feedback inhibition* (Fig. 14.4). Control of this type is economical of material and energy. If, by chance, the cell is provided with an external supply of an amino acid, or has made more than it can use, the biosynthetic pathway to the amino acid, which is no longer needed, is shut down. When the intracellular level of the end product amino acid falls, the inhibition of the first enzyme in the pathway is released and the whole reaction series starts up again. Several variations on this theme exist, particularly when two amino acids are made from the same branch-point intermediate. In this case the first enzyme of the pathway must be sensitive to the concentrations of both final products (Fig. 14.4).

The third way in which the activity of an enzyme may be controlled is simply by mass-action effects. In equilibrium reactions, there is a fixed ratio of the concentrations of reactants to products, and any change in the concentration of either will force the reaction back toward equilibrium by adjusting the reaction in one direction or the other. Since all biological reactions are equilibrium reactions, if the product of a reaction begins to accumulate (because it is not being used), and exceeds the equilibrium concentration, the reaction may be reversed, effectively shutting off the forward reaction. For example, in the phosphorylation of a substrate with ATP:

$$\text{substrate} + \text{ATP} \xrightarrow{\text{Kinase}} \text{substrate-P} + \text{ADP}$$

If substrate-P and ADP accumulate, the reaction will reverse itself and regenerate nonphosphorylated substrate and ATP, until equilibrium concentrations are reached once again.

$$\text{substrate-P} + \text{ADP} \xrightarrow{\text{Kinase}} \text{substrate} + \text{ATP}$$

This reversal is not possible in all cases and is probably not very common, since fairly high concentrations of substrate would be involved. Indeed, in certain dephosphorylation reactions, the energetics are unfavorable for the reverse reaction. An example is the conversion of phosphoenolpyruvate to pyruvate; this reaction is irreversible and has to be bypassed for the purpose of gluconeogenesis.

14.4 Gene amplification: more genes mean more gene products

We have discussed the ways by which gene expression can be regulated by controlling transcription, or translation, or by controlling the activity of gene products (enzymes). One more control mechanism exists; cells can, *in a few specific instances*, control gene expression by varying the number of copies of the gene itself. During oogenesis in certain amphibians, there is a large increase in the number of copies of the genes coding for ribosomal RNA. More copies of the gene means more copies of the gene product; this is necessary because the organism needs many ribosomes to synthesize large quantities of protein during early stages of development. At present, we do not understand the events that trigger the amplification of such a small portion of the genome.

In bacteria, selective amplification of certain genes that exist on transmissible resistance plasmids (Sec. 16.7) can occur when plasmid-carrying strains are exposed to high concentrations of antibiotic; the amplification of resistance-determining genes permits the strains to be resistant to very high concentrations of drug. In this case, the gene products are enzymes that detoxify the antibiotics.

14.5 Hormones—chemical coordinators synchronize your cells

Thus far we have discussed how events are controlled inside individual cells; in complex multicellular organisms, cells are organized into differentiated tissues and the metabolic activities of these tissues have to be sychronized. This is the function of hormones, a large and diverse group of chemical messengers, including peptides, steroids, and amino acid derivatives. Hormones are responsible for most aspects of human physiology and behavior. Many of the warlike characters of history may have been the victims of their own hormone imbalances!

Table 14.2 lists some human hormones and the reactions they control; this table illustrates the diversity of hormones with respect to their chemistry, their sources and targets, and the aspects of metabolism that they control. Plants, like animals, consist of differentiated tissues whose metabolic activities must be coordinated; some examples of plant hormones are given in Table 14.3.

Hormones are produced by one organ and then pass through the circulatory system to exert controlling effects on other organs. The target-cell response usually occurs by means of one of the intracellular control mechanisms already described. In other words, hormonal control is superimposed upon, and works in concert with, the intracellular transcriptional, translational, and post-translational regulatory systems.

In general, hormones do not enter their target organs, but bind to receptors on the cell membrane and trigger responses inside the cell. Hormones of different chemical structures can act on the same target organ through different receptors. It is believed that hormone receptors are membrane-associated proteins; the receptors must have regions with very specific three-dimensional shapes, analogous to the active sites of enzymes, to distinguish between the various hormones.

Binding of the hormone to its membrane receptor triggers a chain of intracellular activities. The bound hormone may stimulate the production of a second chemical messenger, which is, in turn, responsible for the intracellular hormonal effects. In several cases, the binding of hormone stimulates the activity of the enzyme adenyl cyclase (which is present on the inside of the membrane) and hence increases the

TABLE 14.2 Human Hormones

Our body chemistry is coordinated by the activities of a large number of compounds with extremely varied responsibilities. This table includes only a few examples of each of the three chemical classes; a more complete list may be found in E. Frieden et al., *Biochemical Endocrinology of the Vertebrates* (Prentice-Hall, Inc., Englewood Cliffs, N.J., 1971).

Hormone	Chemistry (size or precursor)	Producing organ	Target organ	Physiological response
PROTEIN HORMONES				
Growth hormone	Contains 188 amino acids	Pituitary	Most cells	Increased protein synthesis; increased fat utilization; growth of tissues (easily seen in long bones)
Adrenocorticotropin (ACTH)	39 amino acids	Pituitary	Adrenal cortex	Increased synthesis and release of glucocorticoids; growth of adrenal cortex
Vasopressin	9 amino acids	Pituitary	Kidneys, arteries	Increased reabsorption of water by renal tubules (antidiuretic action); contraction of smooth muscles of blood vessels
Insulin	51 amino acids (Fig. 2.3)	Pancreas	Most cells	Increased glucose uptake into cells, glycolysis and lipid breakdown; inhibition of glycogen synthesis and gluconeogenesis
Glucagon	29 amino acids	Pancreas	Liver	Increased glycogen and lipid breakdown; increased blood sugar
AMINO ACID DERIVATIVES				
Thyroxine	Derived from tyrosine	Thyroid	Most cells	Increased oxygen consumption and metabolic rate; increased growth and development; mechanism for heat regulation
Epinephrine	From tyrosine	Adrenal medulla	Most cells	Increased glycogen and lipid breakdown; increased cardiac activity (in response to sudden environmental change)

LIPID HORMONES[a]

Androgens (testosterone)	Derived from cholesterol	Testes, adrenal cortex	Most cells	Development of accessory reproductive organs in male; promote muscle and skeletal growth at puberty by stimulating protein anabolism
Estrogens (estradiol)	From cholesterol	Ovary, placenta	Most cells	Maintains secondary reproductive organs in female; in uterus, promotes growth of endometrium, including vascularity after menstruation
Progestogens (progesterone)	From cholesterol	Ovary, placenta	Uterus, mammary glands	Development of mammary glands; prepares uterus for reception of embryo after ovulation; suppresses ovulation; levels fall if fertilization does not occur, which stimulates sloughing of uterine wall (i.e., a new menstrual cycle begins)
Mineralocorticoids (aldosterone)	From cholesterol	Adrenal cortex	Kidney; sodium-excreting glands	Increased reabsorption of Na^+ by kidney; decreased Na^+ excretion by sweat glands, salivary glands, gastrointestinal tract
Glucocorticoids (cortisol)	From cholesterol	Adrenal cortex	Most cells	Increased liver gluconeogenesis and glycogen synthesis, increased blood sugar; decreased glucose and glycogen breakdown; anti-inflammatory
Prostaglandins (PGE)	From unsaturated fatty acids	Seminal vesicles, nerves, perhaps most cells	Most cells	Stimulate contraction of smooth muscle (uterus, duodenum); many other functions postulated

[a] Each group of lipid hormones consists of several chemically related compounds (Table 10.4); a metabolically important example from each group is given in parentheses.

TABLE 14.3 Plant Hormones

Like their counterparts in the animal kingdom, plant hormones are of different chemical types and their effects depend on the metabolism of the organ being regulated.

Hormone	Chemical nature	Action
Indole acetic acid	Derived from tryptophan	Stimulates cell extension, cell division, growth; stimulates rooting
Gibberellic acid	Lipid; derived from a precursor of squalene	Stimulates plant growth; in (barley) seeds, induces synthesis of α-amylase, an enzyme needed to use starch reserves; promotes seed germination; releases (potato) bud dormacy
Abscisic acid	Lipid; derived from a precursor of squalene	Inhibits cell extension, cell division; inhibits seed germination; induces dormacy
Ethylene	$H_2C=CH_2$	Stimulates ripening of mature fruit
Cytokinins	Modified adenine ribosides	Stimulate growth

intracellular level of cAMP, which acts as a second messenger. We have already seen how cAMP acts as a second messenger for the hormones epinephrine and glucagon (Fig. 9.2); it is also an intracellular messenger for thyroxine, vasopressin, and several other human hormones. In several of these instances, cAMP activates a protein kinase, which in turn modulates the activities of different proteins by phosphorylating them. Other hormones exert their effects through changes in the concentrations of prostaglandins, lipids produced in many cells.

Steroid hormones differ in that they apparently enter cells and exert a direct effect on transcription, although the mechanism of this action is not clearly known.

Many hormones have opposing functions; for example, a steroid hormone may promote a reaction, whereas a peptide hormone may turn off the same reaction. Because their actions are opposing, the hormones act as checks and balances on each other. Taken as a group, hormones ensure that no one metabolic pathway is over- or understimulated and that the circulating levels of critical metabolites are held within a specified range. This is the *homeostatic function* of hormones, and it requires all the hormones acting when and where they should.

A good example of homeostatic control is the maintenance of blood glucose levels. A high level of blood glucose results in the secretion of insulin, which facilitates the movement of glucose from the blood to the cells. Unchecked, insulin activity would soon reduce blood glucose to near zero. But, as blood glucose levels become critically low, glucagon is secreted, causing a breakdown of liver glycogen and an increase in blood glucose. By acting in sequence, insulin and glucagon maintain blood glucose levels within a precise and consistent range. Likewise, in all organisms, there is a

TABLE 14.4 Human Hormone Imbalances

The symptoms of hormone imbalances vary greatly, depending on the affected individual's sex, age, physiological state (e.g., nonhormonal disorders, pregnancy), and the degree of the imbalance. If the producing organ itself is malfunctioning, imbalances in several hormones may occur simultaneously, leading to the disruption of diverse areas of metabolism.

Hormone	Symptoms of deficiency	Symptoms of excess
Growth hormone	Dwarfism	Giantism (children); acromegaly (adults): large extremities, massive face, protruding jaw
ACTH	Hypotension; hypoglycemia; withstand stress poorly	Cushing's syndrome due to increase in cortisol; muscle wasting; fat deposits on face, neck, trunk; loss of protein from bone causes backache, spinal deformities; lethargy; florid complexion
Vasopressin	Diabetes insipidus: uncontrolled loss of water by kidneys; extreme thirst	Not known
Insulin	Diabetes mellitus: hyperglycemia; glucose in urine; muscle cramps; alterations in eyes; ketosis and eventual coma	Hypoglycemia; reduced activity; disturbed speech, vision, balance; convulsions; coma
Thyroxine	Cretinism (children): subnormal or vegetative intellectual capacity; hypothyroidism (adults): low metabolic rate; low blood pressure; sensitivity to cold; dry, rough skin	Restlessness; fidgeting; flushed, hot skin; "protruding" eyes due to lid retraction and eyeball displacement; rapid pulse, hypertensive heart disease, cardiac failure
Epinephrine	Not known	Hypertension; sweating; central nervous system excitation
Testosterone	Eunuchoidism in boys; infertility; poorly developed muscles and abnormal skeletal growth (i.e., no spurt at puberty)	Virulism in girls; precocious puberty in boys; abnormal skeletal development in both sexes
Estradiol, progesterone	Lack of sexual development in girls; absence of menstruation; sterility	Precocious puberty in girls
Aldosterone	Addison's disease: low blood pressure; loss of Na^+ in urine; pigmentation; hypoglycemia; depression; stress may cause adrenal crisis and death	Aldosteronism: Na^+ retention leads to increased plasma volume; hypertension; K^+ deficiency leads to muscle weakness, paralysis
Cortisol		See ACTH—Cushing's syndrome

constant flow of nitrogen compounds between cell components and the blood, and between the organism and its environment; the positive and negative hormonal controls of nitrogen uptake and excretion must be finely tuned to ensure the required nitrogen balance.

Many pathological conditions are known that are the result of over- or underproduction of hormones (Table 14.4). These imbalances upset the body's homeostasis and hence "throw everything out of kilter." Hormone imbalances can result from many causes, including (1) disease or surgical removal of the hormone-producing organ, (2) an inherited defect in an enzyme needed for hormone synthesis or in a hormone-transporting protein, and (3) a dietary deficiency of a metabolic precursor of the hormone (i.e., iodine needed for thyroxine synthesis) or the ingestion of substances that affect hormone synthesis or activity.

A word of caution: hormones and pheromones (Sec. 14.7) must be distinguished in requirement and function from vitamins. The latter must be provided in the diet and, although deficiencies could affect the production of hormones and pheromones, vitamins are not "messenger" molecules that control and coordinate organ function. Vitamins are usually essential cofactors in enzymatic reactions, or precursors of such cofactors (Table 2.4).

14.6 The mode of action of thyroid hormones

As an example of the way in which a hormone works, we can consider one of the thyroid hormones, thyroxine. Thyroxine is an iodinated derivative of the amino acid tyrosine. This hormone, which is well known and important in man, is even more critical to the development of frogs! Thyroxine and related hormones induce the metamorphosis of tadpoles into frogs, a fact that was discovered in 1912 when a scientist fed thyroid gland extracts to tadpoles and found that this provoked their development into frogs. The characteristic shape changes (legs, etc.) are accompanied by a number of striking biochemical modifications in the tadpole. For example, as a typical aquatic creature, a tadpole excretes excess nitrogen in the form of ammonia. During the thyroxine-triggered metamorphosis, nitrogen metabolism undergoes an extensive modification; frogs excrete urea. Another difference between tapdole and frog is the nature of their blood proteins; a frog lives in an environment (air) that is richer in oxygen than that of the tadpole (water). After thyroid hormone stimulation, the tadpole hemoglobins are replaced by frog hemoglobins; the latter have a lower capacity to bind oxygen. The tadpole–frog metamorphosis is also characterized by the appearance of a number of degradative enzymes, many of which are used in digesting the tail and other unwanted organs. It is not known how the thyroid hormones "direct" all these changes. It is known, however, that administration of thyroid hormone to cells in tissue culture systems promotes extensive RNA and protein synthesis, so that gene expression at the levels of transcription and translation must be stimulated.

14.7 Pheromones

While hormones act between producer and target cells within the same organism, other messenger-type molecules exist in the plant and insect worlds that are produced by one organism to provoke specific changes in a target organism, separate, and often a considerable distance from, the producing organism. The pheromones are produced in small quantities and are extremely potent. Examples of pheromones are the sex attractants of insects, and the mating pheromones of insects, plants, and yeasts. The queen bee produces an inhibitory pheromone that blocks the development of the ovaries in worker bees. Sex responses in moths are believed to be induced by a few molecules of a pheromone that may be transported several miles in the wind. In these cases, specific metabolic and physiological changes are triggered by the active compound, presumably by inducing the expression of one or more genes in the target organism.

Insect sex attractants are often simple organic chemicals that can be commercially synthesized for use in insect control. A number of important crops can be freed of insect predators by using synthetic sex attractants to trap large numbers of the insect pests in order that they may be killed. This may seem like an underhanded way to use sex, but it is an important factor in the economics of crop production.

14.8 Compartmental regulation

When examined under the electron microscope, eucaryotic cells are a complex maze of compartments (Fig. 1.1) that serve different functions: lysosomes contain enzymes for degradation, mitochondria house the oxygen-requiring enzymes and energy-producing reactions, peroxisomes are sacs of enzymes used to carry out reactions involving hydrogen peroxide. Even the cell nucleus can be considered in this fashion, the genetic material being separated from the cytoplasm by the nuclear membrane, to be crossed only when coordinate activity is necessary.

These provide examples of intracellular compartmental control, with such sacs keeping certain key functions "fenced off" from the rest of the cell. The restriction of metabolic reactions to specific intracellular locations increases metabolic efficiency by keeping the substrates and products of related reactions together. Packaging also protects one macromolecule from another, as in the case of the autolytic enzymes which are kept inside lysosomes.

Digging even deeper into the cell, organelles themselves may be compartmentalized. The best-studied example is the mitochondrion, which is subdivided into three regions: a smooth outer membrane, a highly folded inner membrane, and a gel-like internal matrix. Each region contains specific proteins: the inner membrane contains the electron transport system, while the matrix holds most of the Krebs cycle enzymes. Movement from one compartment of the mitochondrion to another is determined by the transport properties of the intermediates, so that required metabolites can be concentrated where they are most needed.

SUMMARY

Cells are capable of adjusting their metabolism to meet immediate metabolic needs, to adapt to changes in their environment, and/or to keep the metabolism of differentiated body organs synchronized. Metabolism is regulated by changing the activities of enzymes catalyzing specific metabolic reactions; this can be done in several ways. The *amount* of an enzyme can be controlled by increasing or decreasing either the transcription of its structural gene or by altering the rate of translation of the appropriate mRNA. The catalytic *activity* of an enzyme can be controlled by effector molecules that bind to the protein and change the shape and hence the functionality of the enzyme's active site.

These forms of intracellular control do not function independently, but act in combination to influence the course of cell metabolism. For example, in bacteria, end-product feedback inhibition of enzyme activity (rapid but energetically wasteful) can combine with end-product repression of enzyme synthesis (slow but energetically cheap) to give both short- and long-term responses to the presence of a required metabolite in the growth medium.

The response of a bacterial cell to a change of carbon source illustrates some of the complexities of intracellular control. A bacterial cell finding itself in the absence of lactose shuts off the transcription of the genes for the enzymatic degradation of lactose (the *lac* operon) with the specific *lac* repressor; conversely, the presence of lactose relieves this inhibition and allows the expression of these genes. cAMP controls sugar metabolism in a more general way; growth of the bacteria in glucose results in low levels of cAMP and a failure to express all operons, including the *lac* operon, involved in the utilization of carbohydrates which are less effective energy sources than glucose.

When cells are components of the differentiated organs in a complex body, extensive intercellular and inter-organ control is necessary. Plants and animals produce hormones which circulate to coordinate the metabolic activities of various organs. Hormones usually act by affecting one or more of the intracellular control mechanisms.

PRACTICE PROBLEMS

1. Every cell contains thousands of structural genes but at the most only three different RNA polymerases. What does this suggest about the base sequences of the promoters associated with the structural genes?

2. Compare and contrast the roles of the inducer and the control protein in positive and negative systems of transcriptional control.

3. The mechanism of negative transcriptional control of gene expression was worked out by studying bacterial mutants with defects in the lactose operon (Fig. 14.3). Strains carrying the following mutations have been isolated:
   (a) A deletion of the *lac* promoter.

(b) A nonsense mutation in the gene for the *lac* repressor.
(c) A mutation in the *lac* repressor gene such that the repressor no longer recognizes lactose.
(d) A mutation in the structural gene for galactosidase which leads to the production of an inactive enzyme.

In which strains could the expression of the lactose operon not be induced by the presence of lactose as the sole carbon source? In which mutants would the lactose operon be turned on all the time? Give brief explanations for your answers.

4. Steroid hormones seem to enter cells and act directly on transcription or translation, while protein hormones bind to the cell membrane and exert their effects by means of a second chemical messenger. How might differences in their chemistry account for the differences in the modes of action of these two classes of hormones?

5. Ingested iodide (I^-) goes primarily to the thyroid, where it is incorporated into thyroxine and other thyroid hormones. Given some iodide tablets, a solution of radioactive iodide, and a (willing) patient with a thyroxine deficiency, how would you determine whether his deficiency was due to (a) a lack of iodide in his diet, (b) a defect in an enzyme of thyroxine biosynthesis or (c) a defect in a thyroxine-transporting protein? If you knew that the patient's sister also had a thyroxine deficiency, could you eliminate any of the possibilities?

6. Under ordinary conditions, intracellular cyclic AMP is destroyed by a phosphodiesterase; this enzyme is inhibited by caffeine. If you were sitting calmly drinking your third cup of after-dinner coffee and someone in the apartment above dropped a bowling ball on the floor, would you be more likely to jump up wildly or to remain sitting quietly? Why?

7. X-chromosome abnormalities are well known in human beings; females with only one X chromosome (XO) and males with two X chromosomes (XXY) are relatively normal people. Individuals with a deficiency in, or an extra copy of, any other chromosome (any autosome) are seriously defective and, in fact, such fetuses account for a large percentage of early spontaneous abortions. By what mechanism can human cells cope with the wrong number of X chromosomes? Why do you think having too many or too few autosomes causes such severe physiological defects?

8. In eucaryotic cells, certain mRNA molecules have lifetimes of many days; they are translated over and over again before being destroyed by ribonucleases in the cytoplasm. Suggest a mechanism by which mRNAs could be protected from unwanted nuclease degradation.

9. Analogs of amino acids often inhibit the growth of bacteria. Suggest three possible reasons for this inhibition.

10. By genetic trickery, the lactose operon and the histidine operon have been fused together; the structural genes of histidine biosynthesis have been joined to the end of the *lac* operon. In this unusual circumstance, suggest what might happen to histidine biosynthesis in the presence of glucose or lactose and what might happen to lactose permease synthesis in the presence of histidine.

SUGGESTED READING

ALLISON, ANTHONY, "Lysosomes and Disease," *Sci. American*, 217, No. 5, p. 62 (1967). A comprehensive discussion of these intracellular packets of hydrolytic enzymes; describes their involvement in silicosis, asbestosis, drug sensitivities, and certain cancers.

BERTRAND, KEVIN, LAURENCE KORN, FRANK LEE, TERRY PLATT, CATHERINE L. SQUIRES, CRAIG SQUIRES, and CHARLES YANOFSKY, "New Features of the Regulation of the Tryptophan Operon," *Science*, 189, p. 22 (1975). Describes how sequence analysis of mutants of the regulatory region controlling the biosynthesis of tryptophan has provided evidence for new regulatory functions.

CHANGEUX, JEAN-PIERRE, "The Control of Biochemical Reactions," *Sci. American*, 212, No. 4, p. 36 (1965). Effectors, small metabolites that increase or decrease enzyme activity, act by changing the three-dimensional structures of the regulated proteins.

DAVIDSON, ERIC H., "Hormones and Genes," *Sci. American*, 212, No. 6, p. 36 (1965). Excellent starting point for learning how hormones control transcription and translation in target cells.

DICKSON, ROBERT C., JOHN ABELSON, WAYNE M. BARNES, and WILLIAM S. REZNIKOFF, "Genetic Regulation: The Lac Control Region," *Science*, 187, p. 27 (1975). Presents the complete nucleotide sequence of the region and discusses its function.

MCEWEN, BRUCE S., "Interactions Between Hormones and Nerve Tissue," *Sci. American*, 235, No. 1, p. 48 (1976). Gonadal steroid hormones act on the brain; these hormones control the (permanent) sexual differentiation of the fetal brain and the (reversible) sexual behavior of adults.

MITTWOCH, URSULA, "Sex Differences in Cells," *Sci. American*, 209, No. 1, p. 54 (1963). Compares the X chromosomes (Barr bodies) found in cells of normal humans with those of humans with X-chromosome abnormalities; good photomicrographs.

NATHANSON, JAMES A., and PAUL GREENGARD, "'Second Messengers' in the Brain," *Sci. American*, 237, No. 2, p. 108 (1977). Both hormones and neurotransmitters activate membrane-bound adenyl cyclase and increase intracellular levels of cAMP.

O'MALLEY, BERT W., and WILLIAM T. SCHRADER, "The Receptors of Steroid Hormones," *Sci. American*, 234, No. 2, p. 32 (1976). Steroid hormones bind to hormone-specific intracellular protein receptors. The hormone–protein complexes then bind to specific sites on the chromosomes and stimulate mRNA synthesis.

PASTAN, IRA, "Cyclic AMP," *Sci. American*, 227, No. 2, p. 97 (1972). Cyclic AMP is the intracellular chemical message for several animal hormones. This nucleotide also regulates transcription in bacteria.

PTASHNE, MARK, and WALTER GILBERT, "Genetic Repressors," *Sci. American*, 222, No. 6, p. 36 (1970). Repressor proteins control gene expression at the transcription level. This article focuses on the negative control of the lactose operon in *E. coli* and the genes of the bacteriophage lambda.

STEIN, GARY S., JANET SWINEHART STEIN, and LEWIS J. KLEINSMITH, "Chromosomal Proteins and Gene Regulation," *Sci. American*, 232, No. 2, p. 46 (1975). In eucaryotes, tissue-specific gene expression is apparently controlled by nonhistone chromosomal proteins.

WILSON, EDWARD O., "Pheromones," *Sci. American*, 208, No. 5, p. 100 (1963). Social insects (ants, bees) communicate by releasing minute amounts of specific organic chemicals. Pheromones have also been implicated in the sexual behavior of other insects and some mammals.

15 Viruses: Cell's Natural Enemies

Viruses are the ultimate parasites; these nonliving entities infect cells and commandeer the metabolic machinery of the host cell to their own ends, to produce progeny viruses. In most cases, the production and release of the viruses leads to the death of the host cell.

Viruses are ubiquitous infectious disease agents that have caused many serious epidemics in humans, animals, and plants throughout history. A list of confirmed viral diseases is given in Table 15.1. There is mounting evidence that many forms of cancer and possibly even such diseases as diabetes are viral in origin; although some of this evidence is only circumstantial, it is often compelling.

Certain viral infections result not in death of the host cell, but rather in alterations in the regulation and metabolism of the host cells. When animal viruses enter cells, part or sometimes all of their genome may be incorporated into the cellular DNA. This can lead to drastic changes in the structure and biochemistry of the host cell [e.g., alterations of the cell surface (membrane) and/or unrestrained growth and division]. These changes are known as transformation events, and many animal viruses, especially those believed to be associated with the development of cancer, are transforming viruses.

The study of viruses—how they infect and take over cells, and how they produce large numbers of progeny during a single infection—has been very profitable. Not only have we learned how viruses are made and how their production is regulated, but the study of viral infections has told us a great deal about the biology of cells in general.

TABLE 15.1(a) Representative Families of Animal Viruses

Eleven of the 15 families of animal viruses are described below. Classification is based on their nucleic acid and on the size and shape of the intact virus.

Family	Nucleic acid[a]	Number of proteins coded for[b]
1. Parvoviridae	ssDNA	3
2. Papovaviridae	dsDNA	6
3. Herpetoviridae	dsDNA	12–24
4. Poxviridae	dsDNA	30
5. Picornaviridae	ssRNA	4
6. Togaviridae	ssRNA	3
7. Coronaviridae	ssRNA	16
8. Orthomyxoviridae	ssRNA	7
9. Paramyxoviridae	ssRNA	6
10. Rhabdoviridae	ssRNA	7
11. Reoviridae	dsRNA	7

[a] ss, single-stranded; ds, double-stranded.
[b] Approximate number of virus-coded proteins; a rough estimate of the relative size of the genome.

One of the most convincing proofs of the role of DNA as the genetic material came from the study of DNA viruses.

Bacterial viruses have also found their place in literature. Sinclair Lewis' fascinating novel *Arrowsmith*, written in 1924, charts the efforts of Martin Arrowsmith, a researcher, to use bacterial viruses to combat human bacterial infections. A good idea but, as we shall see later, doomed to failure.

15.1 Viral structure

All viruses have the same basic structure, with their genetic material surrounded (coated) with a protein or protein–lipid coat. The genetic material of viruses can be DNA or RNA, single- or double-stranded (Tables 15.1 and 15.2). The coat consists of a large number of one or a few kinds of proteins packed neatly into a regular array. The coat of many animal viruses is similar in composition to the membrane of the host in which they were propagated.

Viruses vary in size (Fig. 15.1), but are usually very small compared to the cells they infect. The earliest definition of a virus was that of an infectious agent which could pass through the finest filters known. In general, animal viruses are larger than bacterial viruses; for example, vaccinia virus, which causes cowpox, can be seen in a light microscope and is almost as large as some bacterial cells. Because of their size,

TABLE 15.1(b) Human Viral Diseases

Human beings are susceptible to a multitude of viral diseases; this table lists some of the diseases indigenous to the northern hemisphere. Different viruses, particularly those which infect the respiratory tract, cause similar disease symptoms. The viral infection may spread to many body organs, the one listed here being the most seriously or noticeably affected.

Major affected organ	Disease	Virus family[a,b]	Name(s) of common virus
Respiratory tract	Common cold	5	Rhinoviruses (many types)
		7	—
	Croup	9	Parainfluenza
	Pneumonia		Respiratory syncitial
	Influenza	8	Influenza
Skin	Measles	9	Measles
	Rubella	?6	Rubella
	Chicken pox	3	Varicella-Zoster
	Shingles		
	Cold sores	3	Herpes simplex
	Genital herpes	3	Herpes simplex, type 2
	Smallpox	4	Variola
	Warts	2	Papilloma
Liver	Hepatitis	?1	Hepatitis A
		?	Hepatitis B
Central nervous system; brain	Meningitis	5	Coxsackie
	Encephalitis	6	Many togaviruses
	Rabies	10	Rabies
	Polio	5	Polio, types 1, 2, 3
Gastrointestinal tract	Gastroenteritis	11	Rotavirus
		?1	Norwalk agent
Salivary glands	Mumps	9	Mumps

[a] Numbers correspond to the families in Table 15.1(a).
[b] Some have not yet been classified.

studies of the structure and morphology of viruses have relied heavily on the use of electron microscopes.

Viruses vary in shape from the characteristic head–tail structure of most bacterial viruses to the geometrically exact icosahedral form of many mammalian viruses (Fig. 15.2). In addition, rod-shaped bacterial and plant viruses are known.

15.2 The life cycle of viruses: lysis versus lysogeny

Tadpole-shaped bacterial viruses (also known as bacteriophages or phages) attach to the bacterial cell wall and inject their nucleic acid into the cell through their tail, which acts like a syringe. Infection by mammalian viruses, which do not normally have a tail

TABLE 15.2 Plant Viruses

Plant viruses derive their names from the most obvious symptoms of infection in the plant in which they were originally investigated. The infected plant first studied was usually one of agricultural importance; many plant viruses are now known to have a wide host range.

Since most plant viruses cause changes in leaf coloration and stunt plant growth, names like "mosaic" and "dwarf" are common. Economically, the most damaging features of plant viral infections are stunting and other growth abnormalities, a reduction in the size and number of fruit, and, less frequently, death of part or all of the plant.

Virus	Nucleic acid[a]
Tobacco mosaic	ssRNA
Potato X	ssRNA
Pea enanation[b] mosaic	ssRNA
Tobacco ringspot	ssRNA
Rice dwarf	dsRNA
Cauliflower mosaic	DNA ?ds

[a] ss, single-stranded; ds, double-stranded.
[b] Ridgelike outgrowths from leaves.

structure, commences by envelopment of the entire virus (coat and all) by the cell. In the case of plant infections, the virus enters through injuries or lesions in the cells of the leaf or stem. Once inside the cell, the viral nucleic acid is uncoated and free to direct the synthesis of progeny.

When certain viruses infect cells, there are two possible outcomes. One is that the virus immediately produces many copies of itself, killing the host cell as the progeny viruses are released into the environment; this is called a *lytic infection*. The other possibility is that the viral genome is incorporated into the host chromosome and remains unexpressed until some future event induces a lytic infection; this dormant infection is called a *lysogenic infection*. One of the first events following the entry of viral nucleic acids into host cells is the "decision" between lytic and lysogenic cycles of development. It is worthwhile considering these in a little detail since this will help in our understanding of both viral infections and regulatory processes.

Lytic infections are the most common; they occur with both DNA and RNA viruses. The lytic cycle of all viruses is similar in outline (Fig. 15.3), although it may vary in certain details.

Once inside the cell, the viral nucleic acid takes command of the host cell's biosynthetic machinery. In the case of DNA viruses, the viral nucleic acid first uses host-cell RNA polymerase to transcribe those viral genes required for replication of

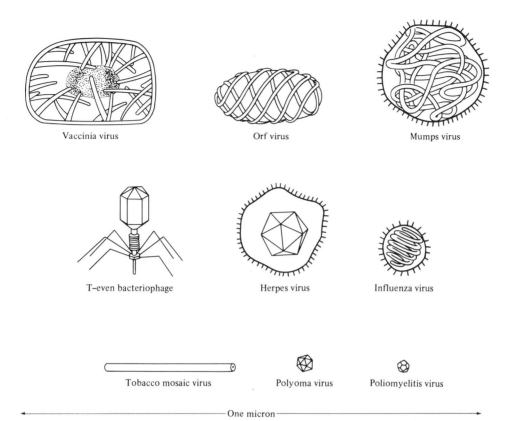

Figure 15.1 Shapes and sizes of viruses. Relative sizes of viruses are shown in this chart. A micron (micrometer), used as a measuring stick, is a thousandth of a millimeter; it is magnified 175,000 times. Other characteristics of these viruses are given in Tables 15.1 and 15.2. Vaccinia and orf are members of the poxviridae family; polyoma is a papovavirus. (After R. W. Horne, *The Structure of Viruses*, Copyright 1963 by Scientific American, Inc. All rights reserved.)

the viral genome and for shutting down host-cell-coded replication and transcription. These are called "early" genes. Host-cell-specific macromolecular synthesis is stopped by the production of a nuclease that specifically breaks down the host chromosome or by the production of inhibitors that prevent host RNA or protein synthesis. The early viral genes may also code for the synthesis of a virus-specific RNA polymerase that is required to interact with viral promoters concerned with the transcription of late genes. The late-acting viral genes include those which code for the viral coat protein(s) that package the viral nucleic acid to form intact progeny viruses inside the cell. The final step is the production of an enzyme that destroys the host membrane and cell wall, causing lysis of the cell. All the progeny viruses are released into the environment,

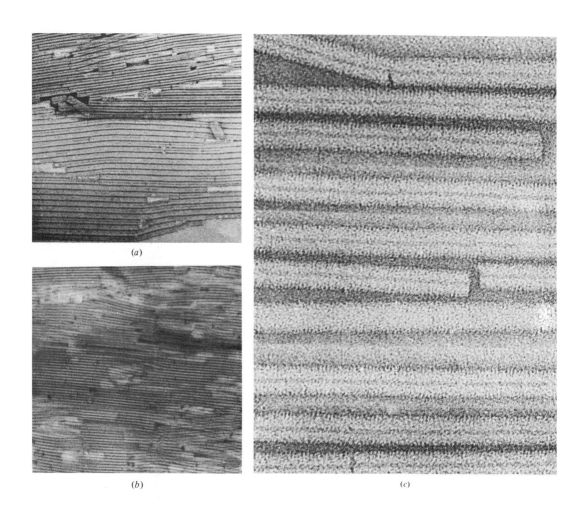
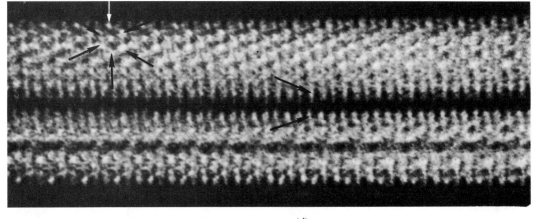

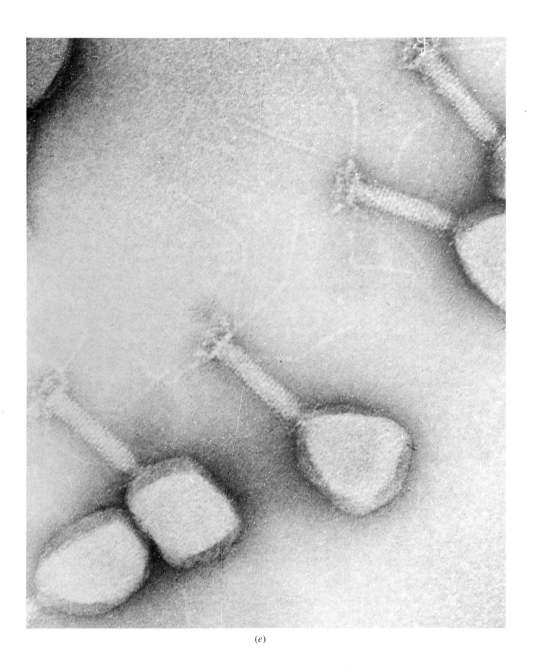

(e)

Figure 15.2 Electron micrographs of viruses. Tobacco mosaic virus (a), (b) and (c) are parallel arrays of the virus at low and high magnification. (d) is an "averaged" image of TMV rods showing the positions of the individual protein structure units in the rod. (e) Bacteriophage T_4 (approx. × 300,000). [(a), (b), and (c) Courtesy Dr. R. W. Horne. (d) After R. W. Horne, J. M. Hobart and R. Markham, "Electron Microscopy of Tobacco Mosaic Virus Prepared with the Aid of Negative Staining-Carbon Film Techniques," *J. Gen. Virol.*, 31, pp. 265–269 (1976), Cambridge University Press; (e) Courtesy Dr. M. F. Moody.]

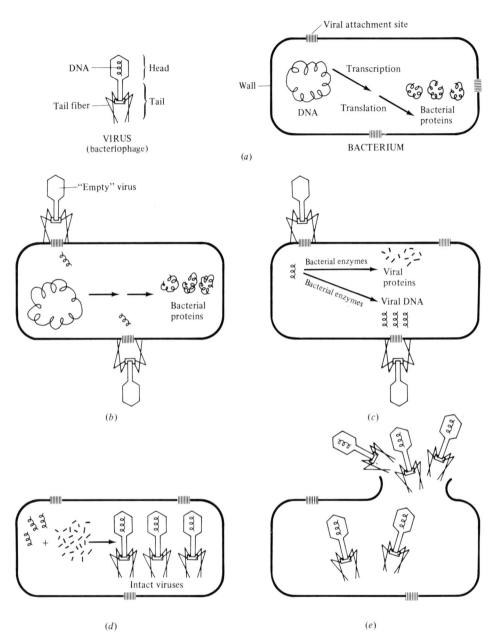

Figure 15.3 Lytic cycle. (*a*) The protagonists. (*b*) Attachment of the virus to the bacterium is followed by injection of the viral DNA. (*c*) Virus expression. The viral DNA directs the production of viral proteins and the replication of viral DNA using bacterial enzymes. The bacterial DNA no longer directs macromolecular synthesis and may even be destroyed. (*d*) Assembly of virus particles. (*e*) Lysis of host bacterium. A virus-coded enzyme, made late in infection, destroys the bacterial wall and the progeny viruses are released.

where they can begin a second round of infection on uninfected cells. In the case of bacterial viruses, the time for a productive infection is about 30–60 minutes and can lead to the formation and release of as many as 1000 progeny viruses from one cell. All this from the infection of one cell by a single viral nucleic acid!

Lysogenic infection is a phenomenon known for only a few viruses, for example, bacteriophage lambda, which infects *Escherichia coli*; it is an important process to understand because of its probable relationship to the transformation of eucaryotic cells by tumor viruses. This process is outlined in Fig. 15.4.

Once inside the host cell, enzymes catalyze the conversion of the linear viral DNA into a circular molecule, which then undergoes recombination with the host chromosome such that the viral DNA becomes integrated into, and part of, the sequence of the chromosome. The integration requires the expression of a limited number of viral genes, in particular those genes whose products promote the integration of the viral DNA into the host chromosome, and those genes coding for the production of a repressor which prevents the expression of the viral genes that trigger the lytic cycle. The repressor must be synthesized continuously to keep the virus in this lysogenic or dormant state. Once inserted into the host chromosome, the viral DNA is replicated like any other normal gene throughout successive cell generations.

How the cell and infecting virus make the decision between lytic and lysogenic infection is a very puzzling problem. Equally puzzling is the question of what releases the viral genome from the lysogenic state and starts the lytic cycle. The viral DNA can remain dormant in the lysogenic state forever, unless and until some external event relieves the inhibition of the lytic genes. Irradiation of a bacterial cell with ultraviolet light, or treatment of the cell with chemicals that interact with DNA provokes the lytic cycle in cells lysogenic for a virus. The exact mechanism is not known, but it is believed that chemical damage of the DNA is one event that can stop repressor production and set off the lytic cycle. It is as if the virus senses that the cell is damaged and will die, and, like a rat, leaves the sinking ship!

15.3 Viruses and cancer

There is now strong circumstantial evidence that viruses can and do cause certain cancers. As mentioned previously, the dormancy of lysogenic viruses is considered to be a reasonable model for the way in which tumor viruses can predispose an animal cell to becoming cancerous. In this case it is not the production or release of virus that damages the host cell; rather, the presence of the viral nucleic acid in the eucaryotic chromosomes may disturb the expression, control, or regulation of normal cellular metabolism and so produce a class of abnormal cells that propagate to produce tumors. One good possibility is that latent cancer-producing or tumor viruses are "hiding" in human chromosomes and that the interaction of DNA with chemicals in the environment leads to cancer cell propagation.

In recent years, a strong statistical correlation has been found between the incidence of cancer in human beings and their exposure to environmental chemicals, such as cigarette smoke, asbestos, insecticides, and hair dyes. In many of these cases, a correlation between the carcinogenic potency (ability to produce cancer) and the ability of the offending chemicals to interact with and damage DNA has been found. There seems to be little doubt that many forms of cancer are peculiar to industrialized modern society, where chemicals are routinely used as preservatives, flavors, dyes, plastics, and so on. We can eliminate most of these hazards, but the principle of freedom of choice makes it difficult to eliminate cancer-producing agents such as cigarette smoke.

Of course, not all carcinogens are by-products of industry. The incidence of skin cancer increases with increased exposure to sunlight (Sec. 12.4). The incidence of cancer of the large intestine is highest among people eating high-meat diets. The daughters of women given DES (Sec. 10.6) show a high incidence of vaginal cancer; thyroid irradiation ("scans") has increased the incidence of thyroid cancer.

15.4 Copying the viral genome

Once inside the host cell, the viral genome, be it RNA or DNA, must be copied enzymatically. In the case of DNA virus infection, the entering virus usually uses the host DNA polymerase to replicate its genome. A different situation exists for RNA viruses, since the host cell does not have an enzyme that can make RNA copies of an RNA template.

Single-stranded bacterial RNA viruses, in a genetic sense, are very simple. They need only to code for a replicase enzyme (to make more copies of the RNA) and a coat protein. The single-stranded viral RNA enters the cell and acts as messenger RNA for the replicase and the coat protein. The replicase then makes more copies of the original viral RNA molecule. (The replicase enzyme is, of course, a polymerase, but it must not be confused with RNA polymerase that makes RNA copies from a DNA template.) RNA viruses in animal cells, flu virus for example, reproduce in essentially the same way, although the viral RNA is double-stranded and the virus has a more complex structure.

Figure 15.4 Lysogenic cycle. (*a*) The protagonists. (*b*) Attachment and injection of viral DNA. The first step in a lysogenic infection is identical to that in a lytic infection. (*c*) Insertion of phage genome into host chromosome. The phage DNA is circularized and inserted into the bacterial chromosome by recombination enzymes. When integrated into the host chromosome, the viral genome is called a *prophage*. (*d*) Replication of the prophage. To the bacterial host, the prophage "looks" like any other gene and is replicated as such when the bacterium divides. The prophage may be replicated for many bacterial generations. The prophage genes are not expressed because a repressor is made that prevents this. (*e*) Release of prophage. Certain *rare* external events inhibit the action of the repressor and cause the replication and release of the prophage from the bacterial chromosome. This sets off the lytic cycle of virus multiplication (Fig. 15.3).

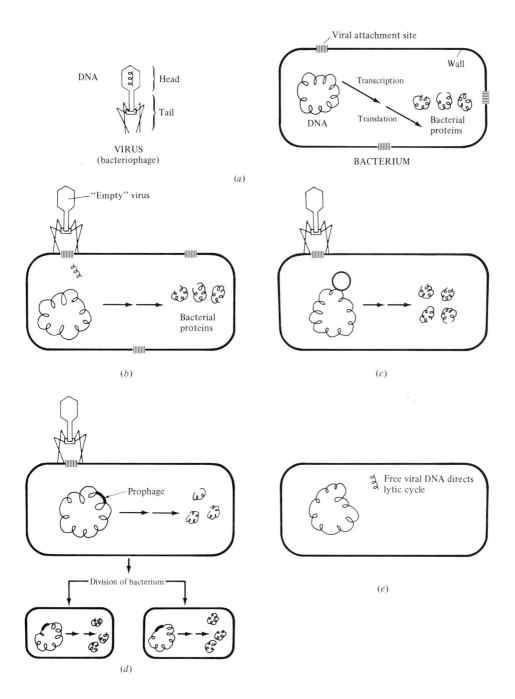

Another group of animal RNA viruses reproduces by means of a DNA copy of the viral RNA. Once the RNA has entered the cell, an enzyme named "reverse transcriptase" [since its activity is opposite that of the normal (DNA → RNA) transcription enzyme] copies the viral RNA into its complementary DNA; the DNA can then be used as a template for the production of viral RNA.

The discovery of reverse transcriptase solved an important biological paradox, for it explained how the genome of RNA tumor viruses often can be incorporated into the host cell's DNA genome. Upon infection, the viral RNA is transcribed into DNA by reverse transcriptase, and this DNA is integrated (by mechanisms unknown) into the DNA of the cell's chromosome. The viral genes will then remain unexpressed until such time as they are transcribed to form the viral RNA genome.

The requirement for reverse transcription by RNA tumor viruses has suggested one approach to the treatment of cancer, namely the isolation of specific inhibitors of reverse transcriptase. Unfortunately, many cells have been found to contain reverse transcriptases, which presumably have essential functions unconnected with their role in viral replication. Therefore, it seems unlikely that this approach will be successful, and we are left with the problem of finding a specific biochemical difference between cancer cells and normal cells, a difference that can be exploited as the point of attack for an inhibitor that will injure tumor cells without affecting normal healthy cells.

15.5 Why we cannot cure viral infections

Infectious diseases of viral origin are among the most significant health problems in our society. The common cold, for example, is the major reason for work time lost in all industry. Many epidemics of viral diseases have been recorded, for example the disastrous flu epidemics of 1918 and more recent years (Fig. 15.5). Herpes simplex is a common infectious agent that can cause venereal infection; these infections are on the upsurge.

Obviously, it is not easy to block the course of a viral infection because these parasites utilize the cell's own biochemical machinery for their propagation. Since there are few specific differences between a normal cell and a viral-infected cell, almost all inhibitors of viral replication also block host-cell metabolism. The best defense against viral epidemics remains vaccination, the establishment of host immunity by stimulating the host organism to produce antibodies against killed or attenuated (noninfective) virus.

Higher animals have an immune system which protects them against invasion by foreign proteins such as the coat proteins of viruses and bacterial toxins. Such foreign proteins, known as *antigens*, stimulate the host's immune system to make specific proteins, *antibodies*, that bind to the invading proteins and lead to their inactivation and destruction. The binding of antigen to antibody is very specific, like the formation of an enzyme–substrate complex, so immunity must be developed independently to each invader. When a virus enters the bloodstream, the viral coat proteins stimulate

Years	Virus antigens	Name of virus
1889–1899	Unknown (? similar to 1957 Asian)	
1900–1917	Unknown (? similar to 1968 Hong Kong)	
1918–1928	Hsw1 N1 (presumed)	Swine
1929–1945	H0 N1	Puerto Rico-8
1946–1956	H1 N1	Fort Monmouth-1
1957–1967	H2 N2	Asian
1968–present	H3 N2	Hong Kong

(a)

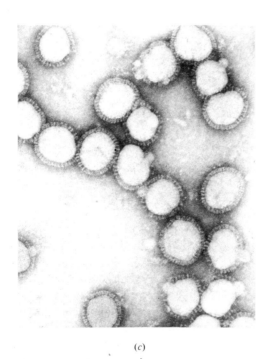

(b)

(c)

Figure 15.5 Causes of major influenza epidemics. (a) The cycles of influenza epidemics, which have occurred approximately every 10 years, have been linked to major changes in virus surface antigens, the proteins hemagglutinin (H) and neuraminidase (N). Because the viruses that caused the pandemics of Asian flu in 1957 and Hong Kong flu in 1968 appear closely related antigenically to the viruses of the epidemics of 1889 and 1900, respectively, some believe the 1918 swine virus will return as the cause of the next pandemic. (b) Schematic diagram of influenza virus. (c) Electron micrograph of influenza virus ($\times$ 160,000). [(c) After Compans et al., *Virology*, 42 pp. 880–889 (1970).]

the host's immune system to produce antibodies against themselves. A complex forms between virus and antibody which is subsequently removed from the blood by scavenger cells known as macrophages.

One of the most effective defenses against anticipated viral (measles, flu) or bacterial (diphtheria) attack is therefore to induce the formation of antibodies, by the administration of virus or bacterial toxin preparations. If subsequent infection occurs, antibody is present in sufficient quantities (or can be made quickly) to neutralize the invading organism. But even this approach is not foolproof, as has been discovered in the case of influenza. The flu vaccine made this year, against this year's virus, may not be effective against next year's virus. The genes for viral coat proteins can mutate, changing the coat structure and negating its interaction with host antibodies. Thus, flu and other viral vaccination programs, although correct in principle, often have only short-term effectiveness.

We know why Martin Arrowsmith's bacteriophages were not effective antibacterial agents: the immune system of the body recognizes the bacteriophages as foreign protein and thus makes antibodies to inactivate them. The search for specific inhibitors of viral functions is a continuing one, being constantly thwarted by the discovery of new aspects of the biochemistry and metabolism of host cells.

15.6 Viroids—science fiction in real life

Viruses are not the only cell parasites nor the smallest agents of infectious disease; in recent years research into several deadly diseases of animals and plants has led to the discovery of *viroids*, small viruslike particles that contain small amounts of nucleic acid.

Studying the course of these infections has proved a difficult task and pure preparations of viroids for biochemical analysis have been obtained only recently. It appears that, in most cases, the nucleic acid is a very small RNA molecule. The molecular weight of the RNA of a viroid causing disease in potatoes is sufficient to code for only 70–80 amino acids, so it is difficult to see how infection (with multiplication of the particle) can occur.

The situation with respect to animal viroids is even more enigmatic, since no detectable nucleic acid has been found associated with these agents. They are resistant to extremely high doses of UV-irradiation, which would normally inactivate any nucleic acid. Animal viroids are associated with a group of diseases of the nervous system known as slow virus diseases, characterized by an insidious onset, chronic and unrelenting progression, and inevitable death. Diseases caused by viroids in animals are scrapie (sheep), mink encephalopathy, and Kuru and Creutzfeldt-Jakob diseases in man. Kuru, found in primitive tribes in New Guinea, was once considered a hereditary disorder, but it was found to be transmissible and its incidence has decreased with the reduction in the practice of cannibalism among these tribes; the infection probably occurred during ritual handling of the highly infective brain prior to cooking.

There may be any number of yet unrecognized viroid infections, particularly among the slow degenerative diseases of man. In spite of our (apparently) great knowledge of the biochemistry of life, we still have to work out how viroids can replicate and make progeny with little or no nucleic acid; this appears to be contrary to all the rules and dogmas of molecular biology. Nature still has surprises up her sleeve!

SUMMARY

Viruses are cell parasites; they infect cells of bacteria, animals, or plants and convert the metabolic machinery of the host cell to their own ends—the production of more viruses. Viruses are simple structures that consist of a nucleic acid genome (RNA or DNA) encapsulated in a protein coat. When a virus infects a cell, the viral nucleic acid reproduces and directs the synthesis of the protein components necessary for making new viruses.

Some viruses are known to infect cells and to remain in a latent state in the host cell chromosome. Sometimes such an infected cell undergoes changes in metabolism (transformation); the multiplication of the transformed cell may lead to the development of tumors. Thus, the study of viruses, in addition to providing much basic information on nucleic acid replication and expression, may also provide clues to the nature of cancer.

PRACTICE PROBLEMS

1. Viruses may contain double- or single-stranded RNA or DNA. Which of these genomes:
   (a) Contains equal amounts of guanine and cytosine?
   (b) Is least sensitive to ultraviolet light?
   (c) Binds to ribosomes?
   (d) Is least sensitive to alkali degradation?

2. Plants do not sneeze, kiss, or engage in intimate relationships, yet viral infections spread to most plants in a field. Suggest two mechanisms by which viruses might be transmitted from one plant to another.

3. The viral genome can be RNA, double-stranded (ds) DNA or single-stranded (ss) DNA. For each nucleic acid, different enzymes are needed to copy the genome during the lytic cycle or to convert the genome into dsDNA which can be incorporated into the host chromosome. Four enzymes involved in these processes are: RNA polymerase, DNA polymerase, replicase, and reverse transcriptase. Fill in the accompanying table by indicating which enzyme(s) is needed in each situation.

	Viral genome		
	dsDNA	ssDNA	RNA
Copying genome in lytic cycle Converting genome to dsDNA			

4. Propose a mechanism by which a predisposition toward cancer could be inherited.

5. Rubella (German measles) is a viral disease which, when contracted during pregnancy, causes birth defects. Before a woman becomes pregnant, many doctors recommend a blood test to determine if she has had this disease. What is being looked for in this blood test? (*Note*: A rubella vaccine is available, but it must be given 3–6 months *before* pregnancy.)

6. Viruses are not "agents of decay"; they do not contribute significantly to the decomposition of dead plant and animal tissue. Why not?

7. About a week after being vaccinated against measles, some children develop a fever and face rash, but these measles symptoms are not contagious. Why not?

8. In general, lifelong immunity results from viral and bacterial infections. We have all had flu. Why do we still get it?

9. Justify the statement "Viruses are not living things."

SUGGESTED READING

CAIRNS, JOHN, "The Cancer Problem," *Sci. American*, 233, No. 5, p. 64 (1975). Most common kinds of cancer are caused by environmental factors and are potentially preventable. The article contains a particularly interesting description of how epidemiological data are used to identify conditions that predispose populations to certain cancers.

CAMPBELL, ALLAN M., "How Viruses Insert Their DNA into the DNA of the Host Cell," *Sci. American*, 235, No. 6, p. 102 (1976). This article includes a good step-by-step description of the lysogenic cycle.

CAPRA, J. DONALD, "The Antibody Combining Site," *Sci. American*, 236, No. 1, p. 50 (1977). Their unique three-dimensional shape allows each of these proteins to bind and inactivate a specific foreign protein.

DALTON, ALBERT J., and FRANCOISE HAGUENAU (eds.), *Ultrastructure of Animal Viruses and Bacteriophage: An Atlas*, Academic Press, Inc., New York, 1973. Electron micrographs of a wide variety of viruses.

DULBECCO, RENATO, "The Induction of Cancer by Viruses," *Sci. American*, 216, No. 4, p. 28 (1967). Mammalian cells, growing in tissue culture, can be transformed by added viruses. Photographs contrast normal and transformed cells.

GAJDUSEK, D. CARLETON, "Unconventional Viruses and the Origin and Disappearance of Kuru," *Science*, 197, p. 943 (1977). An excellent review of current knowledge about the chronic degenerative diseases of man.

HOLLAND, JOHN J., "Slow, Inapparent and Recurrent Viruses," *Sci. American*, 230, No. 2, p. 32 (1974). Evidence is mounting that latent viruses cause many of the chronic degenerative diseases of man.

HORNE, R. W., "The Structure of Viruses," *Sci. American*, 208, No. 1, p. 48 (1963). Excellent diagrams emphasize the symmetrical arrangements of viral coat proteins.

JERNE, NIELS KAJ, "The Immune System," *Sci. American*, 229, No. 1, p. 52 (1973). Describes the biology and the chemistry of the system which inactivates foreign proteins; includes many other references.

LANGAR, WILLIAM L., "Immunization Against Smallpox Before Jenner," *Sci. American*, 234, No. 1, p. 112 (1976). An interesting history of the techniques of smallpox prevention.

MELNICK, JOSEPH L., GORDON R. DREESMAN, and F. BLAINE HOLLINGER, "Viral Hepatitis," *Sci. American*, 237, No. 1, p. 44 (1977). Recent investigations of the epidemiology and immunology of this disease have reduced its incidence and may lead to a vaccine against hepatitis B.

SPECTOR, DEBORAH H., and DAVID BALTIMORE, "The Molecular Biology of Poliovirus," *Sci. American*, 232, No. 5, p. 24 (1975). A step-by-step description of the lytic cycle of this RNA virus.

TEMIN, HOWARD M., "RNA-Directed DNA Synthesis," *Sci. American*, 226, No. 1, p. 24 (1972). RNA tumor viruses use reverse transcriptase to transfer genetic information from RNA to DNA.

WOOD, WILLIAM B., and R. S. EDGAR, "Building a Bacterial Virus," *Sci. American*, 217, No. 1, p. 60 (1967). A combination of genetic and biochemical experiments revealed how a bacteriophage is assembled during the lytic cycle.

16 Antimicrobial Agents

Historical records indicate that man has used chemicals extracted from plants, animals, and microorganisms for the treatment of a variety of his many ailments for a long time. This largely empirical form of chemotherapy* was replaced in the early twentieth century by a more systematic approach. Paul Ehrlich first recognized the need for chemicals that would specifically inhibit the growth of bacteria and other parasites without harming the host. Ehrlich and his collaborators synthesized hundreds of compounds and eventually succeeded in making a "silver bullet": their 666th compound, salvarsan, an organic derivative of arsenic, was an effective, if unpleasant, cure for syphilis. The sulfonamides, discovered in the mid-1930s, were the forerunners of a long line of synthetic antimicrobial agents that have proven to be extremely efficacious. Not until a little later did scientists rediscover the antimicrobial potential of naturally occurring compounds and begin to capitalize on the earlier observations of Fleming that certain molds produce metabolites that inhibit the growth of bacteria. This lead was exploited at the beginning of World War II by a major combined effort between British and American chemists and microbiologists that culminated in the discovery of penicillin. With the subsequent isolation of other antibiotics, such as streptomycin and neomycin by Selman Waksman in the mid-1940s, and their successful application in the treatment of tuberculosis, the antibiotic era was underway.

---

* Chemotherapy is the treatment of disease with chemical agents.

16.1 What are antibiotics?

In most popular science books and encyclopedias, antibiotics are considered to be, along with Scotch tape, one of the "wonders" of modern civilization. It is true that antibiotics have been extraordinarily effective in reducing man's suffering from a variety of infectious diseases caused by bacteria; however, as usual, there is a dark side to the picture, and we shall see later that the use of antibiotics has brought its own problems.

Antibiotics are chemical substances produced by microorganisms that have inhibitory effects on the growth of other organisms. Most antibiotics that are in common clinical use have been isolated from bacterial strains of the actinomycetes group; a few, including penicillin, are produced by fungi. All antibiotics are organic compounds. Their structures range from simple aromatic molecules such as chloramphenicol (Fig. 16.1) to complex molecules such as actinomycin (Fig. 16.2) and peptides such as bacitracin (Fig. 16.3).

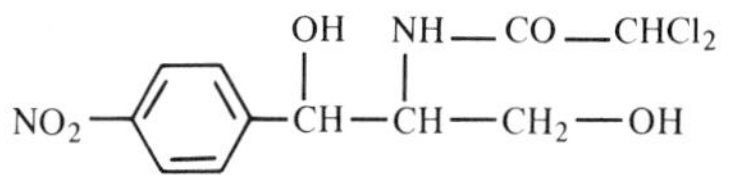

Figure 16.1 Structure of chloramphenicol. This is an unusual antibiotic that contains an aromatic nitro group and chlorine atoms.

Only a handful of the thousands of antibiotics that have been isolated and tested have both the low toxicity to the host and the broad effectiveness required to warrant their introduction into medical practice. There is no perfect antibiotic; each has its own desirable and undesirable characteristics that demand fairly specific and careful applications. It is the use of antibiotics outside the range of their effectiveness that has

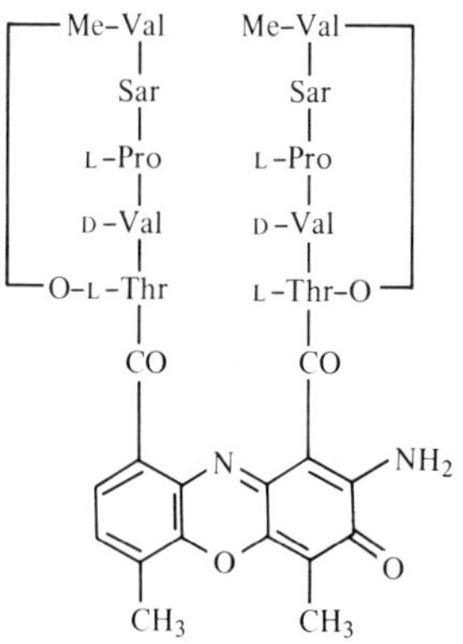

Figure 16.2 Structure of actinomycin D. This antibiotic consists of two parts, the ring system and the attached peptide chains. The latter contain some unusual amino acids that are not found in proteins, such as methyl valine (Me-Val) and methyl glycine (sarcosine, Sar).

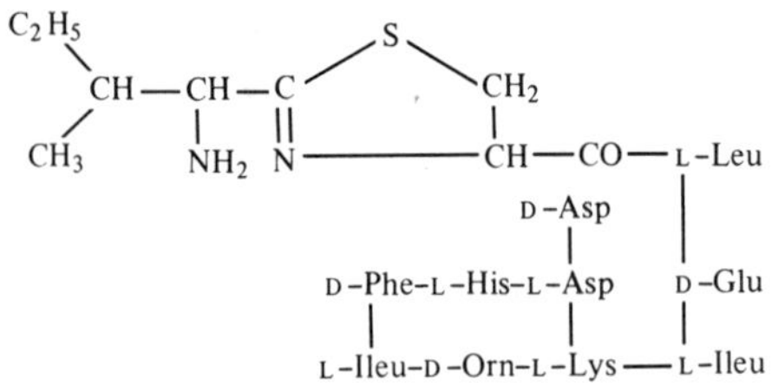

$$\begin{array}{c} \text{C}_2\text{H}_5 \\ \diagdown \\ \text{CH}-\text{CH}-\text{C} \\ \diagup \quad | \quad \| \\ \text{CH}_3 \quad \text{NH}_2 \; \text{N} \end{array} \begin{array}{c} \text{S} \\ \diagdown \\ \quad \diagdown \text{CH}_2 \\ \quad \diagup \\ -\text{CH}-\text{CO}-\text{L-Leu} \end{array}$$

with side chain: D-Asp — D-Glu — L-Ileu, and D-Phe-L-His-L-Asp — L-Ileu-D-Orn-L-Lys — L-Ileu

Figure 16.3 Bacitracin—a peptide antibiotic. This antibiotic also contains amino acids not found in proteins, such as ornithine (Orn). Peptides such as bacitracin are not made on ribosomes using mRNA, and so on; the formation of the peptide bonds between the amino acids is catalyzed by cytoplasmic enzymes.

created problems. For example, the use of antibacterial agents in the treatment of colds and flu, which are caused by viruses, is clearly inappropriate and may have contributed to the development of resistant strains of bacteria (Sec. 16.7).

The use of antibiotics is not limited to the treatment of infectious disease in man; 70% of all antibiotics produced are used in agriculture. Enormous amounts of these drugs are used in the treatment of disease in farm animals and in the prevention of disease on crops such as fruit. The diets of cattle, pigs, and poultry are routinely supplemented with antibiotics, since these feed additives promote a more efficient weight gain per unit of feed in these animals; this has become a vital factor in the economics of farming.

In addition to their therapeutic value, antibiotics have been of considerable interest to biochemists and geneticists, who, from studying the modes of action of these drugs, have learned much useful information about certain metabolic pathways in cells. The isolation and study of antibiotic resistant mutants has also contributed to our knowledge of how antibiotics work and has provided important insights into the way in which cells make protein, RNA, and form cell walls.

16.2 Why are antibiotics effective?

The most important concept in chemotherapy is that the therapeutic agent should be strongly selective in its action: it should inhibit the infectious agent without any effect on the host. As we have mentioned before (Chap. 1), the majority of metabolic reactions are common to all cells. Still, there are several possible sites for the selective action of antibacterial agents based on metabolic differences between the procaryotic bacterial cell and the eucaryotic cells of the host (Table 16.1). Animal cells have no wall, so bacterial cell walls can be sites for the selective action of inhibitors; the penicillins are good examples of this selectivity (Sec. 16.3). In addition, the protein-synthesizing apparatus of bacteria differs substantially from that of animal cells (Chap. 13) and there are a number of antibiotics that bind to and interfere with the activity of bacterial ribosomes, but do not affect the activity of eucaryotic ribosomes. There are

Antimicrobial Agents 305

TABLE 16.1 Some Common Antibiotics and Their Modes of Action Against Bacteria

Antibiotic	Mode of action
Penicillin, cephalosporin, bacitracin	Inhibit wall cell synthesis
Chloramphenicol, erythromycin, lincomycin	Inhibit the peptide-bond-forming step in protein synthesis
Tetracycline	Inhibits protein synthesis by preventing the binding of aminoacyl tRNA to the ribosome
Streptomycin, kanamycin, gentamicin	Inhibit both initiation and elongation steps of protein synthesis; also cause mistakes in reading of genetic code
Rifampicin	Inhibits RNA synthesis by preventing the initiation of transcription

also subtle but significant differences in the enzymes of RNA synthesis and intermediary metabolism that are specific enough for antibiotics to inhibit the bacteria without affecting the host.

In spite of these apparently clear-cut differences between bacterial and animal cell metabolism, most antibiotics do possess toxic side effects, as a result of their interference with some aspect of human metabolism. For example, penicillin produces a severe shock reaction in a small percentage of patients who produce antibodies against the drug. Treatment with chloramphenicol can damage bone marrow, leading to anemia; streptomycin and kanamycin can lead to hearing loss by an effect on the eighth nerve of the inner ear, and also cause liver damage. In spite of a good understanding of the way antibiotics work against bacteria, we know almost nothing about the biochemical basis of their side effects.

16.3 Antibiotics that inhibit bacterial cell wall synthesis

Very few of us have not had a painful shot of penicillin. The consequence of this unfriendly act is that penicillin enters the body and kills sensitive bacteria at high efficiency. It does this by preventing cell wall synthesis; subsequently, the bacteria burst, probably as a result of lysing enzyme action. Penicillin is believed to be an analog of a component of the bacterial cell wall, and to inhibit the action of a cell-wall-synthesizing enzyme by competing with the true substrate. Recent studies have detected several penicillin-binding components in bacterial cells, and any one or all of these may be associated with the inhibitory action of the drug.

The first penicillin to be used was penicillin G (Fig. 16.4); it was a narrow-spectrum antibiotic; that is, it was effective only against certain bacteria (those known as gram-positive). Since that time new penicillins have been isolated from natural sources or chemically synthesized to provide broad-spectrum antibiotics (effective

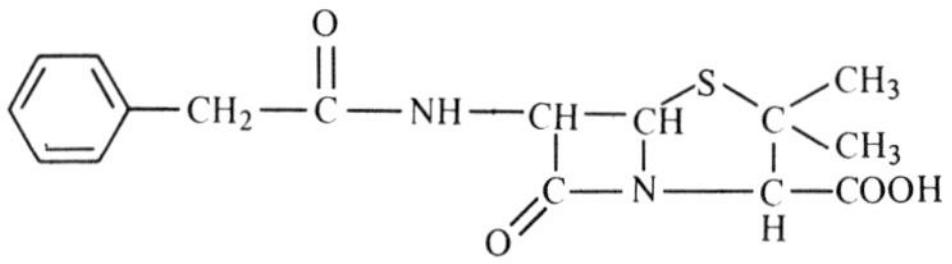

Figure 16.4 Penicillin and cephalosporin. Penicillins are the basic β-lactam antibiotics. The four-membered ring is known as a β-lactam ring; enzymatic hydrolysis of this ring, with concomitant inactivation of the antibiotic, occurs in many resistant bacteria.

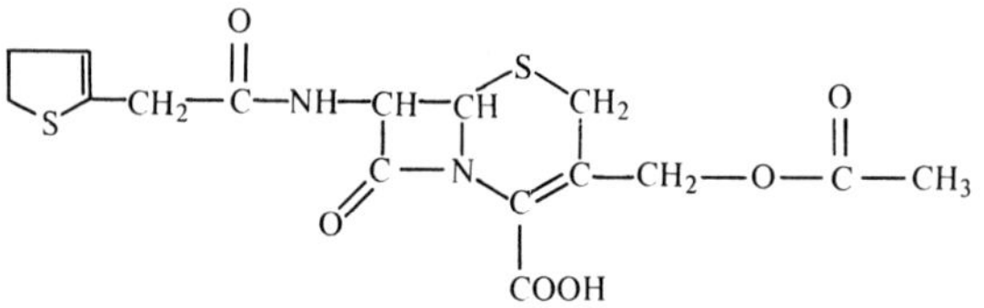

The cephalosporins, such as cephalothin, are chemically similar to the penicillins; they also contain a β-lactam ring. Cephalothin, although less potent than most penicillins, has the advantages of being resistant to certain penicillinases and of not causing allergic reactions.

against both gram-positive and gram-negative bacteria) such as ampicillin. Additional modifications of the penicillin structure make the molecule effective against bacteria that have developed resistance to the earlier forms of the drug.

16.4 Antibiotics that inhibit bacterial protein synthesis

The majority of the antimicrobial agents used in the treatment of human bacterial diseases inhibit protein synthesis by binding to ribosomes. Once the drug is bound to the bacterial ribosomes, it interferes with the normal functioning of these organelles in the synthesis of proteins. This group of antibiotics includes chloramphenicol, tetracycline, erythromycin, streptomycin, kanamycin, gentamicin, and lincomycin. They are used in the treatment of a variety of infections, including typhoid fever, bacillary dysentery, and many forms of sepsis. Puromycin is also an inhibitor of protein synthesis, but it is not a clinically useful antibiotic since it inhibits bacterial and human cells equally (i.e., it lacks the necessary specificity).

In addition to their clinical efficacy, these drugs are of great importance to biochemists studying the mechanism of protein synthesis. They all bind to ribosomes and

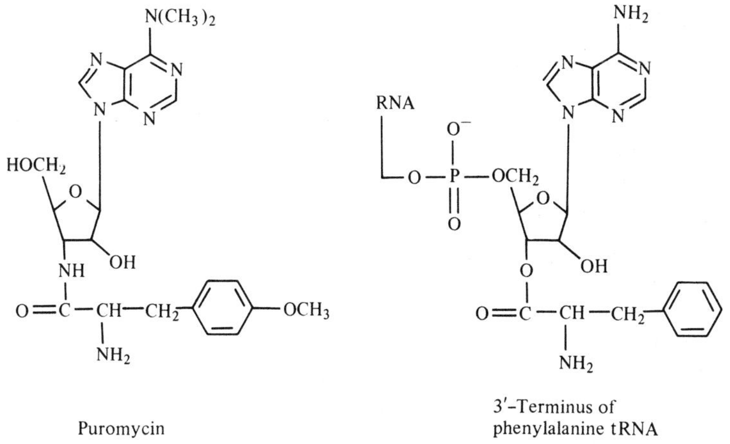

Figure 16.5 Structure of puromycin. As you can see, the structure of puromycin closely resembles that of the 3′-terminal end of amino acyltransfer RNA (Sec. 13.6). The recognition of this similarity suggested how puromycin inhibits protein synthesis by acting as a "false" acceptor of peptide chains.

block various steps in the synthesis of protein; some affect the initiation steps and some inhibit the actual process of peptide-bond formation (Table 16.1). The aminoglycosides, streptomycin, kanamycin, and gentamicin, all inhibit a step in the initiation process in protein synthesis, but, in addition, when bound to the ribosomes, they disturb the recognition between codon and anticodon (Sec. 13.6) and cause the insertion of incorrect amino acids into the growing peptide chain.

The mode of action of puromycin is fairly well understood; it causes the premature release of incomplete peptide chains. As seen in Fig. 16.5, this drug is an analog of aminoacyl-tRNA; it sits on the ribosome and competes with aminoacyl-tRNA as the acceptor of the growing peptide chain. The growing peptide chain is attached to puromycin and is released from the ribosome; polyribosomes are very rapidly degraded when puromycin is added to growing cells. This drug is the only antibiotic known that interferes with protein synthesis because it is similar to, and mimics the function of, one of the components involved in protein synthesis. Much less is known about the specific modes of action of other protein synthesis inhibitors; none of them seem to mimic and replace normal components.

16.5 Antibiotics that inhibit bacterial RNA synthesis

Inhibitors of RNA synthesis are comparatively rare. One of the best known and clinically useful is rifampicin, which has proved effective in the treatment of tuberculosis. TB, once a very serious health problem, has been much reduced in incidence, but it has not been eradicated. It is associated with poverty and poor living conditions and is often difficult to treat since the organism (a bacterium known as mycobacterium) is now resistant to some of the older and previously effective anti-TB drugs such as streptomycin.

Rifampicin inhibits bacterial RNA synthesis by binding to a specific site on RNA polymerase and preventing the initiation of RNA synthesis. Mutant bacterial strains have been isolated in the laboratory that are resistant to rifampicin because they produce an altered RNA polymerase that does not bind the drug; this both proved the mechanism of action of the drug and aided biochemists in their studies of the genetics and biochemistry of RNA polymerase.

Actinomycin D also inhibits RNA synthesis, but it acts by binding directly to DNA. In this way, it prevents the movement of RNA polymerase along the template. Actinomycin D has been a very important tool in studying the properties of messenger RNA in bacteria and animal cells. It is not used for the treatment of bacterial infections because of its toxicity; it is one of the components of the drug "cocktails" that have been of moderate success in the treatment of some forms of cancer.

16.6 Antimetabolites

Not all antibacterial agents are made by bacteria or fungi; some have been man-made by chemists who have tried to design "look-alike" competitive or noncompetitive inhibitors of bacterial enzymes (antimetabolites). Antimetabolites are not classified as antibiotics since they are not naturally occurring compounds. One of the most effective groups of antimetabolites is comprised of the sulfonamides or sulfa drugs; these are simple organic compounds that are obtained by chemical synthesis. As we pointed out in Chapter 4, sulfonamides are analogs of p-aminobenzoic acid, an intermediate in the bacterial formation of the vitamin folic acid.

Sulfa drugs were first made by L. Domagk in 1935 and further developed by P. Trefouel in 1936; they were by-products of the dyestuff industry. Since that time thousands of derivatives of p-aminobenzoic acid have been synthesized in the hope of finding a better sulfa drug. The sulfa drugs remain one of the most effective, nontoxic, and inexpensive antimicrobial agents available. Some newer sulfa drugs are used in the treatment of TB and leprosy.

Other antimetabolites are trimethoprim, which inhibits a step in the biosynthesis of tetrahydrofolic acid (Sec. 11.8), and 5-fluorouracil, which inhibits DNA synthesis (Sec. 11.8). Trimethoprim is a member of a group of structurally related compounds that inhibit the utilization of folic acid by inhibiting the dihydrofolate reductase enzyme. Since this enzyme is present in all organisms, trimethoprim in high concentrations can affect the eucaryotic host. However, the bacterial enzyme is inhibited more strongly by trimethoprim than is the human enzyme. Trimethoprim is generally used in subinhibitory concentrations and in combination with sulfa drugs, since this combination is more effective than either drug alone.

16.7 Bacterial resistance to antibiotics: bacteria fight back

Evolution is a continuous process, and the increasing appearance of bacteria resistant to antibiotics is a good example of the evolution of a life form in response to strong selective environmental pressures.

When antibiotics were first introduced, scientists thought that bacterial resistance to such drugs would not interfere with the treatment of infectious disease, since the frequency of the spontaneous appearance of antibiotic-resistant mutants was too low to be significant. This prediction has not been borne out, and in some instances the development of antibiotic-resistant strains has been both substantial and frightening. Most hospitals harbor large reservoirs of antibiotic-resistant bacteria, probably because the selective force (large-scale use of antibiotics) is most strong in this setting. What has been even more of a surprise is the mechanism by which bacteria become resistant to antibiotics in clinical situations.

It is true that mutations in genes on the bacterial chromosome that alter the cell's response to antibiotics occur at a very low frequency. However, special genes that determine antibiotic resistance have been found on DNA molecules known as infectious resistance plasmids or resistance factors (R-factors). Once one resistant bacterium harboring an R-factor is present in a bacterial population, other bacteria can acquire that resistance at a much higher frequency than is expected from simple inheritance. R-factors are small DNA molecules that can exist apart from the bacterial chromosome. R-factors carry only a small number of genes, but these genes are very significant, since many code for enzymes that inactivate or detoxify antibiotics and thus can protect the cell that harbors them. Upon cell division, R-factors are replicated and passed on to daughter cells. *In addition*, R-factors have the unusual property of being self-transmissible; that is, they can pass from bacterium to bacterium by means of a sexual conjugation process which is independent of cell division. Because of this self-transmissibility, entire bacterial populations can rapidly take up R-factors and hence be converted to antibiotic resistance.

The mechanisms by which genes on R-factors determine resistance to antibiotics are shown in Table 16.2. As can be seen, most of the antibiotics important in the treatment of bacterial diseases can be thwarted by the presence of R-factors. It is common for a single R-factor to code for resistance to several unrelated antibiotics; this is the origin of the term "multiple drug resistance." To make matters worse, bacteria can harbor more than one R-factor, so that some strains are resistant to all antibiotics known!

R-factors were first discovered in Japan in the late 1950s, where they probably evolved in response to the massive use of antimicrobial agents for the treatment of bacillary dysentery. During succeeding years, R-factors have been found in all countries of the world and it is hard not to find resistant strains in hospitals. Fortunately, not all bacteria are resistant and antibiotic therapy is still generally effective.

As new antibiotics are developed and introduced into medical practice, they are followed in time by the appearance of bacterial strains that are resistant by virtue of

TABLE 16.2 Resistance Factors and the Mechanisms by Which They Determine Resistance to Antimicrobial Agents

Genes on R-factors code for (1) enzymes that inactivate antibiotics, (2) altered proteins that retain their biological activity but are no longer sensitive to drug inhibition, or (3) changes in transport mechanisms so that antibiotics cannot enter the cell.

Antimicrobial	Mechanism of bacterial resistance
Tetracycline	Transport of antibiotic into cell is blocked
Penicillins	Activity of antibiotic destroyed by hydrolytic enzyme
Chloramphenicol	Enzymatic acetylation detoxifies the drug
Streptomycin, kanamycin	Enzymatic modifications detoxify the drugs
Erythromycin, lincomycin	Resistant strains have altered ribosomes; drugs can no longer bind to target site
Sulfonamides	Resistant strains have altered dihydrofolate synthetase that is insensitive to drug inhibition

possessing R-factors that have genes coding for resistance to the antibiotics. In addition, multiple drug resistance has become increasingly prevalent in the bacteria present in the domestic livestock population. The use of antibiotics as feed additives to promote animal weight gain, in prophylaxis to prevent infection in farm animals, as fruit sprays, and in many other agricultural areas has provided a strong selective environment for the development of resistant strains. Many people believe that the massive use of antibiotics in agriculture is creating a huge reservoir of resistant organisms that may be transferred to the human population. There are many economic and health factors involved, but some judicious control of antibiotic usage in agriculture and medicine will be necessary to limit the spread of R-factors among the bacteria that infect the animal and human population. There is too much at stake to allow unrestricted use of antibiotics, which could lead to the development of resistant strains that would render antibiotics obsolete.

SUMMARY

There are many substances existing in nature that are toxic to bacteria, plants, or animals. If one of these substances inhibits bacterial growth without adversely influencing the human host, it has the specificity necessary for it to be a useful antibiotic. Antibiotics usually block bacterial functions (protein synthesis, nucleic acid

Antimicrobial Agents

synthesis, cell wall synthesis) without affecting the corresponding but enzymatically different reactions that occur in animal cells. Treatment of viral infections and cancer is difficult, because distinct and exploitable metabolic differences do not exist between host and parasite.

The extensive and often indiscriminate use of antibiotics in the prevention and treatment of infectious disease has led to the selection of strains of bacteria resistant to antibiotics. In the case of hospital-acquired infections, such resistant strains can diminish the effectiveness of antibiotic therapy.

PRACTICE PROBLEMS

1. Several new antibiotics, shown below, have been isolated, tested, and found to be inhibitory to bacteria. Propose a mechanism of action for each of these compounds. Only one was found to be specific enough to be used in the treatment of human bacterial infections; which one?

$$^-O-\overset{O}{\underset{O^-}{\overset{\|}{P}}}-O-CH_2-\overset{O}{\overset{\|}{C}}-COOH$$

$$\text{(structure with } OCH_3, NH-C-CH, S, CH_3, CH_3, C-N, C-COOH, H, OCH_3, O\text{)}$$

$$CH_3-\underset{Cl}{\overset{}{CH}}-\underset{Cl}{\overset{}{CH}}-CH_2-CH_2-COOH$$

$$F-\!\!\!\bigcirc\!\!\!-CH_2-\underset{NH_2}{\overset{}{CH}}-COOH$$

2. A new antibiotic has been discovered that inhibits DNA synthesis but not RNA synthesis. Suggest two possible mechanisms of action for such a drug.

3. Streptomycin and other aminoglycoside antibiotics cause mistakes in the reading of the genetic code. Why is this lethal to the cell? Is all misreading equally disruptive? Compare and contrast the results of misreading with the effects of a missense mutation.

4. (a) Why do polyribosomes disintegrate when puromycin is added to an actively growing bacterial cell? (b) Why is puromycin toxic to human beings?

5. Bacteria growing in a medium containing glucose, ammonium salts, and trace elements are sensitive to sulfa drugs. When this medium is supplemented with adenine, guanine, and thymine, the bacteria are much less sensitive to these antimetabolites; that is, the amount of sulfa drug needed to inhibit bacterial growth becomes much higher. Why?

6. Bacteria can be either streptomycin-sensitive, streptomycin-resistant, or streptomycin-dependent. All three phenotypes are controlled by the same chromosomal gene. Propose a model that would explain the three phenotypes and their genetic control.

7. Give brief explanations for the following:
   (a) In clinical situations, peptide antibiotics are administered topically (for example, on skin) or injected; they are ineffective if given orally.

(b) Erythromycin is not used to combat bacterial diseases of soybean or peanut plants.
   (c) Antibiotic-producing bacteria do not kill themselves.
   (d) Animals given high doses of antibiotics over long periods of time develop deficiencies of thiamine, folic acid, vitamin K, and/or riboflavin.
   (e) Antibiotics used to treat bacterial diseases of man are ineffective against diseases caused by viruses or fungi.

8. Mutants of bacteria resistant to actinomycin D are very rare. Knowing its mode of action, can you explain why this might be the case?

9. Antibiotics are usually classified as bactericidal or bacteriostatic. What are the biochemical explanations for these differences?

10. What experiments might you perform to prove that bacitracin is not synthesized in the cell in the same way that enzymes or transport proteins are made?

SUGGESTED READING

FRANKLIN, T. J., and G. A. SNOW, *The Biochemistry of Antimicrobial Action*, John Wiley & Sons, Inc., New York, 1975. An up-to-date compendium of what antibiotics are, how they work, and how cells have come to resist them.

GORINI, LUIGI, "Antibiotics and the Genetic Code," *Sci. American*, 214, No. 4, p. 102 (1966). Streptomycin and other aminoglycoside antibiotics bind to ribosomes and cause misreading of the coding triplets of mRNA during protein synthesis.

KERMODE, G. O., "Food Additives," *Sci. American*, 226, No. 3, p. 15 (1972). Antibiotics have been used as food preservatives.

WATANABE, TSUTOMU, "Infectious Drug Resistance," *Sci. American*, 217, No. 6, p. 19 (1967). A good description of R-factors (which confer drug resistance) and the mechanism by which they are transferred from one bacterium to another.

Appendix: Important Concepts in Organic Chemistry

Naturally occurring organic compounds consist primarily of the four elements carbon, hydrogen, oxygen, and nitrogen, with lesser amounts of phosphorus and sulfur. The most prevalent element is carbon; to a great extent, the chemistry of the carbon atom determines the structures of these compounds.

A.1 Covalent bonding

Carbon and other elements can form a variety of bonds. Carbon is almost unique among the elements in that it can form long chains of covalent bonds (e.g., in palmitic acid, p. 319).

A covalent bond is formed when two atoms share an electron pair, one electron of the shared pair being donated by each participating atom. Since this electron sharing gives the participating atoms the stable electronic configuration of an (unreactive) inert gas, covalent bonds are very strong and stable; considerable energy is required to break such bonds.

The electronic configuration of the carbon atom is such that each can accommodate four additional electrons. To acquire these four additional electrons, each carbon atom must participate in four covalent bonds; we say that carbon is tetravalent or has a valency of four. (The valences of the other elements in organic compounds

TABLE A.1 Valencies of Elements in Organic Compounds

Element	Symbol	Valency
Carbon	C	4
Hydrogen	H	1
Oxygen	O	2
Nitrogen	N	3
Sulfur	S	2
Phosphorus	P	5

are listed in Table A.1.) The covalent bonds can be carbon–carbon bonds, carbon–hydrogen bonds, and so on. As an example, the carbon atom in methane (CH_4) participates in four covalent carbon–hydrogen bonds:

$$H:\overset{H}{\underset{H}{C}}:H$$
Methane

○ Carbon electron
● Hydrogen electron

The covalent bonds in methane are *single bonds*, since a single electron pair is shared between each hydrogen and the carbon. When there are two shared pairs of electrons, the atoms are said to be linked by a *double bond* (—C=C—); similarly, a *triple bond* is formed when two atoms share three electron pairs (—C≡C—). Both single and double bonds are found in all classes of naturally occurring organic compounds; triple bonds are quite rare. Molecules containing double and triple bonds are called *unsaturated* compounds.

One biologically important organic structure, the benzene ring, contains covalent bonds which are electronically intermediate between single and double bonds. Although the benzene ring is written

the ring does not consist simply of alternating single and double bonds. Rather, all the bonds in the ring are electronically equivalent, with properties between those of a single and of a double bond; the electrons are equally distributed over all the carbon atoms in the ring. Molecules containing one or more benzene rings are classified as *aromatic* compounds.

A.2 The shapes of organic compounds

Organic compounds are not planar, but have distinct three-dimensional shapes. One reason for this is the arrangement of the covalent bonds surrounding each carbon

Important Concepts in Organic Chemistry

atom. The four valence bonds of carbon are directed at the four corners of an equilateral tetrahedron (or four-sided pyramid), with the carbon atom at the center.

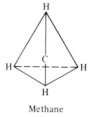
Methane

A carbon atom to which four different atoms or groups of atoms are attached is known as an *asymmetric carbon atom*. Such a carbon atom is found in lactic acid (Figure A.1).

We can draw two different forms of lactic acid:

(+)-Lactic acid   (−)-Lactic acid

The atoms and groups are the same, but their arrangement in space is different; the molecules are mirror images. Such spatially different molecules are known as *isomers*. Since shape and fit are so important in biology, the two isomers would have different properties.

As we examine larger organic compounds, their shapes become more bent, twisted, and irregular, and they are capable of a large number of different spatial arrangements. As you study the more complex biologically important molecules (many

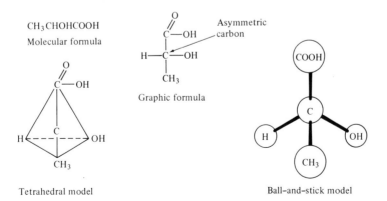

Figure A.1 Structure of lactic acid.

containing hundreds of atoms), you will see that the function of each organic compound in living systems is related to and dependent upon its unique three-dimensional shape.

A.3 The functional groups in organic compounds are responsible for their chemical reactivity

The chemical reactivity of different classes of organic compounds is due to the reactions of their functional groups (Table A.2); compounds can contain more than one functional group.

When writing the reactions of organic molecules, the nonreactive part(s) of the molecule—usually combinations of methyl (—CH_3) and methylene (—CH_2—) groups—is abbreviated as R. Often R is used at a position in the molecule where different groups can be substituted with little effect on the chemical reactivity of the compound.

The major classes of organic compounds and their more biologically important reactions are listed below. Most of these reactions are not spontaneous; nonphysiological conditions (acid or base, heat and/or inorganic catalysts) are required to make them occur. These conditions can be met in the analytical or research laboratory but are seldom present in living cells. Cells avoid the need for extreme conditions partly by carrying out reactions in a series of steps. For example, cells make esters by a two-step process: first the carboxylic acid is chemically activated, and second, the activated acid reacts with the alcohol. In addition, cells use specialized proteins, called enzymes, to catalyze reactions.

Table A.2 Functional Groups in Organic Compounds

Functional group		Classes of compounds containing group
Name	Symbol	
Hydroxyl	—C—OH	Alcohols
Carbonyl	C=O	Aldehydes, ketones
Carboxyl	C=O, HO	Carboxylic acids
Amino	—C—NH_2	Amines
Sulfhydryl	—C—SH	Thiols

(a) ALCOHOLS

Alcohols are compounds that contain one or more hydroxyl groups. Naturally occurring alcohols include

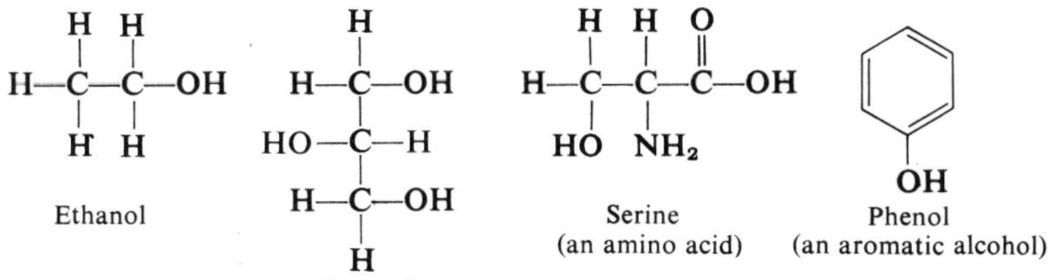

$$\begin{array}{c} H\ H \\ | \ \ | \\ H-C-C-OH \\ | \ \ | \\ H\ H \end{array}$$
Ethanol

$$\begin{array}{c} H \\ | \\ H-C-OH \\ | \\ HO-C-H \\ | \\ H-C-OH \\ | \\ H \end{array}$$
Glycerol
(a trihydroxy alcohol)

$$\begin{array}{c} H\ H\ O \\ | \ \ | \ \ \| \\ H-C-C-C-OH \\ | \ \ | \\ HO\ NH_2 \end{array}$$
Serine
(an amino acid)

Phenol
(an aromatic alcohol)
(benzene ring with OH)

Reactions of Alcohols

1. *Ester formation.* An ester is formed when an alcohol and an acid combine with the release of water. The acid can be organic or inorganic.

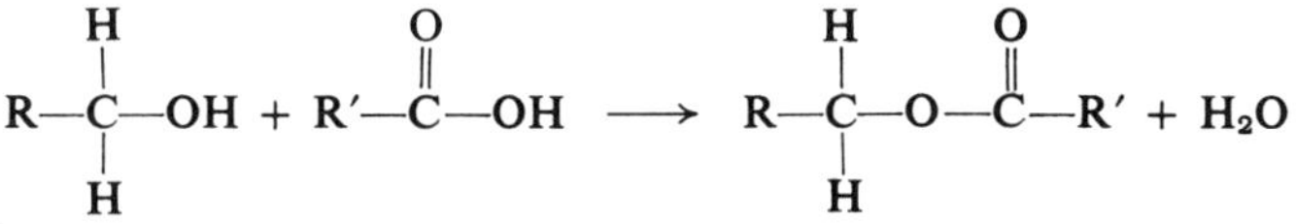

$$\underset{\text{Alcohol}}{R-\underset{\underset{H}{|}}{\overset{\overset{H}{|}}{C}}-OH} + \underset{\substack{\text{Carboxylic} \\ \text{acid} \\ \text{(organic)}}}{R'-\overset{\overset{O}{\|}}{C}-OH} \longrightarrow \underset{\text{Ester}}{R-\underset{\underset{H}{|}}{\overset{\overset{H}{|}}{C}}-O-\overset{\overset{O}{\|}}{C}-R'} + H_2O$$

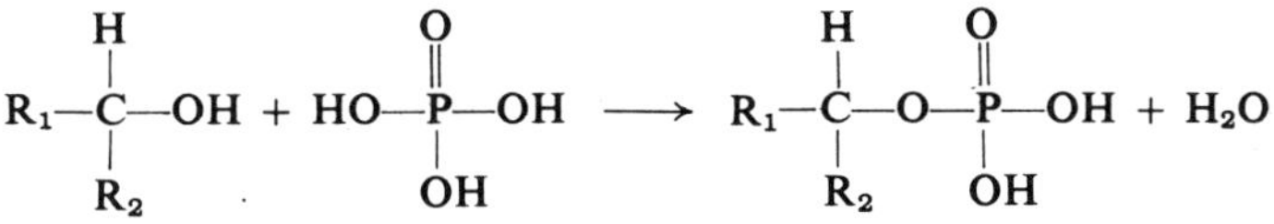

$$\underset{\text{Alcohol}}{R_1-\underset{\underset{R_2}{|}}{\overset{\overset{H}{|}}{C}}-OH} + \underset{\substack{\text{Phosphoric} \\ \text{acid} \\ \text{(inorganic)}}}{HO-\underset{\underset{OH}{|}}{\overset{\overset{O}{\|}}{P}}-OH} \longrightarrow \underset{\text{Phosphate ester}}{R_1-\underset{\underset{R_2}{|}}{\overset{\overset{H}{|}}{C}}-O-\underset{\underset{OH}{|}}{\overset{\overset{O}{\|}}{P}}-OH} + H_2O$$

2. *Oxidation.* Depending on its structure, the removal of two hydrogen atoms from an alcohol produces an aldehyde or a ketone.

$$\underset{\substack{\text{Primary} \\ \text{alcohol}}}{R-\underset{\underset{H}{|}}{\overset{\overset{H}{|}}{C}}-OH} \xrightarrow{-2[H]} \underset{\text{Aldehyde}}{R-C\overset{\nearrow O}{\underset{\searrow H}{}}}$$

$$\underset{\substack{\text{Secondary} \\ \text{alcohol}}}{R_1-\underset{\underset{R_2}{|}}{\overset{\overset{H}{|}}{C}}-OH} \xrightarrow{-2[H]} \underset{\text{Ketone}}{\underset{R_2}{\overset{R_1}{}}\!\!\!\!>\!\!C=O}$$

(b) CARBONYL COMPOUNDS: ALDEHYDES AND KETONES

Both aldehydes and ketones contain a carbonyl group; their general formulas are

$$R-\overset{\displaystyle O}{\underset{\displaystyle H}{C}} \qquad R_1-\overset{\displaystyle O}{\underset{\displaystyle \|}{C}}-R_2$$

Aldehyde   Ketone

Naturally occurring aldehydes and ketones include

$$H-\overset{H}{\underset{H}{C}}-\overset{O}{\underset{}{\overset{\|}{C}}}-\overset{H}{\underset{H}{C}}-\overset{O}{\underset{}{\overset{\|}{C}}}-OH \qquad H-\overset{H}{\underset{H}{C}}-\overset{O}{\underset{H}{C}} \qquad H_3C-\overset{O}{\underset{}{\overset{\|}{C}}}-\overset{O}{\underset{}{\overset{\|}{C}}}-OH$$

Acetoacetic acid   Acetaldehyde   Pyruvic acid (a keto acid)

Reactions of Aldehydes and Ketones

1. *Oxidation.* The addition of oxygen to an aldehyde produces a carboxylic acid.

$$R-\overset{O}{\overset{\|}{C}}-H \xrightarrow{+[O]} R-\overset{O}{\overset{\|}{C}}-OH$$

Aldehyde   Carboxylic acid

2. *Reduction.* The carbonyl group in both aldehydes and ketones can be hydrogenated to produce an alcohol (see alcohols, reaction 2).

3. *Hemiacetal formation.* The reaction between an aldehyde and an alcohol produces a hemiacetal.

$$R-\overset{O}{\overset{\|}{C}}-H + R'-\overset{H}{\underset{H}{C}}-OH \longrightarrow R-\overset{OH}{\underset{H}{C}}-O-\overset{H}{\underset{H}{C}}-R'$$

Aldehyde   Alcohol   Hemiacetal

4. *Aldol condensation.* Under basic conditions, aldehydes can react with each other (condense) to form aldols.

$$R-\overset{O}{\overset{\|}{C}}-H + H-\overset{H}{\underset{H}{C}}-\overset{H}{C}=O \longrightarrow R-\overset{H}{\underset{HO}{C}}-\overset{H}{\underset{H}{C}}-\overset{H}{C}=O$$

Aldehyde   Aldehyde   Aldol

5. *Conversion to enol form.* Aldehydes and ketones are converted to their enol form when a hydrogen atom moves to the oxygen atom and the position of the double bond shifts.

$$R_1-\underset{H}{\overset{H}{C}}-\overset{O}{\underset{}{C}}-R_2 \rightleftharpoons R_1-\underset{}{\overset{H}{C}}=\underset{OH}{\overset{}{C}}-R_2$$

Ketone   Enol

(c) CARBOXYLIC ACIDS

The functional group in carboxylic acids is the carbonyl group. Biologically important carboxylic acids include

Acetic acid   Alanine (an amino acid)   Palmitic acid (a fatty acid)   Succinic acid (a dicarboxylic acid)

Reactions of Carboxylic Acids

1. *Dissociation.* Monocarboxylic acids dissociate in water to give one proton and the negatively charged anion.

$$R-\overset{O}{\underset{}{C}}-OH \rightleftharpoons R-\overset{O}{\underset{}{C}}-O^- + H^+$$

Carboxylic acid   Anion   Proton

This dissociation is readily reversible; the extent of dissociation is proportional to the concentration of hydrogen ions (H^+) in the solution (i.e., the pH of the solution). At pH 7 and above, carboxylic acids exist almost entirely in their dissociated forms. As the pH of the solution is lowered (i.e., the concentration of H^+ is increased), a greater percentage of the acid molecules are in the uncharged form.

2. *Ester formation.* See alcohols, reaction 2.

3. *Amide formation.* When a carboxylic acid combines with an amine with the release of water, an amide is formed.

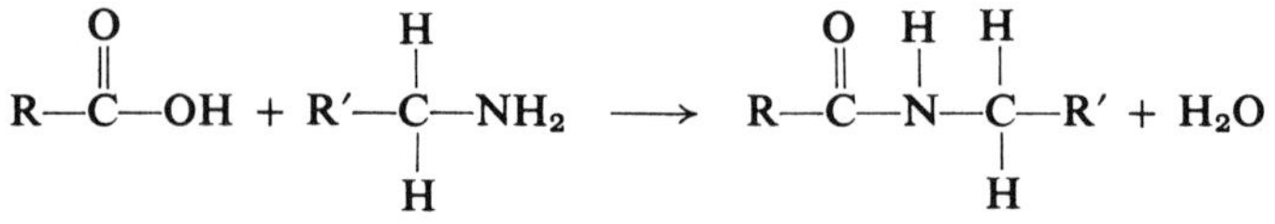

$$\underset{\text{Carboxylic acid}}{R-\overset{O}{\underset{\|}{C}}-OH} + \underset{\text{Amine}}{R'-\overset{H}{\underset{H}{\overset{|}{C}}}-NH_2} \longrightarrow \underset{\text{Amide}}{R-\overset{O}{\underset{\|}{C}}-\overset{H}{\underset{|}{N}}-\overset{H}{\underset{H}{\overset{|}{C}}}-R'} + H_2O$$

(d) AMINES

Amines are compounds containing an amino group. Naturally occurring amines include

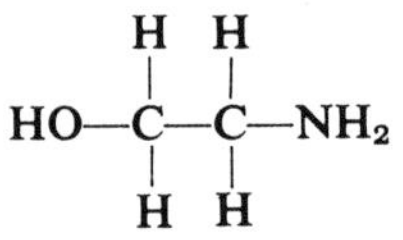

$$\underset{\text{Ethanolamine}}{HO-\overset{H}{\underset{H}{\overset{|}{C}}}-\overset{H}{\underset{H}{\overset{|}{C}}}-NH_2} \qquad \underset{\text{Epinephrine}}{HO-\underset{HO}{\bigcirc}-\overset{H}{\underset{HO}{\overset{|}{C}}}-\overset{H}{\underset{H}{\overset{|}{C}}}-NH_2}$$

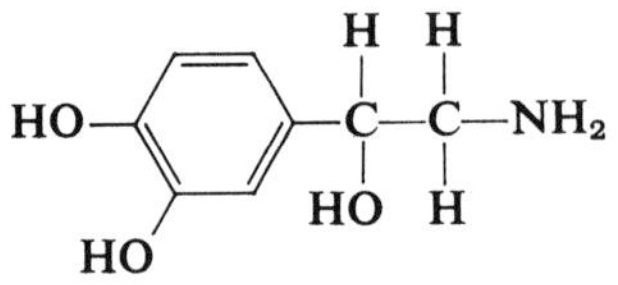

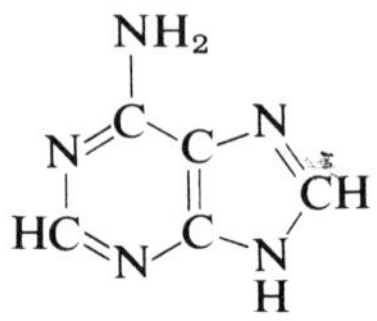

Adenine

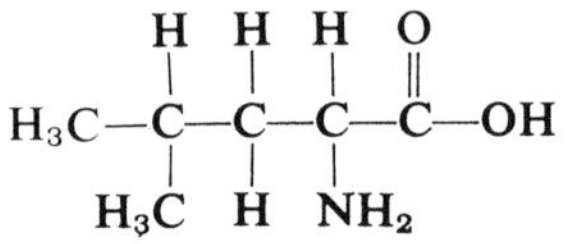

$$\underset{\substack{\text{Leucine} \\ \text{(an amino acid)}}}{H_3C-\overset{H}{\underset{H_3C}{\overset{|}{C}}}-\overset{H}{\underset{H}{\overset{|}{C}}}-\overset{H}{\underset{NH_2}{\overset{|}{C}}}-\overset{O}{\underset{\|}{C}}-OH}$$

Reactions of amines

1. *Dissociation.* In solution, amines exist in both a charged and an uncharged form.

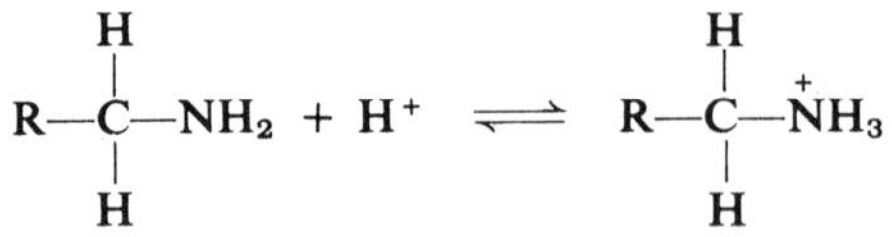

$$\underset{\text{Amine}}{R-\overset{H}{\underset{H}{\overset{|}{C}}}-NH_2} + \underset{\text{Proton}}{H^+} \rightleftharpoons \underset{\text{Cation}}{R-\overset{H}{\underset{H}{\overset{|}{C}}}-\overset{+}{N}H_3}$$

As with the dissociation of carboxylic acids, this reaction is readily reversible and the direction of the reaction depends upon the pH of the solution. At pH 7 and below, the amine is in its cationic form; as the pH increases (i.e., the concentration of H^+ decreases), the amine is converted to its uncharged form.

2. *Amide formation.* See carboxylic acids, reaction 3.

(e) ESTERS

Esters are formed when organic or inorganic acids react with alcohols. Naturally occurring esters include

$$\underset{\text{Acetylcholine}}{H_3C-\overset{O}{\underset{\|}{C}}-O-\overset{H}{\underset{H}{C}}-\overset{H}{\underset{H}{C}}-\overset{+}{N}(CH_3)_3}$$

$$\underset{\substack{\text{Tripalmitate}\\\text{(a triglyceride)}}}{\begin{array}{c}H-C-O-C-(CH_2)_{14}CH_3\\|\\CH_3(CH_2)_{14}-C-O-C-H\\|\\H-C-O-C-(CH_2)_{14}CH_3\\|\\H\end{array}}$$

$$\underset{\text{Glycerol phosphate}}{\begin{array}{c}H-C-OH\\|\\HO-C-H\\|\\H-C-O-P-OH\\|\quad\quad|\\H\quad\quad OH\end{array}}$$

Reaction of esters

1. *Hydrolysis.* Esters are cleaved into their constituent alcohol and acid by the addition of water across the ester bond. This reaction is the reverse of their formation (see alcohols, reaction 2).

(f) AMIDES

Amides are the product of a reaction between an amine and a carboxylic acid. Naturally occurring amides include

$$\underset{\text{Glutamine}}{\begin{array}{c}C-NH_2\\\|\\O\\H-C-H\\|\\H-C-H\\|\\H-C-NH_2\\|\\C-OH\\\|\\O\end{array}}\quad\quad\underset{\text{A protein}}{-N-\underset{R_1}{C}-C-N-\underset{R_2}{C}-C-N-\underset{R_3}{C}-C-}$$

Reactions of Amides

1. *Hydrolysis.* Amides are similar to esters and are cleaved into their constituent acid and amine by the addition of water across the amide bond. This reaction is the reverse of their formation (see carboxylic acids, reaction 3).

(g) SULFUR-CONTAINING COMPOUNDS

Sulfur, which is in the same group of the periodic table as oxygen, can form similar types of compounds. Thus, there are the following compounds, analogous to oxygen-containing compounds

$$\underset{\substack{\text{Thioalcohols}\\ \text{(mercaptans)}}}{\text{R}-\underset{\underset{\text{H}}{|}}{\overset{\overset{\text{H}}{|}}{\text{C}}}-\text{SH}} \quad \text{and} \quad \underset{\text{Thioesters}}{\text{R}-\overset{\overset{\text{O}}{\|}}{\text{C}}-\text{S}-\text{R}'}$$

Although they are relatively few, there are some important naturally occurring sulfur compounds, including

$$\underset{\substack{\text{Cysteine}\\ \text{(an amino acid)}}}{\text{H}_2\text{N}-\overset{\overset{\text{H}}{|}}{\underset{\underset{\text{SH}}{|}}{\text{C}}}-\overset{\overset{\text{O}}{\|}}{\text{C}}-\text{OH}} \qquad \underset{\substack{\text{Acetyl coenzyme A}\\ \text{(a thioester)}}}{\text{H}-\overset{\overset{\text{H}}{|}}{\underset{\underset{\text{H}}{|}}{\text{C}}}-\overset{\overset{\text{O}}{\|}}{\text{C}}-\text{S}-\text{CoA}} \qquad \underset{\substack{\text{Methionine}\\ \text{(an amino acid)}}}{\text{H}_2\text{N}-\overset{\overset{\text{H}}{|}}{\text{C}}-\overset{\overset{\text{O}}{\|}}{\text{C}}-\text{OH}}$$

Reactions of Sulfur Compounds

1. *Reduction.* When two thiols react with the removal of hydrogen, a disulfide is formed.

$$\underset{\text{Thiol}}{\text{R}-\overset{\overset{\text{H}}{|}}{\underset{\underset{\text{H}}{|}}{\text{C}}}-\text{SH}} + \underset{\text{Thiol}}{\text{R}'-\overset{\overset{\text{H}}{|}}{\underset{\underset{\text{H}}{|}}{\text{C}}}-\text{SH}} \xrightarrow{-2[\text{H}]} \underset{\text{Disulfide}}{\text{R}-\overset{\overset{\text{H}}{|}}{\underset{\underset{\text{H}}{|}}{\text{C}}}-\text{S}-\text{S}-\overset{\overset{\text{H}}{|}}{\underset{\underset{\text{H}}{|}}{\text{C}}}-\text{R}'}$$

A.4 Oxidation–reduction reactions

Oxidation–reduction reactions are very important in biological systems. These are coupled reactions: every time one compound is oxidized, another molecule is reduced simultaneously.

There are three possible ways for coupled oxidation–reduction reactions to occur.

Oxidation	Reduction
1. Addition of oxygen	Removal of oxygen
2. Removal of hydrogen atom	Addition of hydrogen atom
3. Removal of an electron	Addition of an electron

A hydrogen atom consists of an electron (e^-) and a proton (H^+).

Oxidation–reduction reactions in biological systems are usually type 2 or type 3. The most common hydrogen donor is NADH; the most common hydrogen acceptor is NAD^+. An example of a type 2 oxidation–reduction reaction is the oxidation of an alcohol to an aldehyde with the concomitant reduction of NAD^+ to NADH.

$$\underset{\text{Alcohol}}{R-\underset{H}{\overset{H}{C}}-OH} + NAD^+ \rightleftharpoons \underset{\text{Aldehyde}}{R-\underset{H}{C}=O} + NADH + H^+$$

In this reaction, the NAD^+ has received two electrons and one proton. An example of a type 3 oxidation–reduction reaction is the reduction of Fe^{3+} to Fe^{2+}. In biological systems, this would occur as follows:

$$\underset{\substack{\text{Ferric}\\\text{ion}}}{Fe^{3+}} + NADH \rightleftharpoons \underset{\substack{\text{Ferrous}\\\text{ion}}}{Fe^{2+}} + NAD^+ + H^+$$

TEXTBOOK REFERENCES

COMPREHENSIVE BIOCHEMISTRY TEXTBOOKS

CONN, ERIC E., and P. K. STUMPF, *Outlines of Biochemistry*, 4th ed., John Wiley & Sons, Inc., New York (1976).

DEBEY, HAROLD J., *Introduction to the Chemistry of Life: Biochemistry*, 2nd ed., Addison-Wesley Publishing Company, Inc., Reading, Mass. (1976).

HARPER, HAROLD A., *Review of Physiological Chemistry*, Lange Medical Publications, Los Altos, Calif. (1973).

LEHNINGER, ALBERT L., *Biochemistry*, 2nd ed., Worth Publishers, Inc., New York (1975).

MAHLER, HENRY R., and EUGENE H. CORDES, *Biological Chemistry*, 2nd ed., Harper & Row, Publishers, Inc., New York (1971).

STRYER, LUBERT, *Biochemistry*, W. H. Freeman and Company, San Francisco (1975).

SUTTIE, JOHN W., *Introduction to Biochemistry*, 2nd ed., Holt, Rinehart and Winston, Inc., New York (1977).

YUDKIN, MICHAEL, and ROBIN OFFORD, *Comprehensible Biochemistry*, Longman Group Ltd., London (1973).

TEXTBOOKS ON SPECIAL TOPICS

BARKER, ROBERT, *Organic Chemistry of Biological Compounds*, Prentice-Hall, Inc., Englewood Cliffs, N.J. (1971).

BREWER, J. M., A. J. PESCE, and R. B. ASHWORTH, *Experimental Techniques in Biochemistry*, Prentice-Hall, Inc., Englewood Cliffs, N.J. (1974).

BROCK, THOMAS D., *Biology of Microorganisms*, 2nd ed., Prentice-Hall, Inc., Englewood Cliffs, N.J. (1974).

FRIEDEN, EARL, et al. *Biochemical Endocrinology of the Vertebrates*, Prentice-Hall, Inc., Englewood Cliffs, N.J. (1971).

JAWETZ, E., J. L. MELNICK, and E. A. ADELBERG, *Review of Medical Microbiology*, Lange Medical Publications, Los Altos, Calif. (1972).

KROGMANN, DAVID W., *The Biochemistry of Green Plants*, Prentice-Hall, Inc., Englewood Cliffs, N.J. (1973).

LARNER, JOSEPH, *Intermediary Metabolism and Its Regulation*, Prentice-Hall, Inc., Englewood Cliffs, N.J. (1971).

LERNER, I. MICHAEL, and W. J. LIBBY, *Heredity, Evolution and Society*, 2nd ed., W. H. Freeman and Company, San Francisco (1976).

MEYERS, F. H., E. JAWETZ, and A. GOLDFEIN, *Review of Medical Pharmacology*, Lange Medical Publications, Los Altos, Calif. (1972).

VAN HOLDE, KENSAL EDWARD, *Physical Biochemistry*, Prentice-Hall, Inc., Englewood Cliffs, N.J. (1971).

WATSON, J. D, *Molecular Biology of the Gene*, 3rd ed., W. A. Benjamin, Inc., Menlo Park, Calif. (1976).

WOLD, FINN, *Macromolecules: Structure and Function*, Prentice-Hall, Inc., Englewood Cliffs, N.J. (1971).

Index

The page numbers in boldface indicate the location of the chemical structures of the compounds.

A

Abscisic acid:
  chemistry, 278
  physiological response, 278
Acetaldehyde, **126**
  in anaerobic glycolysis, 126
Acetoacetic acid, **318**
  in ketosis, 172
Acetone, in ketosis, 172
Acetylcholine, **321**
Acetyl coenzyme A, **127**
  from amino acids, 194–97
  as branch-point compound, 92, 131
  in cholesterol synthesis, 172–74
  control of pyruvate carboxylase, 138
  from fatty acid degradation, 170–72
  in fatty acid synthesis, 168–70
  in glyoxylate cycle, 138
  in Krebs cycle, 131, 133, 137
  from pyruvate, 127–28
Acetyl coenzyme A carboxylase:
  control by citrate, 274
  in fatty acid synthesis, 169

N-acetyl hexosaminidase, in Tay-Sachs disease, 230
Aconitase, inhibition, 82
cis-Aconitic acid, **133**
  in Krebs cycle, 133
Acromegaly, hormone imbalances in, 279
Actin, 14
  in skeletal muscle, 107–9
Actinomycin D, **303**
  mode of action, 308
Activation energy, 68
Activator proteins, in transcriptional control, 268
Active site of enzymes, 70
Active transport, 104–6
  Na^+-K^+ pump, 109–11
  in plant roots, 111
Adaptation, control of, 264
Addison's disease, hormone deficiencies in, 279
Adenine, **202**
  degradation, 208
  in DNA, 212–13
  in RNA, 242–43
  synthesis, 203–4

Index

Adenine ribosides, precursors of cytokinins, 278
Adenosine diphosphate, **103**
  in cellular energy cycle, 102
  concentration in cell, 112
  control of glutamic dehydrogenase, 194
  control of phosphofructokinase, 119–20
  formation from ATP, 101–4
  in glycolysis, 117
  in oxidative phosphorylation, 141
Adenosine monophosphate, **103, 204** (*see also* Cyclic AMP)
  concentration in cell, 112
  control of phosphofructokinase, 119–20
Adenosine triphosphate, **103**
  in cellular energy cycle, 101–6
  in coenzyme A synthesis, 127
  concentration in cell, 112
  control of glutamic dehydrogenase, 194
  control of phosphofructokinase, 119–20
  energy of hydrolysis, 103
  as energy middleman, 101–4
  in fatty acid metabolism, 168–72
  from glycolysis, 117, 143
  from Krebs cycle, 143
  in muscle contraction, 106
  in Na^+-K^+ pump, 109–11
  in nitrogen fixation, 189
  from oxidative phosphorylation, 139–42
  from photosynthesis, 157–60
  in urea cycle, 196
Adenyl cyclase, in cAMP production, 275, 278
ADP (*see* Adenosine diphosphate)
Adrenalin (*see* Epinephrine)
Adrenal steroids (*see also* Glucocorticoids; Mineralocorticoids)
  chemistry, 179–80
  physiological function, 179–81
  synthesis, 173
Adrenocorticotropin (ACTH):
  chemistry, 276
  deficiency, 279
  excess, 279
  physiological response, 276
Adrenogenital syndrome, 230
Afibrinogenaemia:
  clinical features, 230
  metabolic block, 230
Agar, 151
Alanine, **16**
  degradation, 195
  and Krebs cycle, 137
  synthesis, 192
  in transamination reactions, 136

Albinism, 201–2
Alcaptonuria, 201
Alcohol dehydrogenase, in anaerobic glycolysis, 126
Alcoholic beverages, production of, 79, 125
Alcoholism, fatty liver in, 177
Alcohols:
  ester formation, 317–18
  oxidation, 318
  primary, 318
  secondary, 318
Aldehydes (*see* Carbonyl compounds)
Aldolase, in glycolysis, 121
Aldosterone, **180**
  chemistry, 277
  deficiency, 279
  excess, 279
  physiological response, 277
Aldosteronism, 279
Alkaline phosphatase:
  assay, 73
  in medical diagnosis, 78
Alkaloids, from amino acids, 198
Alpha-helix, **34**
  destruction of, 39–42
  in proteins, 29–34
Amethopterin, 206
Amides, hydrolysis, 322
Amines:
  amide formation, 320
  dissociation, 320
Amino acid analyzer, 51
Amino acids (*see also* individual amino acids)
  acidic, 19
  aliphatic, 16
  aromatic, 17
  basic, 18
  charge properties, 20–22
  chromatography, 51–52
  deficiency, 190
  degradation, 137, 194–97
  derived, 19, 190–91
  electrophoresis, 51–53
  essential, 8, 189–91
  genetic code for, 231
  in gluconeogenesis, 153
  isomers, 16
  linkage to tRNA, 254
  metabolic roles, 187–88
  neutral, 16
  in nitrogen cycle, 187–89
  nonessential, 189–91
  peptide bonds between, 22–24
  reactions, 20

Amino acids (cont.):
  structures, **16–19**
  sulfur-containing, 17
  synthesis, 137, 192–93
  three-letter symbols, 16–19, 25
Aminoacyl tRNAs:
  formation, 254–55
  in translation, 254–55, 259
Aminoacyl tRNA synthetase, in protein synthesis, 254
p-Aminobenzoic acid, **84**
Amino group, **317**
δ-Aminolevulinate synthetase:
  in porphyrias, 199
  in porphyrin synthesis, 198–99
δ-Aminolevulinic acid, in porphyrin synthesis, 197–99
Aminopterin, 206
Ammonia:
  from amino acid degradation, 194–96
  in amino acid synthesis, 192–93
  in nitrogen cycle, 187–89
  from pyrimidine degradation, 208
  in urea cycle, 196
Ammonium sulfate, 50
AMP (see Adenosine monophosphate)
Ampicillin, mode of action, 83, 306
Amylase:
  industrial uses, 79
  in maltose production, 152
  pharmacologic uses, 78
  in seed germination, 278
Anabolic pathways:
  definition, 90
  energy required, 90, 101–4
Androgens, 179–80, 277
Antibiotics, 303–9
  from amino acids, 198
  bacterial resistance, 309–10
  chemistry, 303–9
  definition, 303
  as enzyme inhibitors, 83
  inhibitors of cell wall synthesis, 305–6
  inhibitors of protein synthesis, 306–7
  inhibitors of RNA synthesis, 307–8
  selective action, 304–5
  toxicity, 305
  uses, 304
Antibodies, 296–98
  Anti-Rh, 14
  definition, 13
  γG-immunoglobulins, 14
Anticodons, 255

Antigens, 296–98
Antimetabolites, 308
Antimicrobial agents, 302–10
Antimycin A, mode of action, 82
Arachidonic acid, **165**
Arginine, **19**
  degradation, 195
  in ionic bonds, 35
  and Krebs cycle, 137
  in urea cycle, 196
Argininosuccinic acid, **196**
  in urea cycle, 196–97
Aromatic compounds, 314
Arsenic, as enzyme inhibitor, 82
Ascorbic acid:
  in collagen synthesis, 193
  deficiency, 193
  as enzyme cofactor, 38
  metabolic role, 8
Asparagine, 19, **191**
Aspartic acid, **18**
  in asparagine synthesis, 190–91
  degradation, 195
  in ionic bonds, 35
  nitrogen-containing derivatives, 198
  in purine synthesis, 203–4
  in pyrimidine synthesis, 205
  synthesis, 192
  in urea cycle, 196
Assays:
  for enzymes, 71–75
  for proteins, 47
Asymmetric carbon atom, 315
Atherosclerosis, 182–83
ATP (see Adenosine triphosphate)
AT pair, **215**
AU pair, 245

B

Bacitracin, **304**
  mode of action, 305–6
Bacteriophages, 287
Bacteriophage T$_4$, **291**
Barr body, 271
Base pairing (see Watson-Crick base pairing)
Benzene ring, 314
Beriberi, 128
Bile acids:
  metabolic role, 174
  synthesis, 173–74
Bile pigments, 199
Bile salts, 174

Bilirubin:
  in jaundice, 199–200
  from porphyrins, 199–200
Biochemistry, unity of, 1
Biosynthesis, 104–6
Biotin:
  as enzyme cofactor, 38
  in fatty acid synthesis, 169
  metabolic role, 8
Boron, metabolic role, 9
Botulism toxin, mode of action, 82
Branching enzyme, in glycogen synthesis, 150
Branch-point compounds, 90–93
  in feedback inhibition, 273
  in metabolic control, 94
Bread making, 126

C

Caeruloplasmin, in Wilson's disease, 230
Calcium:
  in human body, 5
  metabolic role, 9
  role in muscle, 109
Calvin cycle, 159
Cancer:
  chemotherapy, 83, 308
  environmental causes, 294
  skin, 222
  viral origin, 285, 293–94
Carbamoyl phosphate, **196**
  in pyrimidine synthesis, 205
  in urea cycle, 196–97
Carbohydrates, definition, 115 (see also specific carbohydrates)
Carbon:
  covalent bonds, 15–16, 313–14
  isotopes, 95
  valency, 314
Carbon dioxide:
  from Krebs cycle, 133
  from pentose phosphate pathway, 153
  in purine synthesis, 203–4
  in urea cycle, 196
Carbon dioxide fixation:
  in fatty acid synthesis, 168–70
  in photosynthesis, 158–59
Carbon monoxide, inhibitor of hemoglobin, 82
Carbonyl compounds:
  aldehydes, 318–19
  aldol condensation, 319
  enol form, 319
  hemiacetal formation, 318–19

Carbonyl compounds (cont):
  ketones, 318–19
  oxidation, 318
  reduction, 318
Carbonyl group, 317
Carboxyl group, 317
Carboxylic acids:
  amide formation, 320
  dissociation, 319–20
  ester formation, 320
Carboxypeptidase, in protein sequencing, 54
Carcinogens, 294
  effect on DNA, 221–22
Carriers of metabolic disorders, 234–35
Catabolic pathway:
  definition, 90
  energy released, 90, 101–4
Cauliflower mosaic virus, 288
Cell division, chromosomes during, 220–21
Cell free extracts, 48
  use in determining metabolic pathways, 98
Cell membrane, 4
  active transport, 105
  in DNA replication, 225–26
  hormone receptors, 275
  structure, 174–77
Cells, 1–5
  differences in proteins, 13
  elements in, 4–5
  eucaryotic, 2–4
  procaryotic, 2–4
  requirements for organic compounds, 7
  structure, 2–4
Cell wall, 4
  inhibitors of synthesis, 83, 305–6
Cellulose, 151
  digestion by ruminants, 160
  digestion by termites, 160
Central nervous system, viral diseases, 287
Centrifugation, Svedberg units, 256
Cephalosporin, mode of action, 305–6
Cephalothin, **306**
Cheese production, 40–41
Chemotherapy, 302
Chicken pox, viral origin, 287
Chitin, 151
Chloramphenicol, **303**
  mode of action, 83, 305–7
  R-factor resistance, 310
  toxicity, 305
Chlorine:
  in human body, 5
  metabolic role, 9

Chlorophyll
  in photosynthesis, 157–60
  synthesis, 197–99
Chloroplasts, 4
  photosynthesis in, 157–60
Cholera toxin, mode of action, 82
Cholesterol, **164, 173**
  esterification, 183
  excretion, 174
  heart disease and, 182–83
  hormones from, 277
  metabolic role, 167
  synthesis, 172–74
Cholic acid, **173** (*see also* Bile acids)
Choline:
  methionine in, 198
  in phosphoglycerides, 167
Chondroitin sulfate, 151
Chromatography:
  ion-exchange, 49
  molecular sieve, 48
Chromium, metabolic role, 9
Chromosomes:
  Barr body, 271
  condensation, 270–71
  histones in, 216–17
  human, 216
  replication, 220–21
  in sexual cycle, 235
  structure, 214–20
  in transcriptional control, 270–71
  viruses integrated, 293–94
Chymotrypsin, 14
  amino acid sequence, 30
  formation, 272
  mechanism of action, 69
  pH dependence, 75
  in protein sequencing, 55, 57
  three-dimensional structure, 31
Cirrhosis of liver, 177
Citric acid, **133**
  control of phosphofructokinase, 119–20, 132–34
  in Krebs cycle, 132–33, 137
Citric acid cycle (*see* Krebs cycle)
Citrulline, **196**
  in urea cycle, 196–97
Cobalamin, 199
  as enzyme cofactor, 38
  metabolic role, 8
Cobalt:
  in human body, 5
  metabolic role, 9

Cocaine, 198
Codeine, 198
Codons, 252
Coenzyme A, **127**
Colds, viral origin, 287
Cold sores, viral origin, 287
Colinearity of gene and protein, 232
Collagen, 14
  structure, 37
  synthesis, 193
  three-dimensional structure, 28
Compartmental regulation, 281–82
Complex lipids, 166–67
  amino acids in, 198
  function, 181–82
  structure, 181–82
Constitutive enzymes, 267
Contraceptives, oral, 181
Contractile proteins, 13 (*see also* Actin; Muscles; Myosin)
Control, compartmental, 281–82
Control, metabolic, 264–82 (*see also* Transcriptional control; Translational control; Gene amplification)
Control proteins:
  activators, 268
  repressors, 268
  in transcription, 266–70
Copper:
  in human body, 5
  metabolic role, 9
Cori cycle, 154
Corn syrup, 79
Coronaviridae, 286–87
Cortisol, **180**
  chemistry, 277
  deficiency, 279
  excess, 279
  physiological response, 277
Covalent bonds, 313–14
Coxsackie virus, 287
Creatine phosphate:
  energy of hydrolysis, 103
  formation, 106
  in muscles, 106
Cretinism, 202
  in thyroxine deficiency, 279
Creutzfeldt-Jakob disease, viroid origin, 298
Croup, viral origin, 287
Cushing's syndrome, 279
  hormone imbalances in, 279
Cyanide, as enzyme inhibitor, 82

Cyanogen bromide, in protein sequencing, 55, 57
Cyclic AMP:
  control of RNA polymerase, 267, 269
  control of transcription, 250–51
  in hormone action, 156, 278
  overproduction in cholera, 82
Cysteine, **17**
  in coenzyme A, 127
  degradation, 195
  disulfide bridge, **35**
  and Krebs cycle, 137
  nitrogen-containing derivatives, 198
  reaction with heavy metals, 82
  synthesis, 193
Cytidine monophosphate, **205**
Cytochrome c, 140–45
  evolution, 142–45
Cytochrome oxidase, 141–42
  inhibition, 142
Cytochromes:
  in electron transport chain, 140–41
  porphyrins in, 197, 199
Cytokinins:
  chemistry, 278
  physiological response, 278
Cytosine, **202**
  degradation, 208
  in DNA, 212–13
  in RNA, 242–43
  synthesis, 205
Cytosol, 4

D

Dairy products, commercial production, 126
Dark reaction of photosynthesis, 157–60
Deamination, in nitrogen cycle, 187–89
Debranching enzyme, in glycogen breakdown, 150
Decarboxylases, reactions catalyzed, 66
Dehydrogenases:
  cofactors for, 135
  reactions catalyzed, 66
Deletion mutations, 232–33
Denaturation:
  of proteins, 39–42
  reversible, 41
Denitrification, in nitrogen cycle, 187–89
5'-Deoxyadenosyl cobalamine, as enzyme cofactor, 38
Deoxyribonucleases, reaction catalyzed, 222

Deoxyribonucleic acid:
  actinomycin binding, 308
  base composition, 215
  chromosomal, 214–20
  control of replication, 265
  direction of replication, 224–25
  double helix, 212–13
  effects of radiation, 221–22
  as genetic material, 226–27
  primer for synthesis, 224
  promoter sequence, 250–51
  recombinant, 222–24
  recombination, 221–24
  repair, 221–24
  replication, 217–27
  sense strand, 250
  sensitivity to alkali, 244
  size, 213
  structure, 211–13
  as template, 248
  terminator sequence, 251
  viral, 286, 288, 294–96
Deoxyribonucleotides, formation, 206
Deoxyribose, in DNA, 211–13
Development, control of, 264
Dextran, 151
Diabetes insipidus, 279
Diabetes mellitus, 154–55, 279
  ketosis in, 172
  treatment, 155
Dialysis, 50
Diethylstilbesterol, 181
Differentiation, control of, 264
Digestion:
  bile salts in, 168, 174
  of polysaccharides, 115, 152
  of proteins, 190
  of triglycerides, 170
Diglycerides, 166
Dihydrofolate reductase, trimethoprim inhibition, 308
Dihydroxyacetone phosphate, **121**
  in glycolysis, 117, 121
  in lipid synthesis, 121
Diisopropylfluorophosphate, as enzyme inhibitor, 82
1, 3-Diphosphoglyceric acid, **122**
  energy from, 143
  energy of hydrolysis, 103
  in glycolysis, 117, 122
2, 3-Diphosphoglyceric acid, control of hemoglobin, 122
Diptheria toxin, mode of action, 82

Disaccharides, 151–53
Disulfide bridges:
  destruction of, 39–42
  in proteins, **35**
Disulfides, 322
DNA (*see* Deoxyribonucleic acid)
DNA ligase:
  reaction catalyzed, 222
  in replication, 224–25
DNA polymerase:
  in DNA replication, 218–19, 248
  reaction catalyzed, 222
DNA polymerase I, in DNA repair, 224–26
DNA polymerase II, 224–26
DNA polymerase III, in DNA synthesis, 224–26
DNP-amino acids, 20
Double bond, 314
Double helix, in DNA, 212–13
Drugs:
  antibiotics, 302–8
  as enzyme inhibitors, 81–84
Dwarfism, hormone deficiency, 279

E

Ecdysone, 181
Edman degradation, 54–56
Effectors, control of enzyme activity, 94, 272–74
Electron transport:
  cytochromes in, 140–41
  inhibitors, 82
  mitochondrial electron transport chain, **141**, 130–45
  oxygen requirement, 142
  in photosynthesis, 157–60
Electrophoresis, 51
  of amino acid mixtures, 53
  of isoenzymes, 76
Elements in cells, 5 (*see also* Trace elements)
Elongation factors, in translation, 256–59
Encephalitis, viral origin, 287
Endergonic reactions, 67
Endonucleases, reaction catalyzed, 222
Endoplasmic reticulum, 4
Energy cycle:
  in cells, 102
  quantitation, 111–12
Energy-releasing processes, 101–4
Energy-requiring processes (*see* Work, of cells)
Enolase:
  in glycolysis, 123
  inhibition by fluoride, 123

Enzymatic reactions:
  energy changes during, 67–68
  reversibility, 64
Enzyme activators, function in metabolism, 94
Enzyme activity, control of, 93–94
Enzyme inhibitors, 80–84
  clinically useful, 83
  competitive, 80
  function in metabolism, 94
  irreversible, 80
  noncompetitive, 80
  reversible, 80
  use in determining metabolic pathways, 96–98
Enzymes, 12, 14, 63–84 (*see also* specific enzymes)
  active site, 70
  affect of pH on, 72–75
  affect of temperature on, 72–75
  assays, 71–75
  chemical modification, 272
  cofactors, 8, 37–39, 198
  complexes, 127
  constitutive, 267
  control of activity, 94, 272–74
  effectors, 272–74
  feedback inhibition, 273
  mechanism of action, 68–71
  in medical diagnosis, 77–78
  in metabolic pathways, 90
  nomenclature, 65–66
  one gene/one enzyme hypothesis, 228–31
  pharmacologic uses, 78
  specificity, 65, 71
Enzyme-substrate complex, 68, 70
Epinephrine, 320
  chemistry, 198, 276
  control of carbohydrate metabolism, 155–57
  control of lipid metabolism, 166–67
  excess, 279
  physiological response, 276
Erythromycin:
  mode of action, 83, 305–7
  R-factor resistance, 310
Ester bond, in lipids, 6, 166
Esters, hydrolysis, 321
Estradiol, **180**
  chemistry, 277
  deficiency, 279
  excess, 279
  physiological response, 277
Estrogens, 179–80, 277
Ethanol, **126**

Ethanol (cont.):
  alcoholic beverages, 125–26
  from pyruvic acid, 126
Ethanolamine, **320**
Ethylene, **278**
  in fruit ripening, 278
Eucaryotic cells (*see* Cells)
Eunuchoidism, in testosterone deficiency, 279
Evolution:
  antibiotic resistance, 309–10
  of cytochrome *c,* 142–45
  of DNA base composition, 215
  of DNA processing enzymes, 226
  elements, 4
  evolutionary trees, 144–45
  molecular oxygen and, 159
  natural selection, 236
  of proteins, 142–45
Exergonic reactions, 67
Exonucleases, reaction catalyzed, 222

F

Factors:
  elongation, 256–59
  initiation, 257–59
  termination, 256–59
  in translation, 257, 259
  in translational control, 271
FAD (*see* Flavin adenine dinucleotide)
Fats, (*see* Lipids)
Fat-soluble hormones, (*see* Prostaglandins; Steroids)
Fat-soluble vitamins, 177–79
Fatty acids, **164**
  in complex lipids, 166–67
  degradation, 170–72
  essential, 165
  saturated, 165
  synthesis, 168–70
  unsaturated, 165
Fatty acyl CoA, in glyceride synthesis, 170
Fatty liver, 177
Feedback inhibition, 273
Fertilizers, nitrogen-containing, 189
Fibrinogen, in afibrinogenaemia, 230
Flavin adenine dinucleotide:
  dehydrogenase cofactor, 38, 135
  in electron transport chain, 139–42
  in fatty acid degradation, 169–72
  in Krebs cycle, 135
  in pyruvate dehydrogenase complex, 127
Flavin mononucleotide, as enzyme cofactor, 38

Fluoride:
  inhibition of glycolysis, 123
  metabolic role, 9
Fluoroacetic acid, 132
  as enzyme inhibitor, 82
Fluoro-2, 4-dinitrobenzene, **20**
  in protein sequencing, 55
  reaction with amino acids, 20
5-Fluorouracil:
  mode of action, 206
  structure, 206
Folic acid, conversion to tetrahydrofolic acid, 203
N-Formylmethionine, in translation, 258
Fructose-1, 6-diphosphate, **119**
  in glycolysis, 117, 119
Fructose-6-phosphate, **119**
  control of phosphofructokinase, 119–20
  energy from, 143
  energy of hydrolysis, 103
  in glycolysis, 117, 119
Fruit juice production, 79
Fruiting, hormonal control, 278
Fumaric acid, **133**
  in Krebs cycle, 133, 137
  in urea cycle, 196

G

Galactose, **152**
  in disaccharides, 152–53
  galactosemia, 153, 230
Galactosemia, 153
  clinical features, 230
  metabolic block, 230
Galactose-1-phosphate uridyl transferase, in galactosemia, 230
Gallstones, 174
Gastroenteritis, viral origin, 287
Gastrointestinal tract, viral disease, 287
GC pair, **214**
Gene amplification, 274–75
Genes (*see also* Deoxyribonucleic acid)
  colinearity with proteins, 232
  control of expression, 264–75
  definition, 226
  genetic engineering, 222–24
  inheritance in sexual cycle, 235
  mutations in, 226
  one gene/one enzyme hypothesis, 228–31
  regulatory, 266–67
  for RNA, 251
  transcription of, 248–52
  viral, 288–89

Genetic code, 231–33, **253**
   reading, 232
Genetic engineering, 222–24
Genital herpes, viral origin, 287
Gentamicin, mode of action, 305–7
Giantism, growth hormone excess, 279
Gibberellic acid:
   chemistry, 278
   physiological response, 278
Glucagon:
   chemistry, 276
   control of carbohydrate metabolism, 155–57
   control of lipid metabolism, 166–67
   mode of action, 278
   physiological response, 276
Glucocorticoids, 179–80, 277
Gluconeogenesis, 138, 149, 153–54
   from amino acids, 194–96
   hormonal control, 179, 276–77
   from Krebs cycle intermediates, 137
Glucose, **116**
   blood levels, 149, 153, 179, 278
   as branch-point compound, 149
   concentration in cell, 112
   degradation, 115–28
   in diabetes, 154–55
   energy from, 116, 118, 142
   in glycogen, 149–51
   hormonal control, 156, 179
   in pentose production, 153
   in polysaccharides, 151
Glucose oxidase, in prepared foods, 80
Glucose-1-phosphate, energy of hydrolysis, 103
Glucose-6-phosphate, **119**
   energy of hydrolysis, 103
   from glycogen, 150
   in glycolysis, 117, 119
   in pentose phosphate pathway, 119, 153
   in photosynthesis, 158
Glucose-6-phosphate dehydrogenase, in primaquine sensitivity, 230
Glutamic acid, **18**
   degradation, 195
   in glutamine synthesis, 190–91
   in ionic bonds, 35
   and Krebs cycle, 137
   nitrogen-containing derivatives, 198
   in purine synthesis, 203–4
   synthesis, 192–93
Glutamic dehydrogenase:
   in amino acid synthesis, 192–93
   control of, 194
   in nitrogen cycle, 187–89
Glutamine, 19, **191**

Glyceraldehyde-3-phosphate, **121**
   energy from, 143
   in glycolysis, 117, 121
   in photosynthesis, 158
Glyceraldehyde phosphate dehydrogenase:
   in glycolysis, 122
   inhibition by iodoacetate, 123
Glycerides, 166–67 (*see also* Phosphoglycerides; Triglycerides)
Glycerol, in complex lipids, 166–67
Glycerol phosphate, in glyceride synthesis, 170
Glycine, **17**
   degradation, 195
   and Krebs cycle, 137
   nitrogen-containing derivatives, 198
   in purine synthesis, 203–4
   synthesis, 192
Glycogen, **150**
   bonds in, 149–50
   hormonal control, 156, 179, 276–77
   metabolism, 150
Glycogenic amino acids, 153
Glycogen phosphorylase:
   in glycogen breakdown, 150
   hormonal control, 156
   in McArdle's disease, 230
Glycogen-storage diseases, 150–51
   McArdle's disease, 230
Glycogen synthetase:
   in glycogen synthesis, 150
   hormonal control, 156
Glycolysis, 115–28
   anaerobic vs. aerobic, 143
   energy-producing reactions, 143
   energy yield, 117, 124
   hormonal control, 276–77
   inhibitors, 123
   in muscle, 106
   regulation of, 119–20
Glycosidic bonds:
   in glycogen, 149–50
   in polysaccharides, 6
Glyoxylate, **138**
Glyoxylate cycle, 132
   relation to Krebs cycle, 138–39
Gout, 208
Growth hormone:
   chemistry, 276
   deficiency, 279
   excess, 279
   physiological response, 276
Guanine, **202**
   degradation, 208
   in DNA, 212–13

Guanine (cont.):
  in RNA, 242–43
  synthesis, 203–4
Guanosine diphosphate, control glutamic dehydrogenase, 194
Guanosine monophosphate, **204**
Guanosine triphosphate:
  control glutamic dehydrogenase, 194
  formation in Krebs cycle, 134–35
  in translation, 258–59

H

Heart disease, lipids and, 182–83
Heavy isotopes, 95
Heme, **197**
  in cytochromes, 140–41
  degradation, 199–200
  in hemoglobin, 32, 33
  as protein cofactor, 38
  synthesis, 197–99
Hemoglobin, 14
  control by 2,3-phosphoglyceric acid, 122
  degradation, 199–200
  inhibitors, 82
  in sickle-cell anemia, 230
  three-dimensional structure, 33–34
Heparin, 151
Hepatitis, viral origin, 287
Hepatitis virus, 287
Herpes simplex virus, 287
Herpetoviridae, 286–87
Heterozygotes, 234
Hexokinase:
  affect of insulin, 119
  in glycolysis, 119
Hexosaminidase A, in Tay-Sachs disease, 182
High-energy compounds:
  hydrolysis of, 102
  phosphate bonds, 102
Histamine, 198
Histidine, **19**
  degradation, 195
  and Krebs cycle, 137
  nitrogen-containing derivatives, 198
Histones, 216–17
Homeostasis, of hormone action, 278
Homogentisic acid:
  in alcaptonuria, 201
  from tyrosine, 201
Homozygotes, 234
Hormones, 13, 275–80 (*see also* specific hormones)
  amino acid derivatives, 198, 276

Hormones (cont.):
  chemistry, 276–77
  control of carbohydrate metabolism, 154–57
  control of lipid metabolism, 166–67
  and cyclic AMP, 156
  homeostatic function, 278
  human, 275–80
  imbalances, 279–80
  lipid, 179–81, 277
  physiological response, 276–77
  plant, 275, 278
  protein, 276
  receptors, 275
  second messengers for, 156, 275, 278
  target cells, 275–77
Hyaluronic acid, 151
Hydrogen:
  isotopes, 95
  valency, 314
Hydrogen bonds, **29**
  in DNA, 213–15
  in proteins, 29
Hydrolases, reactions catalyzed, 66
Hydrophilic bonds, 36
  destruction of, 39–42
  in proteins, 36
Hydrophobic bonds, 36
  in cell membranes, 174–77
  destruction of, 39–42
  in lipoproteins, 174–77
  in proteins, 36
β-Hydroxybutyric acid, in ketosis, 172
Hydroxyl group, **317**
Hydroxylysine, 19, **191**
  synthesis, 193
Hydroxyproline, 19, **191**
  synthesis, 193
Hyperglycemia, 155
  causes, 156
  hormone imbalances in, 279
Hypertension, hormone imbalances in, 279
Hypoglycemia, 155
  causes, 156
  hormone imbalances in, 279
Hypotension, hormone imbalances in, 279
Hypothyroidism, 202
Hypoxanthine-guanine phosphoribosyl transferase, in Lesch-Nyhan disease, 230

I

Imino acids, 19 (*see also* Hydroxyproline, Proline)

Immune system, 296–98
Indole acetic acid:
　chemistry, 198, 278
　physiological response, 278
Inducer, in transcriptional control, 268
Influenza:
　epidemics, 296, 297
　viral origin, 287
Influenza virus, 287, **297**
Inherited metabolic disorders, 229–31 (*see also* specific disorders)
　carriers, 234–35
　natural selection, 236
　recessive, 234
　X-linked, 236
Inhibitors (*see* Enzyme inhibitors)
Initiation factors, in translation, 256–59
Inosine monophosphate, **204**
Inositol, in phosphoglycerides, 167
Insects, sex attractants, 281
Insertional mutations, 232–33
Insulin, 14
　amino acid sequence, 27
　chemistry, 276
　control of carbohydrate metabolism, 154–55
　control of lipid metabolism, 166–67
　deficiency, 279
　excess, 279
　mode of action, 278
　physiological response, 276
Invertase, in candy making, 80
Iodine:
　isotopes, 95
　metabolic role, 9
Iodoacetic acid, inhibition of glycolysis, 123
Ion-exchange chromatography, 49, 50
　of amino acid mixture, 52
Ionic bonds:
　destruction of, 39–42
　in proteins, 35
Iron:
　in hemoglobin metabolism, 197–99
　in human body, 5
　metabolic role, 9
Isocitric acid, **133**
　energy from, 143
　in glyoxylate cycle, 139
　in Krebs cycle, 133
Isoenzymes, 75–77
　in medical diagnosis, 77–78
Isoleucine, **16**
　degradation, 195
　and Krebs cycle, 137
Isomerases, reactions catalyzed, 66

Isomers, 315
Isotopes:
　heavy, 95
　radioactive, 95
　use in determining metabolic pathways, 95–96

J

Jaundice, 199–200

K

Kanamycin:
　mode of action, 83, 305–7
　R-factor resistance, 310
　toxicity, 305
Karyotype, human, **216**
Keratin, 14
　denaturation, 42
α-Ketoglutaric acid, **133**
　energy from, 143
　in Krebs cycle, 133, 137
　in transamination reactions, 136
Ketone bodies, 168, 172
　from amino acids, 194
Ketones (*see* Carbonyl compounds)
Ketosis, 172
Kidneys, hormones controlling, 276–77
Kinases, reaction catalyzed, 66, 102
Krebs cycle, **133**, 130–42
　and amino acid degradation, 136–39, 194–96
　and amino acid synthesis, 136–39, 192–93
　energy-producing reactions, 143
　and gluconeogenesis, 136–39, 153
　and glyoxylate cycle, 138
　regulation of, 120
　and urea cycle, 196–97
Kuru, viroid origin, 298
Kwashiorkor, 190

L

Lactase, in lactose breakdown, 151–53
Lactate dehydrogenase, 14
　in anaerobic glycolysis, 124
　electrophoresis, 76
　in humans, 75–77
　in medical diagnosis, 77–78
Lactic acid, **124**
　concentration in cell, 112
　in dairy industry, 40–41
　in gluconeogenesis, 153–54
　from glycolysis, 124
　in muscle, 125–54

Lactose, **152**
  dietary source, 152
  synthesis, 152
Lactose operon:
  genes of, 269
  promoter sequence, **244**
  transcriptional control, 269
Lanolin, 174
Lanosterol, **173**
  in cholesterol synthesis, 173–74
Lecithin, 167
Lesch-Nyhan disease, 207–8
  clinical features, 230
  metabolic block, 230
Leucine, **16**
  degradation, 195
  and Krebs cycle, 137
Life:
  definition, 1
  elements, 4
  evolution of, 4
Ligases, reactions catalyzed, 66
Light reaction of photosynthesis, 157–60
Lincomycin:
  mode of action, 305–7
  R-factor resistance, 310
Linoleic acid, **165**
Linolenic acid, **165**
Lipase:
  pharmacologic uses, 78
  triglyceride digestion, 170
Lipids, 6, 163–83 (*see also* Complex lipids; Fatty acids; Glycerides; Sterols)
  absorption, 167–68, 174
  in adipose tissue, 167–68
  heart disease and, 182–83
  hormonal control, 166, 276–77
  hydrophobic bonding, 175
  in liver, 167–68
  transport, 175–77
Lipid-storage diseases, 181–82
  Tay-Sachs disease, 230
Lipoic acid, in pyruvate dehydrogenase complex, 127
Lipolysis, 166–68
Lipoproteins:
  roles in animals, 168
  structure, 174–77
Liver:
  hormones controlling, 276–77
  viral diseases, 287
Lyases, reactions catalyzed, 66
Lysine, **18**
  degradation, 195

Lysine (cont.):
  in hydroxylysine synthesis, 190–91
  in ionic bonds, 35
  and Krebs cycle, 137
Lysogenic cycle, **295**
  of viruses, 287–93
Lysosomes, 4, 281
Lytic cycle, **292**
  "early" genes, 289
  "late" genes, 289
  of viruses, 287–93

M

Macromolecules (*see also* Lipids; Nucleic acids; Polysaccharides; Proteins)
  bonding in, 6
  definition, 6–7
Magnesium:
  in human body, 5
  metabolic role, 9
Malic acid, **133**
  energy from, 143
  in glyoxylate cycle, 138
  in Krebs cycle, 133
Malonic acid, as enzyme inhibitor, 82
Malonyl CoA, **169**
  in fatty acid synthesis, 168–70
Maltose, **152**
  dietary source, 152
  synthesis, 152
Mammary glands, hormones controlling, 276–77
Manganese:
  in human body, 5
  metabolic role, 9
Mass-action effects, in post-translational control, 274
McArdle's disease:
  clinical features, 230
  metabolic block, 230
Measles, viral origin, 287
Measles virus, 287
Meat tenderizer, 79
Mechanical work, 104–6
  of muscles, 106–9
Melanin:
  in albinism, 201–2
  synthesis, 201–2
  from tyrosine, 198
Meningitis, viral origin, 287
Mercurichrome, 82
Mercury, as enzyme inhibitor, 82
Messenger RNA:
  binding to ribosomes, 256

Messenger RNA (cont.):
　function, 242
　genes for, 251
　genetic code, 253
　size, 242
　in transcriptional control, 266
　in translation, 252, 259
　in translational control, 265, 271
Metabolic pathways, 89–93
　anabolic, 90
　blocks in, 96–98, 229–231
　catabolic, 90
　control of, 93–94, 264–75
　feedback inhibition, 273
　interdependence, 90–93
　studying, 94–98
Metabolism, 88
Methionine, **17**
　degradation, 195
　and Krebs cycle, 137
　nitrogen-containing derivatives, 198
Microtubules, 105
Mineralocorticoids, 179–80, 277
Mink encephalopathy, viroid origin, 298
Misreading, antibiotics causing, 307
Missense mutations, 232–33
Mitochondria, 4
　cells' metabolic furnace, 130
　compartmentation, 281
　electron transport in, 142
　fatty acid degradation in, 171
Molecular-sieve chromatography, 48, 50
Monoglycerides, 166
Monomers:
　definition, 6, 7
　elemental composition, 6
Monosaccharides, 151–53
Monounsaturated fatty acids, 165
Morphine, 198
Mucopolysaccharides, 151
Multimeric proteins, 36–37
　denaturation, 41
　regulation of, 120
　synthesis, 260
Multiple drug resistance, 309
Mumps, viral origin, 287
Mumps virus, 287
Muscles:
　ATP in, 106
　contraction, 106–9
　glycolysis in, 106
　lactic acid production, 125
　red vs. white, 106
　role of calcium, 109

Muscles (cont.):
　sliding filament model, 107
　structure of skeletal, 107
Mutants:
　definition, 226
　in genetic/biochemical studies, 228–29
　induction of, 228
Mutations:
　deletion, 232–33
　effect on phenotype, 229–31
　effect on protein structure, 260
　and evolution, 145
　inheritance, 235
　insertion, 232–33
　missense, 232–33
　nonsense, 232–33
　use in determining metabolic pathways, 96–98
Myosin, in skeletal muscle, 107–9

N

Na^+-K^+ pump, 109–11
　energy requirement, 109–11
　pump protein, 109–10
NAD^+ (see Nicotinamide adenine dinucleotide)
NADH, in electron transport chain, 139–42 (see also Nicotinamide adenine dinucleotide)
NADPH:
　in deoxynucleotide synthesis, 206
　in fatty acid synthesis, 169–70
　formation from $NADP^+$, 135–36
　from pentose phosphate pathway, 119, 153
　in photosynthesis, 158–59
Nerve cells:
　active transport, 105
　lipids in, 181–82
Nerve gas, 82
Neurospora crassa, in genetic studies, 228–29
Niacin:
　deficiency, 207
　as enzyme cofactor, 38
　metabolic role, 8
　synthesis from tryptophan, 198, 207
Nicotinamide, in NAD^+, 135, 207
Nicotinamide adenine dinucleotide, **207**
　anaerobic regeneration, 124–26
　cytoplasmic, 143
　dehydrogenase cofactor, 38, 135
　and electron transport chain, 139–42
　in fatty acid degradation, 169–72
　in glycolysis, 117
　in Krebs cycle, 133, 135
　mitochondrial, 143
　synthesis, 207

Nicotine, 198
Night-blindness, 178
Ninhydrin, reaction with amino acids, 20, 51
Nitrate, in nitrogen cycle, 187–89
Nitrate reductase, in nitrogen cycle, 187–89
Nitrification, in nitrogen cycle, 187–89
Nitrogen:
   isotopes, 95
   in nitrogen cycle, 187–89
   valency, 314
Nitrogenase, in nitrogen fixation, 189
Nitrogen cycle, 187–89
Nitrogen fixation:
   as cell work, 105
   in nitrogen cycle, 187–89
Nonsense mutations, 232–33
   effect on translation, 260
Norwalk agent, 287
Nucleic acids, 6 (*see also* Deoxyribonucleic acid; Ribonucleic acid)
Nucleoside, 202
Nucleotides, 202
   metabolic roles, 203
   phosphate exchange, 206
   salvage pathways, 207
Nucleus, 4
Nutrition (*see* Amino acids, essential; Fatty acids, essential; Trace elements; Vitamins)

O

Oleic acid, **164**
One gene/one enzyme hypothesis, 228–31
One gene/one protein hypothesis, 231
Operator, in transcriptional control, 244, 267
Operon, 267
Organelles, 2–4 (*see also* individual organelles)
   functions, 4
   isolation, 3
   macromolecules in, 6–7
Organic compounds:
   chemistry of, 313–23
   elements in, 313
   functional groups, 316–17
   shapes of, 314–15
Ornithine, **196**
   in urea cycle, 196–97
Orotic acid, in pyrimidine synthesis, 205
Orthomyxoviridae, 286–87
Oxaloacetic acid, **133**
   in gluconeogenesis, 153–54
   in Krebs cycle, 132–33, 136–37
   from pyruvate, 138
   in transamination reactions, 136

Oxidases:
   and molecular oxygen, 174
   reactions catalyzed, 66
Oxidation-reduction reactions, 322–23
Oxidative phosphorylation, 139–42
Oxidoreductases, reactions catalyzed, 66
Oxygen:
   in electron transport, 139
   from photosynthesis, 158–59
   valency, 314
Oxygen debt, 125

P

Palmitic acid, **164**
Palmitoleic acid, **170**
   synthesis, 170
Pantothenic acid:
   in coenzyme A, 127
   metabolic role, 8
Papain, 79
Papilloma virus, 287
Papovaviridae, 286–87
Parainfluenza virus, 287
Paramyxoviridae, 286–87
Parvoviridae, 286–87
Pea enanation mosaic virus, 288
Pectinases, industrial uses, 79
Pellagra, 207
Penicillin, **306**
   mode of action, 83, 305–6
   R-factor resistance, 310
   synthesis, 198
   toxicity, 305
Pentose phosphate pathway, ribose production, 153
Peptidases:
   industrial uses, 79
   reactions catalyzed, 66
Peptide antibiotics, 304
Peptide bond, 22–23
   in proteins, 6
   formation of, 23, 257–60
   stability, 23
Peptidyl transferase, on ribosomes, 257
Pernicious anemia, 199
Peroxisomes, 281
pH:
   definition, 7
   effect on amino acids, 20–22
Phages (*see* Bacteriophages)
Phenol, 317
Phenylalanine, **17**
   degradation, 195, 200–202

Phenylalanine (cont.):
  and Krebs cycle, 137
  metabolic disorders, 200–202
  in PKU, 200–201
Phenylalanine hydroxylase:
  in PKU, 200–201, 230
  in tyrosine synthesis, 200–201
Phenylisothiocyanate, 55–56
Phenylketonuria, 200–201
  clinical features, 230
  detection of carriers, 200
  metabolic block, 230
Phenylpyruvic acid, in PKU, 200
Pheromones, 281
Phosphatases, reactions catalyzed, 66
Phosphate, inorganic:
  concentration in cell, 112
  in glycolysis, 117, 122
  in oxidative phosphorylation, 141
Phosphatidyl choline, **164**
Phosphocreatine (*see* Creatine phosphate)
Phosphodiester bonds:
  in DNA, 6, 211–13
  in RNA, 6, 242–43
Phosphoenolpyruvate carboxykinase, in gluconeogenesis, 154
Phosphoenolpyruvic acid, **123**
  concentration in cell, 112
  energy from, 143
  energy of hydrolysis, 103
  in gluconeogenesis, 153–54
  in glycolysis, 117, 123
Phosphofructokinase:
  in glycolysis, 119–20
  regulation of, 119–20
  regulation by citrate, 132–34, 274
Phosphoglucose isomerase, in glycolysis, 119
Phosphoglycerate kinase, in glycolysis, 122
2-Phosphoglyceric acid, **123**
  in glycolysis, 117, 123
3-Phosphoglyceric acid, **122**
  in glycolysis, 117, 122
  in photosynthesis, 159
Phosphoglycerides, **164**, 166
  in cell membranes, 175–77
  in lipoproteins, 175–77
  metabolic role, 167
  synthesis, 170
Phospholipid bilayer, 175
Phosphoribosyl pyrophosphate, **204**
  degradation, 207–8
  in nucleotide synthesis, 203–6
Phosphorus:
  isotopes, 95

Phosphorus (cont.):
  valency, 314
Phosphorylase kinase, 156
  hormonal control, 156
Phosphorylases, reactions catalyzed, 66
Photosynthesis, 157–60
Pickling, 126
Picornaviridae, 286–87
Pigments, from amino acids, 198
Plants:
  hormones, 278
  viruses, 288
Plasmids, in recombinant DNA, 222–24
  (*see also* Resistance factors)
Pneumonia, viral origin, 287
Poisons (*see* Enzyme inhibitors)
Polio, viral origin, 287
Polio virus, 287
Polyribosomes, 256, 258
Polysaccharides, 6, 149–51 (*see also* Cellulose; Glycogen; Starch)
  digestion of, 115
  storage, 149–51
  structural, 151
Polysomes, 256
Polyunsaturated fatty acids, 165
  essential, 8, 165
  and heart disease, 182–83
Porphyrias, 199
Porphyrins:
  degradation, 199–200
  in metabolic diseases, 199
  as protein cofactors, 199
  from succinic acid, 137–38
  synthesis, 198–99
Post-translational control, 265, 272–74
  enzyme modification, 272
  feedback inhibition, 273
  mass-action effects, 274
Potassium:
  in human body, 5
  metabolic role, 9
  Na^+-K^+ pump, 109–11
Potato X virus, 288
Poxviridae, 286–87
Primaquine sensitivity, 230
Procaryotic cells (*see* Cells)
Progesterone, **180**
  chemistry, 277
  deficiency, 279
  excess, 279
  physiological response, 277
Progestogens, 179–81, 277
Proline, **19**

Proline (cont.):
  degradation, 195
  disrupts a-helix, 33
  in hydroxyproline synthesis, 190–91
  and Krebs cycle, 137
  nitrogen-containing derivatives, 198
  synthesis, 192
Promoters, 244
  RNA polymerase binding, 250–51, 266, 270
  role in transcription, 250–51
Prophage, 293, 295
Prostaglandin E_1, **180**
Prostaglandins, 277
  metabolic roles, 181
  synthesis, 179–81
Protein cofactors, 37–39
  function, 37–39
  metal ions, 38
  organic molecules, 38
  porphyrins, 199
Protein malnutrition, 190
Proteins, 6
  amino acid sequence, 25
  assays, 47
  backbone of peptide bonds, 22–23
  chromosomal, 216–17
  colinearity with genes, 232
  denaturation, 39–42
  determining amino acid composition, 50–53
  determining amino acid sequence, 54–59
  determining three-dimensional shape, 59–60
  digestion, 190
  disulfide bridges, 35
  extracellular, 13
  genetic control, 231
  a-helix, 29–34
  hydrolysis, 50
  hydrophilic bonds, 36
  hydrophobic bonds, 36
  ionic bonds, 35
  isolation, 13
  metabolic roles, 12–15
  multimeric, 36–37
  purification, 47–50
  R-group interactions, 28–37
  synthesis of, 252–60 (*see also* Translation)
  three-dimensional shape, 27–37
Purines (*see also* Adenine; Guanine)
  base pairing, 213–15
  degradation, 207–8
  in DNA, 212
  metabolic diseases, 207–8
  nomenclature, 202
  in RNA, 243

Purines (cont.):
  synthesis, 203–4
Puromycin, **307**
  mode of action, 306–7
Pyridoxal phosphate:
  as enzyme cofactor, 38
  metabolic role, 8, 38
  in transaminase reactions, 136–38
Pyrimidines: (*see also* Cytosine; Thymine; Uracil)
  base pairing, 213–15
  degradation, 207–8
  in DNA, 212
  nomenclature, 202
  in RNA, 243
  synthesis, 205
Pyrroles, **198**
Pyruvate carboxylase, 138
  in anaerobic glycolysis, 126
  control by acetyl CoA, 138
Pyruvate dehydrogenase complex, 127
Pyruvate kinase, in glycolysis, 123
Pyruvic acid, **123**
  from amino acids, 194–96
  concentration in cell, 112
  conversion to acetyl CoA, 127–28
  conversion to ethanol, 126
  conversion to lactic acid, 124–25
  conversion to oxaloacetate, 138
  energy from, 143
  from glycolysis, 117, 123
  in transamination reactions, 136

Q

Quinine, 198

R

Rabies, viral origin, 287
Rabies virus, 287
Radiation:
  effect on DNA, 221–22
  release of prophage, 293
Radioactive isotopes, 95
Recombinant DNA, 222–24
Recombination, of DNA, 221–24
Reducing agents, 41
Reductases, reactions catalyzed, 66
Reduction, 41
Regulation (*see* Control, metabolic)
Regulatory genes:
  operator, 267
  promoter, 267
Reoviridae, 286–87
Repair, of DNA, 221–24
Repair enzymes, DNA polymerase I, 224–26

Replicase, in RNA synthesis, 294
Repressor proteins:
  in transcriptional control, 268
  in viral infection, 293
Resistance factors:
  antibiotic resistance, 309
  gene amplification, 275
  genes on, 309
  transmission, 309
Respiratory syncitial virus, 287
Respiratory tract, viral diseases, 287
Reverse transcriptase, 296
R-factors (see Resistance factors)
Rhabdoviridae, 286–87
Rhinoviruses, 287
Riboflavin:
  as enzyme cofactor, 38
  in FAD, 135
  role in metabolism, 8
Ribonuclease, 14
  amino acid sequence, 27
  denaturation, 41
  reaction catalyzed, 244–45
  sequencing RNA, 244–45
Ribonuclease, pancreatic, base specificity, 244
Ribonuclease T_1, base specificity, 244
Ribonuclease U_2, base specificity, 244
Ribonucleic acid (see also Messenger RNA; Ribosomal RNA; Transfer RNA)
  alkali sensitivity, 244
  antibiotics inhibiting synthesis, 307–8
  DNA template, 248
  metabolic roles, 241–42
  as primer for DNA synthesis, 224
  rare bases in, 242
  replicase, 294
  reverse transcriptase, 296
  sequencing, 244–45
  structure, 242–47
  synthesis, 248–52
  viral, 242, 286, 288, 294–96
Riboses, 153
  formation, 153
  in RNA, 242
Ribosomal RNA:
  function, 242
  genes for, 251
  procaryotic, 256–57
  size, 242
Ribosomes, 4
  antibiotic binding, 306
  binding sites, 256
  enzymatic activity, 257
  eucaryotic, 255

Ribosomes (cont.):
  polyribosomes, 256
  procaryotic, 255
  structure, 255–57
  subunits, 256–57
  in translation, 255–57, 259
  in translational control, 272
Ribulose-1, 5-diphosphate, 159
  in photosynthesis, 158–59
Ribulose diphosphate carboxylase, in photosynthesis, 159
Rice dwarf virus, 288
Rickets, 178
Rifampicin, mode of action, 83, 305, 307–8
RNA (see Ribonucleic acid)
RNA polymerase:
  in bacteria, 250
  binding to promoter, 266–70
  control by cAMP, 267, 269
  in eucaryotes, 250
  inhibitors, 83
  isoenzymes, 250
  rifampicin binding, 308
  in transcription, 248–52
  in transcriptional control, 265–70
Roots, active transport in, 111
Rotavirus, 287
Rubella, viral origin, 287
Rubella virus, 287
Ruminants, digestion of cellulose, 160

S

Salivary glands, viral disease, 287
Salvarsan, 302
Saturated fatty acids, 165 (see also Fatty Acids)
  and heart disease, 182–83
Scrapie, viroid origin, 298
Scurvy, 193
Seeds:
  energy for germination, 138
  hormones in germination, 278
Selenium, metabolic role, 9
Semiconservative replication of DNA, 217–21
  direction of replication, 224–25
Sequenator, 55
Serine, 17
  degradation, 194–95
  and Krebs cycle, 137
  nitrogen-containing derivatives, 198
  in phosphoglycerides, 167
  reaction with diisopropylfluorophosphate, 82
  synthesis, 192

Sex hormones:
  chemistry, 179–80
  function, 179–81
  synthesis, 173
Shingles, viral origin, 287
Sickle-cell anemia:
  clinical features, 230
  inheritance, 233
  natural selection, 236
  metabolic defect, 230, 233–34
Silage, 126
Silicon, metabolic role, 9
Single bond, 314
Skin, viral diseases, 287
Smallpox, viral origin, 287
Sodium:
  in human body, 5
  metabolic role, 9
  Na^+-K^+ pump, 109–11
Spectrophotometer, 72
Sphingomyelin, 181
Sphingosine:
  in complex lipids, 181–82
  serine in, 198
Squalene, in cholesterol biosynthesis, 172–74
Starch:
  chemistry, 151
  digestion, 152
  from photosynthesis, 158–59
  synthesis, 151
Starvation:
  fatty liver in, 177
  ketosis in, 172
Steroid hormones:
  mechanism of action, 278
  metabolic roles, 179–80
  synthesis, 179–80
Sterols, **164**, 167
  synthesis, 172–74
Storage proteins, 13–14
Streptodornase, pharmacologic uses, 78
Streptomycin:
  mode of action, 83, 305–7
  R-factor resistance, 310
  toxicity, 305
Structural proteins:
  definition, 13–14
  three-dimensional structure, 37
Strychnine, 198
Substrate-level phosphorylation:
  definition, 134
  in glycolysis, 117, 143
  in Krebs cycle, 134–35, 143
Succinic acid, **133**

Succinic (cont.):
  energy from, 143
  in glyoxylate cycle, 139
  in Krebs cycle, 133, 137
  in porphyrin synthesis, 197–99
  from pyrimidine degradation, 208
Succinic dehydrogenase, inhibition by malonic acid, 82
Succinyl CoA, **133** (*see also* Succinic acid)
Sucrose, **152**
  dietary source, 152
  synthesis, 152
Sulfa drugs:
  mode of action, 83–84, 203, 308
  R-factor resistance, 310
  sulfonamides, 84
Sulfhydryl group, **317**
Sulfonamides, **84** (*see also* Sulfa drugs)
Sulfur:
  isotopes, 95
  valency, 314
Svedberg units of centrifugation, 256
Synthetase kinase, hormonal control, 156
Synthetases, reactions catalyzed, 66

T

Target cell, of hormone action, 275
Tay-Sachs disease, 182
  clinical features, 230
  diagnosis, 78
  metabolic block, 230
Termination factors, in translation, 256–59
Termites, digestion of cellulose, 160
Testosterone, **180**
  chemistry, 277
  deficiency, 279
  excess, 279
  physiological response, 277
Tetracycline:
  mode of action, 83, 305–7
  R-factor resistance, 310
Tetrahydrofolic acid, in purine synthesis, 203–4
β-Thalassemia, 260
Thiamine pyrophosphate:
  as enzyme cofactor, 38
  deficiency disease, 128
  in pyruvate dehydrogenase complex, 127
  role in metabolism, 8
Thioalcohols, 322
Thioesters, 322
Thiols, reduction, 322

Threonine, **17**
  degradation, 194–95
  and Krebs cycle, 137
Thymidine monophosphate, **205**
Thymine, **202**
  degradation, 208
  in DNA, 212–13
  effect of UV light, 222
  synthesis, 205–6
Thyroxine:
  cAMP and, 278
  chemistry, 198, 276
  deficiency, 279
  excess, 279
  mode of action, 280
  physiological response, 276
Tin, metabolic role, 9
Tobacco mosaic virus, 288, **291**
Tobacco ringspot virus, 288
Togaviridae, 286–87
Toxins:
  definition, 13–14
  denaturation, 39
  inhibitors of protein function, 82
Trace elements:
  cellular requirements, 8–9
  in human body, 5
  metabolic roles, 8–9
  uptake by plants, 111
  as protein cofactors, 9, 38
Transaminases, 136–38
  cofactors for, 136–38
  reactions catalyzed, 66
Transamination, 136–38
  in amino acid synthesis, 193
Transcription, 248–52 (*see also* Transcriptional control)
  antibiotics inhibiting, 305
  "start" and "stop" signals, 250–51
Transcriptional control, 94, 265–70
  chromosome condensation, 265, 270–71
  inducer, 268
  of lactose operon, 269
  negative, 267–70
  at operator, 267
  positive, 270
  at promoter, 250–51, 266
Transferases, reactions catalyzed, 66
Transfer RNA, **246**, 247
  binding to ribosomes, 256
  function, 242
  genes for, 251
  initiator, 255

Transfer RNA (cont.):
  linkage to amino acids, 254
  rare bases in, 246–47
  size, 242
  in translation, 254–55, 259
  in translational control, 272
  triplet recognition, 255
Transformation, in bacteria, 227
Translation, **259**, 252–60 (*see also* Translational control)
  antibiotics inhibiting, 305
  energy for, 258–59
  outline, 252
  speed of, 258
Translational control, 94, 265, 271–72
Transport proteins:
  cofactors for, 37–39
  defects in, 230
  definition, 13–14
  lipoproteins, 175–177
  Na^+-K^+ pump, 109–11
Tricarboxylic acid cycle (*see* Krebs cycle)
Triglycerides, **164**, 166
  metabolic role, 166
  synthesis, 170
Trimethoprim, mode of action, 308
Tripalmitate, **164**
Triple bond, 314
Triplet code (*see* Genetic code)
Trypsin:
  pharmacologic uses, 78
  in protein sequencing, 55, 57
Tryptophan, **18**
  conversion to niacin, 207
  deficiency, 207
  degradation, 195
  and Krebs cycle, 137
  nitrogen-containing derivatives, 198
  precursor indole acetic acid, 278
Tyrosine, **18**
  degradation, 195, 200–202
  and Krebs cycle, 137
  metabolic disorders, 200–202
  nitrogen-containing derivatives, 198
  synthesis, 193

U

UDP-glucose, in glycogen synthesis, 150
Ultraviolet light, effect on DNA, 222
Unsaturated compounds, 314

Unsaturated fatty acids, 165
  essential, 165
  hormones from, 277
  synthesis, 170
Unwinding protein, in DNA replication, 225–26
Uracil, **202**
  degradation, 208
  in RNA, 242–43
  synthesis, 205
Urea, **196**
  synthesis, 196–97
Urea cycle, **196**, 196–97
  energy required, 197
  and Krebs cycle, 196–97
Uric acid, 196, **208**
  in gout, 208
  from purines, 208
Uridine diphosphate:
  energy of hydrolysis, 103
  UDP-glucose, 150
Uridine monophosphate, **205**
Uterus, hormones controlling, 276–77

V

Vaccination, 296–98
Vaccines, preparation, 39
Valency, 313–14
Valine, **16**
  degradation, 195
  nitrogen-containing derivatives, 198
  and Krebs cycle, 137
Vanadium, metabolic role, 9
Varicella-Zoster virus, 287
Variola virus, 287
Vasodilators, from amino acids, 198
Vasopressin:
  cAMP and, 278
  chemistry, 276
  deficiency, 279
  physiological response, 276
Vinegar production, 126
Viroids, 298–99
Viruses, 285–96
  and cancer, 293–94
  classification, 286
  genetic material, 286, 288
  in human disease, 287
  lysogenic cycle, 287–93
  lytic cycle, 287–93
  plant, 288
  replicating genome, 294–96

Viruses (cont.):
  RNA in, 242
  size, 286–87, 289
  structure, 286–87, 289
  tumor, 293
Vitamin A, **178**
  deficiency, 178
  metabolic role, 8, 178
  overdose, 178
Vitamin B_1 (*see* Thiamine pyrophosphate)
Vitamin B_2 (*see* Riboflavin)
Vitamin B_6 (*see* Pyridoxal phosphate)
Vitamin B_{12} (*see* Cobalamin)
Vitamin C (*see* Ascorbic acid)
Vitamin D, **178**
  deficiency, 178
  metabolic role, 8, 178
  overdose, 178
  synthesis, 177
Vitamin D_3, **173**
  synthesis, 173
Vitamin E, **178**
  deficiency, 178
  metabolic role, 178
Vitamin K, 8, **178**
  deficiency, 178
  metabolic role, 178
  overdose, 178
Vitamins: (*see also* specific vitamins)
  fat-soluble, 177–79
  metabolic roles, 8
  as protein cofactors, 38

W

Warts, viral origin, 287
Water:
  in cells, 5
  ionization, 5
  role in cell pH, 7
Watson-Crick base pairing:
  codon:anticodon recognition, 255
  in DNA, 213–15
  in DNA replication, 218–20
  in RNA, 245
  in RNA synthesis, 250
Waxes, 182
Wild-type genes, inheritance, 235
Wilson's disease:
  clinical features, 230
  metabolic block, 230

Work, of cells, 101–13
   active transport, 104
   ATP required, 111–12
   biosynthesis, 104
   mechanical, 104

X

Xeroderma pigmentosum, 222
X-linked genes, inheritance of, 236
X-ray crystallography, of proteins, 59–60

Y

Yeast:
   alcoholic beverage production, 125
   in bread making, 126

Z

Zinc:
   in human body, 5
   metabolic role, 9

81   0025